D0165865

WEATHERFORD COLLEGE LIBRARY

Hazardous Materials Handbook

Awareness and Operations Levels

WEATHERFORD COLLEGE LIBRARY

WEATHERFORD COLLEGE LIBRARY

Hazardous Materials Handbook
Awareness and Operations Levels

WEATHERFORD COLLEGE LIBRARY

DELMAR
CENGAGE Learning

Australia • Brazil • Japan • Korea • Mexico • Singapore • Spain • United Kingdom • United States

**Hazardous Materials Handbook:
Awareness and Operations Levels
Delmar, Cengage Learning**

Vice President, Career and Professional
Editorial: Dave Garza

Director of Learning Solutions: Sandy Clark

Product Development Manager: Janet Maker

Managing Editor: Larry Main

Senior Product Manager: Jennifer A. Starr

Editorial Assistant: Maria Conto

Vice President, Career and Professional
Marketing: Jennifer McAvey

Marketing Director: Deborah S. Yarnell

Senior Marketing Manager: Erin Coffin

Marketing Coordinator: Shanna Gibbs

Production Director: Wendy Troeger

Production Manager: Mark Bernard

Senior Content Project Manager:
Jennifer Hanley

Senior Art Director: Bethany Casey

Technology Project Manager:
Christopher Catalina

Production Technology Analyst: Thomas Stover

Cover Image: ©Getty Images, Inc. / Stockbyte
collection

© 2008 Delmar, Cengage Learning

ALL RIGHTS RESERVED. No part of this work covered by the copyright herein
may be reproduced, transmitted, stored, or used in any form or by any means
graphic, electronic, or mechanical, including but not limited to photocopying,
recording, scanning, digitizing, taping, Web distribution, information networks,
or information storage and retrieval systems, except as permitted under Section 107 or 108 of the 1976 United States Copyright Act, without the prior written permission of the publisher.

For product information and technology assistance, contact us at
Cengage Learning Customer & Sales Support, 1-800-354-9706

For permission to use material from this text or product,
submit all requests online at **www.cengage.com/permissions**.
Further permissions questions can be e-mailed to
permissionrequest@cengage.com.

Library of Congress Control Number: 2008924373

ISBN-13: 978-1-4283-1971-4
ISBN-10: 1-4283-1971-9

Delmar
5 Maxwell Drive
Clifton Park, NY 12065-2919
USA

Cengage Learning is a leading provider of customized learning solutions
with office locations around the globe, including Singapore, the United
Kingdom, Australia, Mexico, Brazil and Japan. Locate your local office at:
international.cengage.com/region

Cengage Learning products are represented in Canada by Nelson
Education, Ltd.

For your lifelong learning solutions, visit **delmar.cengage.com**
Visit our corporate Web site at **cengage.com**

Notice to the Reader
Publisher does not warrant or guarantee any of the products described herein
or perform any independent analysis in connection with any of the product
information contained herein. Publisher does not assume, and expressly
disclaims, any obligation to obtain and include information other than that
provided to it by the manufacturer. The reader is expressly warned to consider
and adopt all safety precautions that might be indicated by the activities described herein and to avoid all potential hazards. By following the instructions
contained herein, the reader willingly assumes all risks in connection with such
instructions. The publisher makes no representations or warranties of any kind,
including but not limited to, the warranties of fitness for particular purpose or
merchantability, nor are any such representations implied with respect to the
material set forth herein, and the publisher takes no responsibility with respect
to such material. The publisher shall not be liable for any special, consequen-
tial, or exemplary damages resulting, in whole or part, from the readers' use of,
or reliance upon, this material.

Printed in the United States of America
1 2 3 4 5 12 11 10 09 08

Firefighter's Handbook and *Hazardous Materials Handbook* are dedicated to the greathearted and courageous firefighters and emergency responders who have given of themselves the greatest sacrifice, their lives. Every year, the fire and emergency service community changes forever when the lives of firefighters are ended or damaged in the course of their duties. As we continue on we are left with the scars of these losses and the tears of many loved ones left behind. We share in the heartache of the loss of every firefighter and emergency responder. Let their lives shine on in the dedication and bravery of those left to respond when the tones drop, the bells ring, and the sirens blare.

This text is also dedicated to the driving force behind the continuation of firefighter heritage, the sharing of wisdom and experience, and the art of discovery and learning—trainers and educators. Every classroom session, practical scenario, and review session directly affects the quality of response the fire service provides. Never underestimate the power of positive change the training and education community holds.

And to every firefighter who has touched the life of someone in need and made a positive difference—you are truly the epitome of human compassion and selflessness. Don't ever stop caring.

DEDICATION

Contents

1

Hazardous Materials: Laws, Regulations, and Standards

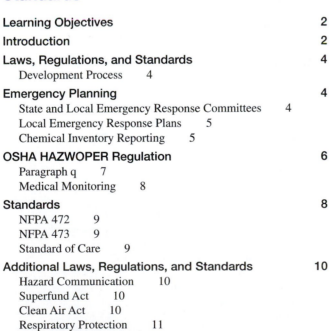

2

Hazardous Materials: Recognition and Identification

viii   Contents

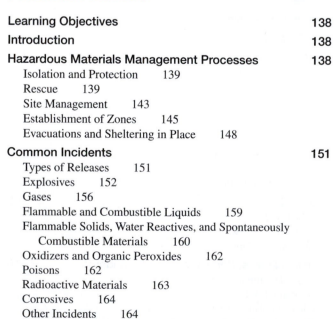

3

Hazardous Materials: Information Resources

4

Hazardous Materials: Personal Protective Equipment

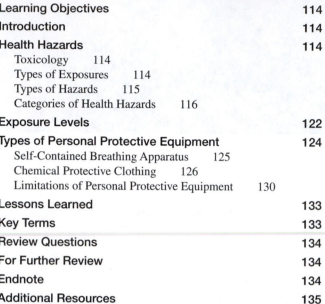

5

Hazardous Materials: Protective Actions

6

Product Control and Air Monitoring

7

Terrorism Awareness

Preface

OUR HISTORY

Proudly serving the fire service for over ten years, Delmar, Cengage Learning has been publishing trade-related training materials since the post–World War II era.

Delmar Publishers began in 1945 immediately following the end of World War II. Soldiers returning from the war were in need of jobs, and the industry needed people educated in the trades to fill available positions. It was an economic boom, and textbooks played a major role in educating these new workers. With this new educational need, Delmar was born as a publisher.

In the early 1980s, the Thomson Corporation bought Delmar Publishers, which quickly became part of the Thomson Learning branch of information providers. It has over the past few decades transitioned our company into what it is today—a provider of a variety of training solutions, including textbooks, CDs, videos and online courses.

In 2006, the growth and potential of Thomson Learning attracted the attention of both Apax Partners, a global company focused on investment, and Omers, a Canadian-based company offering pension services to retired municipal employees. Through a partnership, Apax and Omers bought Thomson Learning with the intention of inspiring further value and growth.

Now known as Cengage Learning ("center of engagement"), Delmar and fellow companies continue to grow globally, working to deliver even more training solutions, highly customized options and specialized content to colleges, universities, professors, students, reference centers, government agencies, corporations and professionals around the world.

As we proceed through the twenty-first century we hope that you will turn to us to meet all your training needs as we continue to *Answer Your Call!*

OUR AUTHORS

We are proud to present this brand-new offering drawn from our third edition of *Firefighter's Handbook—Hazardous Materials Handbook: Awareness and Operations Levels!* As is the tradition in the fire service, this edition has truly become a reality through team effort. These chapters are a result of many hours of collaboration to ensure that we offer a practical approach and deliver the most current information available, in addition to meeting the requirements set forth by NFPA 472 *Standard for Competence of Responders to Hazardous Materials/Weapons of Mass Destruction Incidents Qualifications,* 2008 Edition.

We wish you the best as you enter the world of firefighting, and as you begin your training, we would like to share with you three guiding principles:

1. Build your skill sets.
2. Remain informed.
3. Above all—stay safe!

Firefighter's Handbook and Hazardous Materials Handbook

The hazardous materials information in these pages is drawn from *Firefighter's Handbook: Firefighting and Emergency Response*, Third Edition. We thank Content Editor Andrea Walter and author Chris Hawley for their dedication to ensuring that the content for these hazardous materials chapters remains accurate, current, and compliant with the 2008 Editions of NFPA Standards 1001 and 472.

Andrea A. Walter

Content Editor: Firefighter's Handbook series
Founding Author of Firefighter's Handbook chapters: Emergency Medical Services; Disaster and Large Incident Response

Andrea A. Walter is a firefighter with the Metropolitan Washington Airports Authority at the Washington Dulles International Airport and a life member and former officer of the Sterling Volunteer Rescue Squad, in Sterling, Virginia. Walter has been active in the fire and emergency services community for many years, serving as the Manager of the Commission on Fire Accreditation International for the International Association of Fire Chiefs, and assisting in a variety of projects with the National Volunteer Fire Council, Women in the Fire Service, and the United States Fire Administration. She has over nineteen years of experience in the fire and emergency services. In addition to being an author for this text, Walter serves as the project's content editor/lead author and has been with the text project since its inception. Walter is involved in a variety of other Delmar text projects and instructional supplements. She is also an author and the content editor for Delmar's *First Responder Handbook: Fire Service Edition* and *First Responder Handbook: Law Enforcement Edition,* as well as *Exam Preparation: Firefighter I and II,* and *Exam Preparation: Hazardous Materials Awareness and Operations.*

Chris Hawley

Author of Firefighter's Handbook and Hazardous Materials Handbook chapters: Hazardous Materials: Laws, Regulations, and Standards; Hazardous Materials: Recognition and Identification; Hazardous Materials: Information Resources; Hazardous Materials: Personal Protective Equipment; Hazardous Materials: Protective Actions, Product Control and Air Monitoring; Terrorism Awareness

Chris Hawley is a Deputy Program Manager for Computer Sciences Corporation (CSC) and is responsible for several WMD courses within the DOD/FBI/DHS International Counterproliferation Program. This program provides threat assessment, HazMat and Anti-Terrorism training and full scale exercises worldwide and is focused on Eastern Europe and Central Asia. Previous to the International work, Chris retired as a Fire Specialist with the Baltimore County Fire Department. Prior to this assignment he was as-signed as the Special Operations Coordinator. As the Special Operations Coordinator he reported to the Division Chief of Special Operations, and was responsible for the coordination of the Hazardous Materials Response Team and the Advanced Technical Rescue Team along with two team leaders. Prior to this assignment he was assigned to the Fire Rescue Academy as one of four shift instructors. As a shift instructor he was responsible for all County-wide training on his shift. He has been a hazardous materials responder for over 19 years. Chris has twenty-five years experience in the fire service and prior to Baltimore County he was a Hazardous Response Specialist with the City of Durham, NC Fire Department.

Chris has designed innovative programs in hazardous materials and anti-terrorism and has assisted in the development of many other training programs. He has assisted in the development of programs provided by the National Fire Academy, Federal Bureau of Investigation, U.S. Secret Service, and many others. Chris has presented at numerous local, national, and international conferences, and writes articles on a regular basis. He serves on a variety of pivotal committees and groups at the local, state, and federal levels. He also works with local, state, and federal committees and task forces related to hazardous materials, safety, and terrorism.

Chris has published four texts on hazardous materials and terrorism response for Delmar. He co-authored *Special Operations for Terrorism and HazMat Crimes,* along with Mike Hildebrand and Greg Noll through Red Hat Publishing. He also assists a variety of publishers with the review and development of emergency services texts and publications.

Firefighter's Handbook: Essentials of Firefighting

We encourage you to check out the companion to this book—*Firefighter's Handbook: Essentials of Firefighting,* Third Edition—which contains the firefighting chapters that are drawn from *Firefighter's Handbook: Firefighting and Emergency Response,* Third Edition. Combined with *Hazardous Materials Handbook,* the Essentials book meets the requirements set forth by NFPA Standard 1001 for Firefighter I and II certification.

Our founding authors helped to create *Firefighter's Handbook: Firefighting and Emergency Response* in its first edition, when the fire service was in need of new options for fire training, and some have returned with each edition to keep the tradition of *Firefighter's*

Handbook moving toward the future. Some of the authors listed here are new to the third edition, and we have welcomed them onto our team—they are also committed to the same educational values that were the foundation for the first edition.

Richard Bonnett

Author of Firefighter's Handbook chapter: Emergency Medical Services
Bonnett is an EMS Captain with the Metropolitan Washington Airport's Authority and a life member of the Millwood Volunteer Fire and Rescue Department. He has been actively involved with fire and EMS for over twenty years. He is a Nationally Registered Paramedic and is certified as an Advanced Life Support Coordinator. He has been an active instructor in the State of Virginia for over twelve years teaching at the state and local level. He is training center faculty in Advanced Cardiac Life Support at Winchester Medical Center and also teaches Pediatric Advanced Life Support. Bonnett recently co-founded Comprehensive Safety and Health Training, a company offering basic and advanced life support classes in the tri-state area.

Willis T. Carter

Founding Author of Firefighter's Handbook chapter: Communications and Alarms
Chief Willis Carter has been a member of the fire service for thirty-five years. He began his career in 1972 as a firefighter with the Shreveport (La.) Fire Department. In 1978, Carter performed a lateral transfer into what was then known as "Fire Alarm," where he was a call taker and dispatcher until his promotion to Chief of Communications in 1986. Carter is responsible for the management and operations of the Fire Communications Division, which has a staff of 45 who operate the primary Public Safety Answering Point (PSAP) for the Caddo Parish (Louisiana) 9-1-1 system. In addition to PSAP operation, the Chief is also responsible for the Information Technology section of the Fire Department which supports the network, hardware, and software utilized by the 650-member Fire Department.

Carter is active in public safety communications outreach and education. He is a member of the International Fire Chief's Association and is an assessment-team leader for the Commission on Accreditation for Law Enforcement Agencies, Inc. (CALEA).

Carter led the effort for the Shreveport Fire Department Communications Center to become the first public safety communications center in the state of Louisiana to achieve accredited status through CALEA and the only fire communications center in the nation to receive CALEA accreditation.

Carter currently serves as President Elect of the Association of Public Safety Communications Officials, International (APCO) Board of Officers. APCO is the world's oldest and largest not-for-profit public safety communications association with a membership of over 15,000 members worldwide. He will be installed as President of this association at the close of the APCO annual training conference which will be held in Baltimore in August of 2007.

Dennis Childress

Author of Firefighter's Handbook chapters: Forcible Entry; Ventilation; Fire Suppression
Childress, an author, lecturer and fire services instructor, recently retired from the Orange County Fire Authority in Southern California, and has been in the fire service for just over thirty-nine years. He is a Certified Chief Officer with the State of California, and he holds an Associate of Arts degree in Fire Science and a Bachelor of Science degree in Fire Protection Administration. He holds a seat on the Board of Directors for the Southern California Fire Training Officers Association, and sits on the Statewide Training and Education Advisory Committee for the State Fire Marshal's Office. He is a principal member of the NFPA 1500 Fire Service Occupational Safety and Health Committee, and the NFPA 1561 Standard for Emergency Services Incident Management System Committee. He has authored a number of articles in fire service publications over the years, and he has also been an instructor in Fire Command and Management in the California State Fire Training System for over twenty years. He has also been a major part of the development of the Fire Officer Command curriculum for the State of CA over the last fifteen years.

Ronny J. Coleman

Author of Firefighter's Handbook chapter: Overview of the History, Tradition and Development of the American Fire Service

Chief Coleman is a nationally and internationally recognized member of the fire service who formerly served as the Chief Deputy Director, Department of Forestry and Fire Protection, and as California State Fire Marshal. He has served in the fire service for thirty-eight years. Previously he was Fire Chief for the Cities of Fullerton and San Clemente, California, and was the Operations Chief for the Costa Mesa Fire Department. Chief Coleman possesses a Master of Arts Degree in Vocational Education from Cal State Long Beach, a Bachelor of Science Degree in Political Science from Cal State Fullerton, and an Associate of Arts Degree in Fire Science from Rancho Santiago College. He has served in many elected positions in professional organizations, including President, International Association of Fire Chiefs; Vice President, International Committee for Prevention and Control of Fire; and President, California League of Cities, Fire Chiefs Department.

David W. Dodson

Author of Firefighter's Handbook chapters: Fire Behavior; Firefighter Safety; Building Construction; Firefighter Survival

Dodson is the owner of and lead instructor for Response Solutions, LLC in Eastlake, Colorado. He is an independent contract instructor for fire departments and fire service trade conferences nationwide, and is a regular speaker at the annual Fire Department Instructor's Conference (FDIC) and Firehouse World Expo. He has many years in the fire service and specializes in firefighter safety issues.

He started his fire service career with the U.S. Air Force. He served at Elmendorf AFB in Alaska and spent two years teaching at the USAF Fire School. After the USAF, Dodson spent almost seven years as a Fire Officer and Training/Safety Officer for the Parker Fire District in Parker, Colorado. He became the first Career Training Officer for Loveland Fire and Rescue in Colorado and rose through the ranks, including time as a HAZMAT Technician, Duty Safety Officer, and Emergency Manager for the city. He accepted a Shift Battalion Chief position for the Eagle River Fire District in Colorado before starting his current company, Response Solutions, which is dedicated to teaching firefighter safety and practical incident handling. Chief Dodson has served on numerous national boards including the NFPA Firefighter Occupational Safety Technician Committee and the International Society of Fire Service Instructors (ISFSI). He also served as president of the Fire Department Safety Of-

ficers' Association. In 1997, Dodson was awarded the ISFSI "George D. Post Fire Instructor of the Year."

Robert F. Hancock

Founding Author of Firefighter's Handbook chapters: Ropes and Knots; Rescue Procedures

Hancock is Assistant Chief/Administration with Hillsborough County Fire Rescue in Tampa, Florida, a department with 615 career personnel and 205 volunteers. He was hired in November 1974 as a firefighter and was promoted through the ranks to his present position in October 1993. He was awarded an Associate of Science Degree in Fire Science, with honors, from Hillsborough Community College. He graduated from the Executive Fire Officers Program at the National Fire Academy and has been certified as an instructor with the State of Florida since 1983. Hancock is chairman of the Florida Fire Chiefs' Disaster Response Communications Sub-Committee, charged with identifying short- and long-term solutions to the disaster response communication issue statewide.

T. R. (Ric) Koonce, III

Founding Author of Firefighter's Handbook chapters: Portable Fire Extinguishers; Water Supply; Fire Hose and Appliances; Nozzles, Fire Streams, and Foam; Protective Systems

Koonce is an Assistant Professor and Program Head of Fire Science Technology at J. Sergeant Reynolds Community College in Richmond, Virginia. He is a retired Battalion Chief with the Prince George's County (Maryland) Fire Department and has thirty-five years of fire service experience. He is an adjunct instructor for the Virginia Department of Fire Programs. He holds two associate degrees, a Bachelor of Science degree in Fire Service Management from University College of the University of Maryland, and a Certificate of Public Management from Virginia Commonwealth University.

Frank J. Miale

Founding Author of Firefighter's Handbook chapters: Fire Behavior; Ladders; Ventilation

Miale is a Battalion Chief (ret.) with over thirty years in the FDNY. A twenty-five-year active member in his local Volunteer Lake Carmel Fire Department, he maintains a busy role as treasurer and training instructor. A former high school teacher, he holds two Bachelor of Science degrees with several concentrations in Education, Biology, and Fire Administration. During his career in the FDNY, he taught at the NYC fire academy, participated in the introduction of a communication system using apparatus-mounted computers, and headed a special Emergency Command Unit while an active line officer. Formerly the Training Officer for the 27th Battalion in the FDNY, he taught many ladder company and ventilation courses throughout the country. His career was spent primarily in busy ladder companies in Brooklyn, Harlem, and the South Bronx sections of New York City prior to promotion to Chief Officer. He is the recipient of nine awards for courage and valor, including two department medals from the FDNY, and has been published many times in WNYF, Fire Command, and Fire Service Today.

Geoff Miller

Founding Author of Firefighter's Handbook chapter: Salvage, Overhaul, and Fire Cause Determination
Miller is a thirty-three-year veteran of the fire service and is currently the Deputy Chief of Operations with the Sacramento Metropolitan Fire District in California. Previous assignments have included seven years as an Assistant Chief, six years as a Battalion Chief, four years as the district's Training Officer, ten years as a line Captain, and two years as an Inspector. He is assigned to a CAL FIRE Type 1 Incident Command Team and FEMA Incident Support Team as a Plans Chief. He is also a task Force Leader and Plans Manager for USAR CA TF 7 Sacramento. Through the FEMA assignment he responded to the Pentagon and World Trade Center disasters along with numerous hurricanes and National Security Special Events. He has been involved in several California Fire Fighter I and II curriculum development workshops as well as participating in the rewrite of Fire Command 1A and 1B.

Robert Morris

Founding Author of Firefighter's Handbook chapter: Forcible Entry
Morris is a veteran of the New York City Fire Department and has been assigned to some of the busiest fire companies in New York City, including Ladder Company 42, Engine 60 in the Bronx, and Rescue Company 3 in Manhattan. After serving in the Bronx and Harlem, he served as Company Commander of Ladder Company 28. Captain Morris is currently Company Commander of Rescue Company 1 in Manhattan. Captain Morris is the recipient of seventeen meritorious awards, including three department medals.

Jeff Pindelski

Author of Firefighter's Handbook chapters: Fire Department Organization, Command, and Control; Communications and Alarms; Personal Protective Clothing, Equipment, and Ensembles; Self-Contained Breathing Apparatus; Rescue Procedures

Jeffrey Pindelski is an eighteen-year-plus student of the fire service. Jeff is a Battalion Chief with the Downers Grove Fire Department in Illinois. He previously served for twelve years as a Firefighter and Lieutenant on the Truck and Heavy Rescue Company. In addition to his background in a career position, he has also served on departments in a volunteer and part-time capacity.

Jeff is a staff instructor at the College of Du Page and also instructs courses at the Romeoville Fire Academy. He is a Certified Fire Officer III and Instructor III while also being certified as a Fire Suppression Incident Safety Officer. Chief Pindelski holds a Masters Degree in Public Safety Administration from Lewis University, a Graduate Certificate in Managerial Leadership and a Bachelors Degree from Western Illinois University.

He has been involved with the design of several training programs dedicated to firefighter safety and survival including R.I.C.O. (Rapid Intervention Company Operations) which is a forty-hour Rapid Intervention training program held on a national level. Jeff has served on review committees for Delmar and is coauthor of *Rapid Intervention Company Operations*. Chief Pindelski was a recipient of the State of Illinois Firefighting Medal of Valor in 1998 and has been published regularly in several trade journals on various fire service related topics.

Marty Rutledge

Author of Firefighter's Handbook chapters: Portable Fire Extinguishers; Water Supply; Fire Hose and Appliances; Nozzles, Fire Streams, and Foam; Protective Systems; Ladders;

Ropes and Knots; Salvage, Overhaul, and Fire Cause Determination; Prevention, Public Education, and Pre-Incident Planning

Rutledge resides in Loveland, Colorado. He is a member of the Fire Certification and Advisory Board to the Colorado Division of Fire Safety, as well as serving as the State First Responder program coordinator. He is also a member of the Colorado State Fire Fighter's Association and has over seventeen years of fire and emergency services experience in both volunteer and career ranks, serving as firefighter, engineer, lieutenant, and EMS program quality assurance manager. He is a member of Wind and Fire Motorcycle Club, and has ridden motorcycles with firefighters in different countries around the world. Rutledge has authored and served as technical expert for a supplementary firefighter training package for *Firefighter's Handbook* and has co-authored Delmar's *First Responder Handbook: Fire Service Edition, First Responder Handbook: Law Enforcement Edition,* as well as *Exam Preparation: Firefighter I and II,* and *Exam Preparation: Hazardous Materials Awareness Operations.*

Donald C. Tully

Founding Author of Firefighter's Handbook chapter: Prevention, Public Education, and Pre-Incident Planning
Tully is a member of the Orange County, California, Fire Authority. With over thirty years in the fire service, he has also been a Division Chief/Fire Marshal in Buena Park and Westminster, California, for ten years, and a Fire Technology Instructor at Santa Ana College, California. He is Past President of the Orange County Fire Prevention Officers' Association and was a member of IFSTA's Fire Investigation Committee. He also served as a member of NFPA Committees 1221 (CAD Dispatch and Public Communications) and 72 (Fire Alarms), and as a member of the California State Fire Marshal Committees on Fire Sprinklers and Residential Care Facilities (ad hoc committees). He is a California State Certified Chief Officer, Fire Officer, Fire Investigator, Fire Prevention Officer, and Fire Service Instructor and Technical Rescue Specialist.

Thomas J. Wutz

Founding Author of Firefighter's Handbook chapters: Fire Department Organization, Command, and Control; Self-Contained Breathing Apparatus

As a Chief of the Fire Services Bureau, New York State Office of Fire Prevention and Control, Wutz manages the Fire Services Bureau's twenty full-time staff and 300 part-time instructors as well as develops and delivers fire training outreach programs, training policies and procedures serving 1,850 fire departments (career and volunteer), with 24,000 students annually. In addition, he provides program management for the New York State Fire Mobilization and Mutual Aid Plan. Chief Wutz is also certified as a Type II Incident Commander for the New York State Incident Management Assistance Team, most recently serving as deputy IC for the State IMAT assistance to Jackson County, Mississippi, September 2005.

Chief Wutz retired as base fire chief for the 109th Military Airlift Group, United States Air Force/New York Air National Guard (twenty-eight years) with thirty-one years total military service and has served as a volunteer firefighter in various departments in New York state for thirty-five years.

Author's Corner

Delmar is proud to have our *Firefighter's Handbook* series authors publish other titles with us. If you are interested in learning more about topics in this book, or wish to read other works by these authors, we encourage you to check out the following titles. A comprehensive list of all our fire titles is included at the back of the book.

Jeff Pindelski and Mike Mason

Rapid Intervention Company Operations This book is a one stop reference for learning the procedures and techniques required for firefighters to rescue one of their own. Order#: 978-1-4018-9503-7

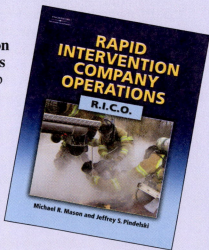

David Dodson

Fire Department Incident Safety Officer, 2ⁿᵈ ed.
This book primes aspiring and current Safety Officers to aggressively pursue the operation of a highly efficient safety program in the department. Based on the 2007 Edition of NFPA Standard 1521, this book focuses on practices and procedures to help reduce firefighter injuries and deaths.
Order#: 978-1-4180-0942-7

Chris Hawley

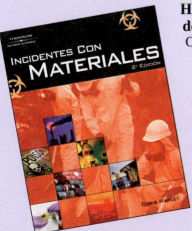

Hazardous Materials Incidents, 3ʳᵈ ed.
This book provides practical knowledge of how to effectively and safely respond to hazardous materials incidents. Based on the 2008 Edition of NFPA Standard 472, this book keeps emergency responders up to date with national concerns and emerging technologies.
Order#: 978-1-4283-1796-3

Hazardous Materials Incidents: Spanish Edition
Order#: 978-1-4180-1156-7

Hazardous Materials Air Monitoring and Detection Devices, 2ⁿᵈ ed.
This book explains technical information in an easy to understand manner and provides hazardous materials teams with a thorough guide to effective air monitoring in emergency response situations. Updated to the proposed 2008 Edition of NFPA 472, this book includes discussion of the latest technology in this field.
Order#: 978-1-4180-3831-1

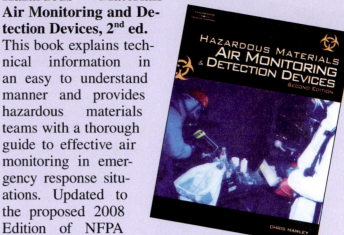

Ron Coleman

Going for Gold
Discover what it takes to be a successful fire chief. This book helps readers understand how to prepare for the job, make the transition to this new position, and what is required for a successful track record.
Order#: 978-0-7668-0868-3

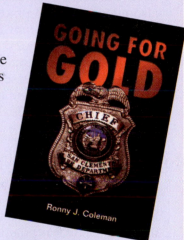

Andrea Walter, Marty Rutledge, and Chris Hawley

Exam Preparation: Firefighter I and II
Order#: 978-1-4018-9923-3

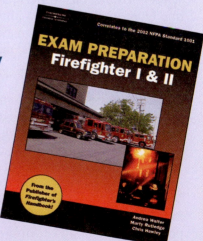

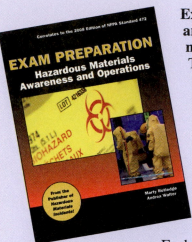

Exam Preparation: Hazardous Materials Awareness and Operations

These books prepare candidates for their certification exams by providing the necessary practice and review critical for success. Including a chapter on test-taking strategies, each book contains 900 test questions, plus a 200-question final exam. Each section of questions increases in difficulty to ensure mastery of the content, and each question provides a correlation to the accompanying Delmar book and NFPA Standard, as well as insightful rationale. An interactive back-of-book CD-ROM containing the questions in the book also allows users to practice computer-based testing as well as offers a *bonus* 200-question final exam! Order#: 978-1-4180-0962-5

Other Exam Preparation titles and online versions are also available! Please refer to the Additional Fire Titles from Delmar section in the back of this book for more information.

ABOUT THIS BOOK

In our tradition of offering quality firefighter training manuals, Delmar is proud to introduce this new offering to the third edition of the *Firefighter's Handbook* series—*Hazardous Materials Handbook: Awareness and Operations Levels*. Within the pages of this book you will find all the information you need to successfully complete hazardous materials awareness and operations training courses which prepare students for certification testing.

Hazardous Materials Handbook is a comprehensive guide to the basic principles and fundamental concepts involved in hazardous materials response, as defined by the 2008 edition of the National Fire Protection Association 472 *Standard for Competence of Responders to Hazardous Materials/Weapons of Mass Destruction Incidents*. With a practical, straightforward approach and step-by-step sequences to explain knowledge and skill criteria, this book can be used by both new and experienced hazardous materials responders to enable them to perform a wide variety of firefighting and emergency service activities.

Drawn from *Firefighter's Handbook: Firefighting and Emergency Response*, Third Edition, this hazardous materials content strives to maintain the quality of the previous editions, while focusing on keeping you and your team safe, explaining new initiatives, and highlighting new technology. This book is intended not only to train firefighters, but also to ensure optimal proficiency on the job.

As a firefighter you will make a difference in the lives of many. Use your knowledge, practice your skills, and above all—be safe.

Organization of This Book

The new offering for the third edition consists of seven chapters, covering hazardous materials and terrorism awareness. The chapters are set up to deliver a straightforward, systematic approach to training, and each includes an outline, objectives, introduction, lessons learned, key terms, review questions, and a list of additional resources.

Also included at the front of the book is an NFPA 472 Correlation Guide that correlates the requirements outlined in the Standard to the content in *Hazaradous Materials Handbook* by chapter and page references. These resources can be used as a quick reference and study guide.

Features of This Book

Hazardous Materials Handbook contains a number of features that make it unique. It offers a realistic approach to emergency response—it is comprehensive in coverage, including essential information, but presents the content in a clear and concise manner, so it is easy to read, follow, and understand. We also recognize that it is essential in this field to not only acquire knowledge, but to put it into practice as well.

The following recommends how you can best utilize the features in this book to gain competence and confidence in learning the essentials of hazardous materials response.

NFPA Standard 472 Correlation Guide

This grid provides a correlation between *Hazardous Materials Handbook* and the requirements for the awareness and operations levels of NFPA Standard 472, 2008 Edition. Each performance requirement is linked to specific book chapters and page numbers for easy reference.

Street Stories

Each chapter opens with a personal experience written by notable contributors from across the nation. These personal accounts engage you in the events and help to remind you of the importance of practicing the knowledge and skills presented in the chapter.

It was the start of what we call a normal day on the job, until later in the morning when an alarm for a chemical leak inside a beverage warehouse was sounded. The dispatch consisted of what we call a hazardous materials box: four engine companies, one truck company, a rescue squad, basic life support unit, hazardous materials company, and command officer. I was working at the hazardous materials company that day. While en route, a radio transmission by the first arriving company advised the Emergency Communications Center of a major ammonia leak inside the warehouse storage area and that they were taking protective actions. This area contained multiple storage of beverages and boxes within an enclosed and secured area that also contained valves and piping for the anhydrous ammonia refrigeration system.

Upon our arrival, the warehouse had been evacuated and a strong odor of ammonia had already consumed the entire area surrounding the warehouse. Once we had performed a hazard risk assessment and ensured that the first responders had taken appropriate protective actions, we then selected our level of protection. Three hazardous materials technicians and I entered the release area to shut off the valve to the leaking pipe. After locating the release area we located the valve and made an attempt to close the valve. While closing the valve a sudden release of gaseous and liquid ammonia covered the personnel working at and around the valve. Visibility was taken from us almost instantaneously because of the gaseous release and communications were lost between all four technicians. I was able to find my way out and noticed that my personnel were still inside the release area. Prior to making another entry to locate my personnel, I noticed a white smoke coming from my chemical boots.

After further investigation, I realized that the oil-based paint from the concrete floor was causing a chemical reaction under the soles of my boots. I reentered the release area, located my personnel, and immediately withdrew from the release area to the decontamination area. Once we were refreshed, a second entry attempt was made into the release area, where we were able to locate another sectional valve and stop the leak. The hardest part of the second attempt was removing our SCBA inside our suits to squeeze past piping and valves to get to the right one.

Prior to leaving the scene we finally determined that prior to our arrival a firefighter had entered the release area and closed the valve without notifying command and/or hazardous materials personnel. When hazardous materials personnel entered the release area thinking that the valve was not closed, they actually reopened it, which caused the valve to freeze in the open position. In this event, a number of factors affected our response. First-arriving crews needed to address isolation and evacuation issues, and the type of release, but one firefighter was endangered by not taking appropriate protective actions. No personnel were injured or exposed to the ammonia, but the incident proved to be very dangerous as a result of personnel freelancing and the lack of training present at an emergency scene.

—*Street Story by Gregory L. Socks, Captain, Montgomery County, MD, HAZMAT Team*

137

Job Performance Requirements

Step-by-step photo sequences illustrating important procedures are integrated throughout the chapters. These are intended to be used as a guide in mastering the job performance skills and to serve as important review.

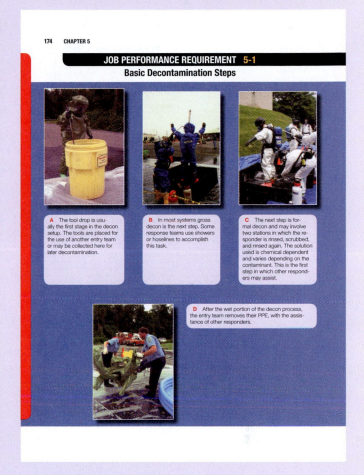

174   CHAPTER 5

JOB PERFORMANCE REQUIREMENT 5-1
Basic Decontamination Steps

A The tool drop is usually the first stage in the decon setup. The tools are placed for the use of another entry team or may be collected here for later decontamination.

B In most systems gross decon is the next step. Some response teams use showers or hoselines to accomplish this task.

C The next step is formal decon and may involve two stations in which the responder is rinsed, scrubbed, and rinsed again. The solution used is chemical dependent and varies depending on the contaminant. This is the first step in which other responders may assist.

D After the wet portion of the decon process, the entry team removes their PPE, with the assistance of other responders.

Articles

These featured articles are set apart from the main flow of the text and include further information on important points. Offering interesting facts, practical advice, and highlighting key initiatives, these articles are a must-read.

NEW! ViewPoints

New for this edition, these reports recognize variances in firefighter practices across the United States, providing a more informed view of the world of firefighting.

SAMPLING AND EVIDENCE COLLECTION

One of the major concerns of local law enforcement and the FBI regarding the prosecution of a terrorism case is the purity of the evidence. In order for the evidence to be used in a successful prosecution, the evidence must be pure beyond all doubts. If there is any suspicion that the evidence has been tampered with, altered, or contaminated, it could affect the outcome of the prosecution. When responders enter the hazard area, there is great potential for evidence to be destroyed. The best way to proceed is to coordinate your efforts with your local FBI WMD coordinator, who can assist you with the preservation of evidence. Evidence collection is very process driven, but procedures can be implemented that will help preserve any potential evidence. The gold standard for

any courtroom is laboratory analysis, so evidence must be available to be analyzed by a laboratory. This doesn't mean that every incident requires that material be sent off to the laboratory, but steps must be taken to preserve a portion of the evidence in the event that laboratory testing is required.

When there is a large amount of potential evidence, it can be overwhelming, but following a process helps eliminate error and preserve evidence. There is great potential to become involved in a court case, and following the proper procedures every time minimizes your risk in the courtroom. The one time that you don't follow evidentiary procedures will be the time that you are grilled in the courtroom about contaminating the evidence. The FBI and local law enforcement have response teams that collect evidence in these situations, and they can provide great assistance when trying to process a potential crime scene.

A cooperative effort is needed to combat a terrorist attack. The primary functions are rescue/life safety by fire and EMS personnel, hazard identification by the hazardous materials team, identification of possible secondary devices by a bomb technician, and incident management. It is important to communicate the hazards to all personnel, and to limit the response to essential personnel. Instead of having the whole alarm assignment report to the front of the building, it is preferable to use one or two companies to investigate while staging the other companies away from dumpsters, mailboxes, or dead-end streets. A secondary device can be hidden almost anywhere, but the key is to look for something out of the ordinary. When dealing with victims it is essential to isolate them until the cause is identified. The victims can have a large amount of information and should be questioned quickly. Questions to ask include these: What did you see, hear, or smell? Was this coming from one area or was it throughout the building? Did you see anything else suspicious? What type of signs and symptoms do you have? In addition, the police will need to conduct interviews. Documentation and preservation of the evidence are essential to the successful prosecution of the terrorist.

GENERAL GROUPINGS OF WARFARE AGENTS

Terrorists could use any of a number of possible warfare agents. They are classified into three broad areas. Weapons of mass destruction are commonly used

by the military. Some of the regulations that prohibit the making, storing, or using of terrorism agents are called WMD laws or regulations. Any item that has the potential to cause significant harm or damage to a community or a large group of people is considered a WMD. The other two classifications are nuclear, biological, and chemical (NBC) and chemical, biological, radiological, nuclear, and explosive (CBRNE). Both of these are descriptions of the types of materials that could be used in a terrorism attack. Although there are slight variations, they are all used to describe the various types of agents that a terrorist could use. Most of the language differences come from funding legislation or a specific federal agency.

The military has devised a naming system for many of these agents, many of which are listed in **Table 7-1.** Responders should become familiar with these names because much of the literature and help guides refer to these agents by these names. For instance, when using a military detection device, the military name is used. When dealing with terrorism, firefighters are entering another world that has its own language. The fire service has to adopt this new language to survive in this new world. The three groupings mentioned earlier are further subdivided into the categories discussed in the following subsections.

Nerve Agents

Nerve agents are related to organophosphorus pesticides and include tabun, sarin, soman, and V agent. They were designed for one purpose and that is to kill people. Although very toxic, their ability to kill large

---

FIGURE 2-67 This is an intermodal tank, commonly referred to as an IM or IMO. This is a bulk tank that carries an average of 3,000 to 5,000 gallons. This tank is an IMO-101, which is an atmospheric tank. The orange panel indicates that it has a United Nations (UN) hazard code of 60 and a UN number of 2572. The code of 60 indicates that the product is a toxic material. The UN number indicates that the product is phenylhydrazine. The name is also stenciled on the sides of the container, and there are toxic placards.

FIGURE 2-68 This is another IMO-101, and the photo shows that the orange panel indicates a hazard code of 66 and a UN number of 2810. The hazard code indicates that the product is a highly toxic material, and there are fifty-two different materials listed for UN 2810. One would have to look at the shipping papers to figure out what is being carried in this tank. The tank is marked "foodstuff only," which means that it is holding a product used in food. It is interesting to note that many of the chemicals listed for UN 2810 are chemical warfare agents such as sarin nerve agent and comparable materials. The most likely candidate, considering the "foodstuff only" label, would be a medical-type product.

specification numbers, IM-101, IM-102, Spec 51 (specification 51). When used internationally they are called IMOs. They are made to be dropped off at a facility and when empty picked up for refilling.

VIEWPOINT

The righting of overturned tank trucks can be controversial at times. The general rule is that anytime a truck is overturned, it must be off-loaded prior to the truck being righted. There is substantial risk of tank failure if a loaded truck is righted by a tow truck. Some tow companies use air bags and have been trained to right fully loaded tank trucks, but it is not advisable to allow them to do so. The towing company may be aggressive in wanting to right the truck, but the incident commander is in charge, and if the tank were to fail and the product entered a water way, the incident commander could be held responsible for violations of the Clean Water Act.

A tank truck is designed to function with its wheels down and when it is wheels up, the strength of the tank is in question. The stresses on a tank during a rollover cannot be seen and when rolled over there are more stresses placed on the tank. Gasoline tankers (MC-306/DOT-406) are especially susceptible to failure during a righting operation. The principal responsible party is required by law to provide response experts, either from within their company or an outside contractor, to mitigate this type of incident. In some cases the local public sector hazardous materials team would remain in command and monitor the scene until the scene is no longer an emergency. In some parts of the country, the local hazardous material team is aggressive in this mitigation and performs the offload. No matter who handles the incident the product needs to be offloaded from the truck.

There are several methods of removing the product from the tank, some easier than others. One method is to remove the product through existing valves and piping, which can sometimes be awkward when the truck is upside down. Another method is to attach a valve assembly to the existing valves and then offload the product. Comparable to this method are flexible bladders that attach to the valves and are connected to piping. Another method is to drill the tank to allow for the offload of the product. Each of the methods has benefits and pitfalls and may not apply in every case. First responders can speed up this process by calling for another compatible tank truck to transfer product into and calling for a vacuum truck or pump to conduct the transfer. These requests should be made in consultation with the responsible party and the hazardous materials team. In some states, box trailers that have any quantity of chemicals, including household commodities, are required to offload prior to righting, as shown in **Figure 2-69.**

Street Smart Tips

As is true in any profession, sometimes experience can be the best teacher. These tips are power-packed with a wide variety of hints and strategies that will help you be a street-smart firefighter.

Firefighter Facts

These boxes offer a detailed snapshot of facts based on firefighting history, experience, and recorded date to provide essential background information.

---

128 CHAPTER 4

STREETSMART TIP

When wearing a Level A ensemble prehydration is highly recommended, and dressing should take place in a cool quiet area, preferably in the shade.

During emergency operations all members must be monitored. Lack of full visibility and heat stress are major concerns when using a Level A ensemble. The Level A ensemble typically consists of the following components:

- Encapsulated suit with attached gastight gloves and boots
- Inner and outer gloves
- Hard hat
- Communication system
- Cooling system
- SCBA
- PBI/Nomex coveralls
- Overboots

To be an NFPA 1991 (Encapsulated Suit Specifications) certified Level A suit, the suit must have some flash fire resistance. To assist in compliance, newer suits use a blended fabric that offers chemical resistance as well as the flash resistance. Older style suits

use a flash suit overgarment that is made of aluminized PBI/Kevlar fabric. This flash protection is not intended for firefighting, but it offers three to thirteen seconds of protection when involved in a flash fire.

Level B Ensembles

Within the Level B ensemble family, there is a lot of variety in suit types, as shown in Figures 4-10A–C. Although the EPA and OSHA acknowledge two basic types, a large number of styles are available. The two basic types are coverall style and encapsulated, but even these have subvarieties. The encapsulated style of Level B suits is similar to the Level A style, but does not have attached gloves, nor is it vapor-tight. It typically has attached booties, and the SCBA is worn on the inside. Some manufacturers provide glove ring assemblies that allow for the gloves to be preattached during storage. A variety of fabrics are available for the Level B ensemble style of suits, and compatibility is important. The Level B encapsulated suit is the workhorse of hazardous materials teams; it is the most common suit used.

The encapsulated Level B suit is sometimes referred to as a Bubble B or a B plus suit. These suits have some of the same heat stress issues associated with them as do the Level A ensembles, even though

FIGURE 4-10 The photos here all represent Level B suits: (A) a coverall style for law-enforcement officers; (B) an encapsulated style, which is not gastight; and (C) a military-designed two-piece garment worn by tactical officers. The respiratory protection is a rebreather style, which provides a four-hour air supply.

---

46 CHAPTER 2

FIGURE 2-59 MC-331 tanks carry liquefied compressed gases such as propane and ammonia. They are made of steel and are designed to carry a variety of products.

to have is 80 percent to allow for expansion when heated. The liquid in the tanks is at atmospheric temperature but on release can go below 0°F (–18°C) and could cause frostbite upon contact. The pressure in these tanks is of concern when responding to incidents involving these tanks.

FIREFIGHTER FACT

When a propane tank is emptied and, hence, the pressure reduced, the temperature of the propane drops below 0°F (–18°C). Temperature, pressure, and volume are interrelated. Think of an SCBA bottle. When it gets filled it becomes hot because the pressure and volume are increasing. When the SCBA bottle is used, it becomes cold because it is losing pressure and volume. Any time one of the parameters is changed, there is a corresponding change in the other properties. When a pressurized gas is pressurized to a point that it becomes a liquid, as is the case with propane, it allows for a lot of propane to be stored in a small container. When released, however, the temperature will drop because the pressure is decreasing in the container. This is known as autorefrigeration.

SAFETY

If the pressure increases at a rate higher than the relief valve can handle, the tank will explode. These explosions have been known to send pieces of the tank up to a mile, with the ends of the tank typically traveling the farthest, although any part is subject to becoming a projectile.

Boiling Liquid Expanding Vapor Explosion (BLEVE)

When tanks, trucks, tank cars, or other containers are involved in a fire situation there are a number of hazards. One very deadly hazard is known as a boiling liquid expanding vapor explosion (BLEVE), Figure 2-60. A large number of firefighters have been killed by propane tank BLEVEs, and when a BLEVE occurs it usually results in more than one firefighter being killed at a single incident. The type of container and the product within the container will dictate how severe a BLEVE may be. The basis of a BLEVE is the fact that the pressure inside the container increases and exceeds the maximum pressure the container was designed to handle. The contents are violently released, and if the material is flammable, an explosion or large fireball occurs. In the recent past there have been several incidents involving BLEVEs that resulted in emergency responder deaths and injuries, thus emphasizing the need to recognize and prevent this event before it occurs.

Another phenomenon that transpires with containers is known as violent tank rupture (VTR), which occurs with nonflammable materials. The concept is the same as a BLEVE, but there is not a characteristic fireball and explosion. With both a BLEVE and a VTR there is some form of heat increase inside the container, typically from a nearby fire. The fire heats the container, which heats the contents. The contents will boil, which creates expanding vapors, which in turn increase the pressure inside the container. In some containers the relief valve will activate, relieving the pressure. In some cases the pressure inside the tank is greater than the relief valve can handle and the

Notes

This feature highlights and outlines important points for you to learn and understand. Based on key concepts, this content is an excellent source for review.

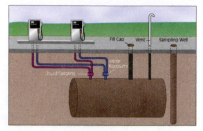

FIGURE 2-77 Piping system of a gas station.

amount in the piping aboveground and in the hose, if everything works properly.

The loading piping is located separately and all of the fill pipes are located in the same area for easy transfer from a tank truck. The fill pipes are usually color coded and marked so that the driver can differentiate between unleaded, unleaded super, and diesel fuel, although the color coding is not standard across the country and varies from company to company. At some location on the property there will be vent pipes for the tanks, generally away from the pumps and near the property line. There will also be other manhole covers approximately 6 to 8 inches (15.2–20 cm) in diameter that have a triangle on the top of the cover. These are inspection wells and typically surround the tank. The holes are drilled at various depths so that leaks can be easily detected by air monitoring of the well, or if water is in the well, by taking a sample. If a facility has had a leak, there may be a large number of these wells on and around the property. Most gas station tanks are 10,000 to 25,000 gallons (37,854–94,635 liters) in size and, if not properly monitored, can slowly release a substantial amount of product over a short period of time.

NOTE

It is not uncommon to find gasoline bubbling up in a basement miles away from the gas station and to find that the release occurred several years ago.

Another common problem arises when farms are redeveloped into housing developments. If unknown USTs are located on the property, they may or may not be discovered during the construction and can eventually leak, causing problems. The tank and environmental industry refers to a **leaking underground storage tank** as a LUST.

SAFETY

A recent incident in Baltimore County, Maryland, demonstrates that regardless of how many protection systems may be in place, a release off the property is still possible. One afternoon a gas station was getting a tank filled with fuel, as per normal procedure, and the tank filling went as expected with no problems. The gas station had installed a state-of-the-art alarm system that would indicate a fault in the system and alert the owners to any potential releases. The system goes through a system check after each sale, if no pump is operated for a period of thirty seconds. If a pump is on, or the period between sales is less than thirty seconds, then the system check does not begin to work. During a busy afternoon, it could be estimated that the system check would not function for a long period of time, because at least one of the pumps is operating all the time.

In addition to the system check of each of the storage tank locations, some piping and the pumps all have electronic monitors that detect the presence of a liquid. When a liquid is detected, an alarm activates—the same type of alarm as if the system check fails the system. One of the potential problems with many self-service gas stations is that there is only one attendant, who for many reasons cannot leave the work area to either check on a problem outside or, depending on the alarm panel location, cannot check any alarms. In some gas stations the alarms are in other rooms and cannot be seen or heard by the only attendant. Another problem is that the liquid alarms will activate when it rains, tripping for rainwater.

Safety and Cautions

As a firefighting professional, you will face situations in which you will need to react immediately in order to ensure your safety and the safety of others. These tips provide this necessary advice.

Lessons Learned

Lessons Learned summarizes the main points presented in the chapter and is ideal for review purposes.

Review Questions

At the end of every chapter, the review questions assist you with the learning process and help you to evaluate your knowledge and ensure mastery of the content.

Additional Resources

These lists of references are found at the end of each chapter and offer additional sources of information for future study of important topics.

NEW! For Further Review

Discover additional resources from Delmar that offer further practice of important concepts.

New To This Edition

The third edition content for *Hazardous Materials Handbook* contains many new updates and additional information to address the needs of the emergency responders of today:

New Initiatives

- Emphasis on the implementation of the National Incident Management System (NIMS) stresses the importance of teamwork and a structured response to incidents.
- New JPR photo sequences covering critical response skills, such as fighting fire for containers under pressure, implementing defensive operations, using air monitoring equipment . . . and more!
- Reports on the latest terrorist and hazardous materials crimes, and how to respond to a criminal/terrorist activity keep responders informed of potential threats to the community.
- Reflects new international placarding system and introduces and defines new D.O.T. terminology so you remain informed.
- Metric conversions accompany all U.S. Customary/English measurements for quick reference.

Keeping You Safe

- New practices to ensure firefighters remain safe on the fireground through proper use of personal protective equipment
- New information on radiation to prepare hazardous materials responders for this potential threat, including radiation detection, radioactive material containers, radiation hazards, and proper protection.

New Technology and Equipment

- Current information on air monitoring helps hazardous materials responders keep informed of new technology in this field.
- New photos reflect current apparatus, equipment, tools and procedures.

OUR CUSTOM OPTIONS

As we look toward the future, Delmar is proud to be the first to offer custom solutions to our firefighters. We recognize that the fire service is a complex and changing world, and it is our duty to provide you with options for training. We are committed to providing you with a package to meet your specific needs. Listed here are the four versions of *Firefighter's Handbook,* Third Edition, that we offer to the fire service:

Firefighter's Handbook: Firefighting and Emergency Response

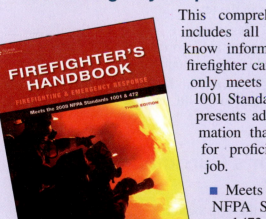

This comprehensive book includes all the need-to-know information for the firefighter candidate. It not only meets 100% of the 1001 Standard, but it also presents additional information that is essential for proficiency on the job.

- Meets 100% of 2008 NFPA Standard 1001 and 472
- Includes basic firefighter training, EMS and hazardous materials operations
- For those who combine training on all Firefighter I and II topics in preparing candidates for certification

Order#: 978-1-4180-7320-6

Firefighter's Handbook: Essentials of Firefighting

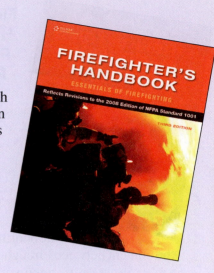

For those who teach hazardous materials in a separate course, this book provides all the need-to-know information for firefighting, exclusive of the hazardous materials content in the third edition of *Firefighter's Handbook*.

- Updated to the 2008 NFPA 1001 Standard
- Includes the same chapters and content as the original third edition, without the hazardous materials response chapters
- For those who conduct hazardous materials training separate from firefighter training

Order#: 978-1-4180-7324-4

Hazardous Materials Handbook: Awareness and Operations Levels

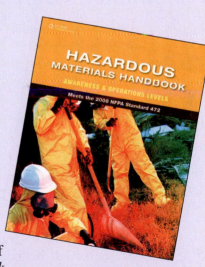

A handy resource for hazardous materials training, this book provides all the information included in the hazardous materials chapters of the third edition of *Firefighter's Handbook*.

- Meets the 2008 NFPA Standard 472
- Includes exclusively the hazardous materials response content
- For those who conduct hazardous materials training separately

Order#: 978-1-4283-1971-4

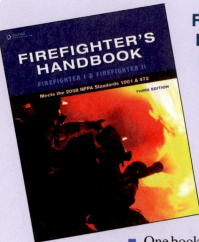

Firefighter's Handbook: Firefighter I and Firefighter II

Firefighter I and II levels are divided into separate sections in this brand-new offering, providing convenience to those who teach these two levels separately.

- One book that actually divides out the content between Firefighter levels I and II
- Meets 100% of the 2008 NFPA Standard 1001 and 472
- Includes basic firefighter training, EMS and hazardous materials response

Order#: 978-1-4283-3982-8

OUR CURRICULUM PACKAGE
Instructor's Curriculum Kit

The Instructor's Curriculum Kit is designed to allow instructors to run programs according to the standards set by the authority having jurisdiction where the course is conducted. It contains the information necessary to conduct Firefighter I, Firefighter II, hazardous materials awareness, and hazardous materials operations courses. It is divided into sections to facilitate its use for training:

- **Administration** provides the instructor with an overview of the various courses, student and instructor materials, and practical advice on how to set up courses and run skill sessions.
- **Modularized Lesson Plans** are ideal for instructors, whether they are teaching at fire departments, academies, or longer-format courses. Each section presents learning objectives, recommended time allotment, equipment and reading assignments for each lesson and outlines key concepts presented in

each chapter of the text with coordinating Power-Point slides, textbook readings, and Job Performance Requirement and Supplement Skill sheets.

- **Equipment Checklist** offers a quick guide for ensuring the necessary equipment is available for hands-on training.
- **Job Performance Requirement and Supplemental Skills Sheets** outline important skills that each candidate must master to meet requirements for certification and provide the instructor with a handy checklist. A Job Performance Requirement Correlation Guide cross-references the Standard with the Job Performance Requirements outlined in *Firefighter's Handbook*. These guides can be used for quick reference when reviewing important skills and for studying for the Firefighter I and II certification exam.

  The Skill Sheets are also available separately as a soft cover book.

- **Progress Log Sheets** provide a system to track the progress of individual candidates as they complete the required skills.
- **Quick Reference Guides** contain valuable information for instructors. Two grids are included: NFPA Standard 1001 and 472 Correlation Guides used to cross-reference *Firefighter's Handbook* with these standards; New Edition Correlation Guide used to cross-reference the revisions between the second and third editions of *Firefighter's Handbook*.
- **Answers to Review Questions** are included for each chapter in the text.
- **Additional Resources** offer supplemental resources for important information on various topics presented in the textbook.

The Instructor's Curriculum Kit also includes the Instructor's Curriculum CD-ROM.
Order#: 978-14180-7321-3

Instructor's Curriculum CD-ROM

Available in the Instructor's Curriculum Kit and as a separate item, the Instructor's Curriculum CD-ROM ensures a complete, electronic teaching solution for the third edition. Designed as an integrated package, it includes the following:

- Complete curriculum available in Word format allows instructors to add their own notes or revise to meet the requirements of the Authority Having Jurisdiction.
- PowerPoint presentations outline key concepts from each chapter, and contain graphics and photos from the text, as well as video clips, to bring the content to life.
- A Test Bank containing hundreds of questions helps instructors prepare candidates to take the written portion of the certification exam for Firefighter I and II.
- A searchable Image Library containing hundreds of graphics and photos from the text offers an additional resource for instructors to enhance their own classroom presentations or to modify the PowerPoint provided on the CD-ROM.

Order#: 978-1-4180-7323-7

Instructor's Curriculum Kit: Firefighter I and Firefighter II

Another option for instructors, this curriculum kit contains all the same information in the third edition curriculum kit, with the Lesson Plans, PowerPoint presentations, Skills Sheets, Test Bank, and Progress Logs conveniently separated out into Firefighter I and II levels. Ideal for those using *Firefighter's Handbook: Firefighter I and Firefighter II*, this curriculum kit also includes an accompanying curriculum CD-ROM.
Order#: 978-1-4180-7325-1

Instructor's Curriculum CD-ROM: Firefighter I and Firefighter II

Order#: 978-14180-7327-5

NEW! Skill Sheets

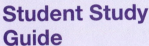

Drawn directly from the curriculum, the Job Performance Requirement and Supplement Skills sheets are also available separately in this soft cover book. Handy for tracking individual candidate progress, this book is three hole punched and perforated for ease of use on the training ground.
Order#: 978-1-4180-7326-8

Student Study Guide

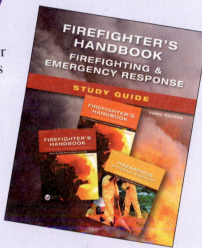

Thoroughly revised for the third edition, this is helpful in the classroom setting as a guide for study and a tool for assessing progress. The study guide consists of questions in multiple formats, including new and revised questions to support the new edition of the book.
Order#: 978-1-4180-7322-0

New! Firefighter Skills DVD Series

Covering all the Skills outlined in the Instructor's Curriculum, this four part series combines live action video, graphics, and animations to illustrate important step-by-step skills essential to the job of a firefighter. Each clip walks the viewer through the steps of the skill, while providing important review information and safety tips.

- Firefighter Skills DVD Series
  Order#: 978-1-4283-1085-8

- Hazardous Materials Awareness and Operations
  Order#: 978-1-4283-1089-6

ACKNOWLEDGMENTS

Hazardous Materials Handbook: Awareness and Operations Levels remains true to the Delmar tradition of remaining dedicated to the individuals we serve—among them, both aspiring and experienced firefighters. However, we would not be able to accomplish this without the contributions of many professionals whose passion, commitment, and hard work have helped shape a book of which we all can be proud.

We wish to thank those that contributed to the previous editions of this book. These founding reviewers and advisory board members offered valuable insight—they provided guidance to the content since inception and were our sounding boards in development. For their assistance, we extend our deepest gratitude.

In addition, we recognize the contributors to the third edition of this book:

Our Focus Group

A group of experts from across the United States gathered in a meeting focused on this *Firefighter's Handbook* series. Our sincere thanks go to those who traveled to meet with us and to provide us with thoughts, ideas, and recommendations for the new edition:

James Dalton, instructor, Chicago Fire Academy, Chicago, Illinois

Gary Fulton, Curriculum Specialist, Office of the State Fire Commissioner, Lewistown, Pennsylvania

Aaron Heller, Captain, Hamilton Township NJ Fire District #9, Township of Hamilton, New Jersey

Jason Lloyd, Instructor/Industrial Fire, Texas Engineering Extension Service (TEEX), College Station, Texas

Gary McCarty, Battalion Chief, Fire Prevention/Community Relations, Salt Lake City Fire Department, Salt Lake City, Utah

Pat McAuliff, Director of Fire Science and EMS, Colin County Community College, McKinney, Texas

Frank Miratsky, Battalion Chief of Training, Omaha Fire & Rescue, Omaha, Nebraska

Howard Sykes, Assistant Chief, Lebanon Volunteer Fire Department/Durham Technical Community College, Durham, North Carolina

Andrea Walter, lead author, Firefighter/Technician, Metropolitan Washington Airports Authority, Dulles Airport Fire Department, Washington D.C.

Our Reviewers

Our respect and appreciation to those who invested time in the review of the manuscript to provide us with insight and advice for the content in *Hazardous Materials Handbook:*

Charlie Brush, Standards Supervisor, State Fire Marshal, Bureau of Fire Standards and Training, Ocala, Florida

Andrew Byrnes, Hazardous Materials Program Coordinator, Institute of Emergency Services and Homeland Security, Utah Fire & Rescue Academy, Provo, Utah

Robert W. Royall, Jr., Chief of Emergency Operations, Harris County Fire Marshall's Office, Crosby, Texas

Tom Ruane, Fire and Life Safety Consultant, Peoria, Arizona

Glen Rudner, Hazardous Materials Officer, Virginia Department of Emergency Management, Dumfries, Virginia

Our Advisory Board Members

To those experts who work behind the scenes and provide us with continuing guidance on this book as well as other learning materials on our emergency services list, we extend our gratitude:

Steve Chikerotis, Battalion Chief, Chicago Fire Department, Chicago, Illinois

Tom Labelle, Executive Director, New York State Association of Fire Chiefs, East Schodack, New York

Jerry Laughlin, Deputy Director of Education Services, Alabama Fire College, Tuscaloosa, Alabama

Jeff Pindelski, Battalion Chief, Downers Grove Fire Department, Downers Grove, Illinois

Peter Sells, District Chief of Operations Training, Toronto Fire Services, Toronto, Canada

Billy Shelton, Executive Director, Virginia Department of Fire Programs, Glen Allen, Virginia

Doug Fry, Fire Chief, City of San Carlos Fire Department, San Carlos, California

Pat McCauliff, Director of Fire Science and EMS, Colin County Community College, McKinney, Texas

Michael Petroff, Western Director for Fire Department Safety Officers Association, St. Louis, Missouri

Adam Piskura, Director of Training, Connecticut Fire Academy, Windsor Locks, Connecticut

Theresa Staples, Program Manager, Firefighter and HazMat Certifications, Centennial, Colorado

Street Stories

Chapter 1: Rob Schnepp, Assistant Chief of Special Operations for the Alameda County (California) Fire Department

Chapter 2: Gregory G. Noll, Emergency Planning and Response Consultant, Hildebrand and Noll Associates, Lancaster, Pennsylvania

Chapter 3: Tom Creamer, Special Operations Coordinator, City of Worcester Fire Department, Worcester, Massachusetts

Chapter 4: Mike Callan, President, Callan and Company, Middlefield, Connecticut

Chapter 5: Gregory L. Socks, Captain, Montgomery County, Maryland, Hazmat Team

Chapter 6: Bill Hand, Houston Fire Department Hazardous Materials Response Team, Houston, Texas

Chapter 7: Christopher Hawley, Baltimore County Fire Department (Ret.)

Delmar Emergency Services Team

And to those we rarely take the time to recognize because this is their job, a special thanks. The Delmar team developed, produced, and marketed the third edition of the *Firefighter's Handbook* series, setting an example not only for getting the job done, but also having the creativity and fortitude to go above and beyond. Our appreciation goes to Alison Pase, Janet Maker, Rich Hall, Jennifer Starr, Jennifer Hanley, Maria Conto, Erin Coffin, and Patti Garrison.

HOW TO CONTACT US

At Delmar, listening to what our customers have to say is the heart of our business. If you have any comments or feedback on *Hazardous Materials Handbook* or any of our products, you can e-mail us at delmar.fire@cengage.com, or fax us at 518-881-1262, Attention: Fire Rescue Editorial.

For additional information on other titles that may be of interest to you, or to request a catalog, refer to the Additional Fire Titles from Delmar section in the back of this book, or visit delmarfire.cengage.com.

NFPA 472, 2008 Edition Correlation Guide

	NFPA 472, Chapter 4 Awareness	Learning Objective(s)	Chapter(s)	Page(s)
4.1	**General.**			
4.1.1	**Introduction.**			
4.1.1.1	Awareness level personnel shall be persons who, in the course of their normal duties, could encounter an emergency involving hazardous materials/weapons of mass destruction (WMD) and who are expected to recognize the presence of the hazardous materials/WMD, protect themselves, call for trained personnel, and secure the area.	1-1, 1-2	1	2–5, 7–9
4.1.1.2	Awareness level personnel shall be trained to meet all competencies of this chapter.	1-2	1	2–12
4.1.1.3	Awareness level personnel shall receive additional training to meet applicable governmental occupational health and safety regulations.	1-6	1	7–12
4.1.2	**Goal.**			
4.1.2.1	The goal of the competencies at the awareness level shall be to provide personnel already on the scene of a hazardous materials/WMD incident with the knowledge and skills to perform the tasks in 4.1.2.2 safely and effectively.	1-1, 1-2	1	2–5, 7–9
4.1.2.2	When already on the scene of a hazardous materials/WMD incident, the awareness level personnel shall be able to perform the following tasks:	1-1, 1-2	1	2–5, 7–9
(1)	Analyze the incident to determine both the hazardous material/WMD present and the basic hazard and response information for each hazardous material/WMD agent by completing the following tasks:	2-1, 2-2, 2-3, 2-4, 2-5, 2-6, 2-7, 2-8	2	17–61
(a)	Detect the presence of hazardous materials/WMD.	2-1, 2-3	2	17–61

Reprinted with permission from NFPA 472, Professional Competence of Responders to Hazardous Materials Incidents, copyright © 2008, National Fire Protection Association. This is not the complete and official position of the NFPA, which is represented solely by the Standard and its entirety.

NFPA 472, Chapter 4 Awareness		Learning Objective(s)	Chapter(s)	Page(s)
(b)	Survey a hazardous materials/WMD incident from a safe location to identify the name, UN/NA identification number, type of placard, or other distinctive marking applied for the hazardous materials/WMD involved.	2-1, 2-2	2	17–34
(c)	Collect hazard information from the current edition of the DOT Emergency Response Guidebook.	3-5	3	78–95
(2)	Implement actions consistent with the emergency response plan, the standard operating procedures, and the current edition of the DOT Emergency Response Guidebook by completing the following tasks:	3-5	3	78–95
(a)	Initiate protective actions.	5-1	5	138–151, 177
(b)	Initiate the notification process.	1-3	1	5
4.2	**Competencies—Analyzing the Incident.**			
4.2.1*	**Detecting the Presence of Hazardous Materials/WMD.** Given examples of various situations, awareness level personnel shall identify those situations where hazardous materials/WMD are present and shall meet the following requirements:	2-1, 2-2, 2-3	2	17–71
(1)*	Identify the definitions of both *hazardous material* (or *dangerous goods*, in Canada) and *WMD*.	2-1	2	17–71
(2)	Identify the UN/DOT hazard classes and divisions of hazardous materials/WMD and identify common examples of materials in each hazard class or division.	2-1, 2-2	2	17–34
(3)*	Identify the primary hazards associated with each UN/DOT hazard class and division.	2-2	2	17–34
(4)	Identify the difference between hazardous materials/WMD incidents and other emergencies.	1-1	1	2–4
(5)	Identify typical occupancies and locations in the community where hazardous materials/WMD are manufactured, transported, stored, used, or disposed of.	2-1, 2-3	2	17–61
(6)	Identify typical container shapes that can indicate the presence of hazardous materials/WMD.	2-4	2	34–61
(7)	Identify facility and transportation markings and colors that indicate hazardous materials/WMD, including the following:	2-5	2	34–61
(a)	Transportation markings, including UN/NA identification number marks, marine pollutant mark, elevated temperature (HOT) mark, commodity marking, and inhalation hazard mark	2-2, 2-5, 2-6, 2-7	2	17–61
(b)	NFPA 704, *Standard System for the Identification of the Hazards of Materials for Emergency Response*, markings	2-5	2	30–31
(c)*	Military hazardous materials/WMD markings	2-5	2	31–32
(d)	Special hazard communication markings for each hazard class	2-2, 2-5	2	30–32
(e)	Pipeline markings	2-5	2	32–33
(f)	Container markings	2-2, 2-6, 2-7	2	34–61
(8)	Given an NFPA 704 marking, describe the significance of the colors, numbers, and special symbols.	2-5	2	30–31
(9)	Identify U.S. and Canadian placards and labels that indicate hazardous materials/WMD.	2-2	2	17–33

Reprinted with permission from NFPA 472, Professional Competence of Responders to Hazardous Materials Incidents, copyright © 2008, National Fire Protection Association. This is not the complete and official position of the NFPA, which is represented solely by the Standard and its entirety.

	NFPA 472, Chapter 4 Awareness	Learning Objective(s)	Chapter(s)	Page(s)
(10)	Identify the following basic information on material safety data sheets (MSDS) and shipping papers for hazardous materials:	3-1, 3-2, 3-3	3	95–102
(a)	Identify where to find MSDS.	3-1, 3-2, 3-3	3	95–102
(b)	Identify major sections of an MSDS.	3-1, 3-2, 3-3	3	95–102
(c)	Identify the entries on shipping papers that indicate the presence of hazardous materials.	3-4	3	100–105
(d)	Match the name of the shipping papers found in transportation (air, highway, rail, and water) with the mode of transportation.	3-4	3	100–105
(e)	Identify the person responsible for having the shipping papers in each mode of transportation.	3-4	3	100–105
(f)	Identify where the shipping papers are found in each mode of transportation.	3-4	3	100–105
(g)	Identify where the papers can be found in an emergency in each mode of transportation.	3-4	3	100–105
(11)*	Identify examples of clues (other than occupancy/ location, container shape, markings/color, placards/ labels, MSDS, and shipping papers) the sight, sound, and odor of which indicate hazardous materials/WMD.	2-1, 5-6	2, 5	17–61, 151–165
(12)	Describe the limitations of using the senses in determining the presence or absence of hazardous materials/WMD.	2-1	2	61
(13)*	Identify at least four types of locations that could be targets for criminal or terrorist activity using hazardous materials/WMD.	7-1	7	217–219
(14)*	Describe the difference between a chemical and a biological incident.	7-7	7	226–232
(15)*	Identify at least four indicators of possible criminal or terrorist activity involving chemical agents.	7-2	7	219–224
(16)*	Identify at least four indicators of possible criminal or terrorist activity involving biological agents.	7-2	7	219–232
(17)	Identify at least four indicators of possible criminal or terrorist activity involving radiological agents.	7-1, 7-2	7	219–232
(18)	Identify at least four indicators of possible criminal or terrorist activity involving illicit laboratories (clandestine laboratories, weapons lab, ricin lab).	7-1, 7-2, 7-4	7	219–232
(19)	Identify at least four indicators of possible criminal or terrorist activity involving explosives.	7-1, 7-2, 7-7	7	219–232
(20)*	Identify at least four indicators of secondary devices.	7-2	7	219–232
4.2.2	**Surveying Hazardous Materials/WMD Incidents.** Given examples of hazardous materials/WMD incidents, awareness level personnel shall, from a safe location, identify the hazardous material(s)/ WMD involved in each situation by name, UN/NA identification number, or type placard applied and shall meet the following requirements:			
(1)	Identify difficulties encountered in determining the specific names of hazardous materials/WMD at facilities and in transportation.	2-2, 3-1	2	17–71, 83
(2)	Identify sources for obtaining the names of, UN/NA identification numbers for, or types of placard associated with hazardous materials/WMD in transportation.	2-1, 2-2	2	17–61
(3)	Identify sources for obtaining the names of hazardous materials/ WMD at a facility.	1-4, 1-5, 3-8	1, 3	95–102, 108

Reprinted with permission from NFPA 472, Professional Competence of Responders to Hazardous Materials Incidents, copyright © 2008, National Fire Protection Association. This is not the complete and official position of the NFPA, which is represented solely by the Standard and its entirety.

NFPA 472, Chapter 4 Awareness	Learning Objective(s)	Chapter(s)	Page(s)
4.2.3* **Collecting Hazard Information.** Given the identity of various hazardous materials/WMD (name, UN/NA identification number, or type placard), awareness level personnel shall identify the fire, explosion, and health hazard information for each material by using the current edition of the DOT *Emergency Response Guidebook* and shall meet the following requirements:			
(1)* Identify the three methods for determining the guidebook page for a hazardous material/WMD.	3-5	3	78–95
(2) Identify the two general types of hazards found on each guidebook page.	3-5	3	78–95
4.3* **Competencies—Planning the Response. (Reserved)**			
4.4 **Competencies—Implementing the Planned Response.**			
4.4.1* **Initiating Protective Actions.** Given examples of hazardous materials/WMD incidents, the emergency response plan, the standard operating procedures, and the current edition of the DOT *Emergency Response Guidebook*, awareness level personnel shall be able to identify the actions to be taken to protect themselves and others and to control access to the scene and shall meet the following requirements:			
(1) Identify the location of both the emergency response plan and/or standard operating procedures.	1-4, 1-5	1	4–6
(2) Identify the role of the awareness level personnel during hazardous materials/WMD incidents.	1-2	1	7–10
(3) Identify the following basic precautions to be taken to protect themselves and others in hazardous materials/WMD incidents:	4-1	4	114–124
(a) Identify the precautions necessary when providing emergency medical care to victims of hazardous materials/WMD incidents.	5-1	5	138–151
(b) Identify typical ignition sources found at the scene of hazardous materials/WMD incidents.	5-6	5	156
(c)* Identify the ways hazardous materials/WMD are harmful to people, the environment, and property.	4-1, 4-2	4	114–124
(d)* Identify the general routes of entry for human exposure to hazardous materials/WMD.	4-1, 4-2	4	114–124
(4)* Given examples of hazardous materials/WMD and the identity of each hazardous material/WMD (name, UN/NA identification number, or type placard), identify the following response information:	2-1, 2-2	2	17–71
(a) Emergency action (fire, spill, or leak and first aid)	3-5	3	78–95
(b) Personal protective equipment necessary	4-4	4	124–132
(c) Initial isolation and protective action distances	3-5	3	78–95
(5) Given the name of a hazardous material, identify the recommended personal protective equipment from the following list:	4-4	4	124–132
(a) Street clothing and work uniforms	4-4	4	124–132
(b) Structural fire-fighting protective clothing	4-4	4	124–132
(c) Positive pressure self-contained breathing apparatus	4-4	4	124–132
(d) Chemical-protective clothing and equipment	4-4	4	124–132

Reprinted with permission from NFPA 472, Professional Competence of Responders to Hazardous Materials Incidents, copyright © 2008, National Fire Protection Association. This is not the complete and official position of the NFPA, which is represented solely by the Standard and its entirety.

NFPA 472, Chapter 4 Awareness		Learning Objective(s)	Chapter(s)	Page(s)
(6)	Identify the definitions for each of the following protective actions:	3-5	3	78–95
(a)	Isolation of the hazard area and denial of entry	3-5	3	78–95
(b)	Evacuation	3-5	3	78–95
(c)*	Sheltering in-place	3-5	3	78–95
(7)	Identify the size and shape of recommended initial isolation and protective action zones.	3-5	3	78–95
(8)	Describe the difference between small and large spills as found in the Table of Initial Isolation and Protective Action Distances in the DOT *Emergency Response Guidebook.*	3-5	3	78–95
(9)	Identify the circumstances under which the following distances are used at a hazardous materials/WMD incidents:	3-5	3	78–95
(a)	Table of Initial Isolation and Protective Action Distances	3-5	3	78–95
(b)	Isolation distances in the numbered guides	3-5, 5-2, 5-3	3, 5	78–95, 145–165
(10)	Describe the difference between the isolation distances on the orange-bordered guidebook pages and the protective action distances on the green-bordered ERG (*Emergency Response Guidebook*) pages.	3-5	3	78–95
(11)	Identify the techniques used to isolate the hazard area and deny entry to unauthorized persons at hazardous materials/WMD Incidents.	3-5	3	78–95
(12)*	Identify at least four specific actions necessary when an incident is suspected to involve criminal or terrorist activity.	7-3, 7-4	7	224–226
4.4.2	**Initiating the Notification Process.** Given scenarios involving hazardous materials/WMD incidents, awareness level personnel shall identify the initial notifications to be made and how to make them, consistent with the emergency response plan and/or standard operating procedures.			
4.5*	**Competencies—Evaluating Progress. (Reserved)**			
4.6*	**Competencies—Terminating the Incident. (Reserved)**			

Operations Level, Chapter 5

NFPA 472, Chapter 5, Operations		Learning Objective(s)	Chapter(s)	Page(s)
5.1	**General.**			
5.1.1	**Introduction.**			
5.1.1.1*	The operations level responder shall be that person who responds to hazardous materials/weapons of mass destruction (WMD) incidents for the purpose of protecting nearby persons, the environment, or property from the effects of the release.	1-1, 1-2, 1-3	1	2–11
5.1.1.2	The operations level responder shall be trained to meet all competencies at the awareness level (Chapter 4) and the competencies of this chapter.	1-1	1	2–11

Reprinted with permission from NFPA 472, Professional Competence of Responders to Hazardous Materials Incidents, copyright © 2008, National Fire Protection Association. This is not the complete and official position of the NFPA, which is represented solely by the Standard and its entirety.

NFPA 472, Chapter 5, Operations		Learning Objective(s)	Chapter(s)	Page(s)
5.1.1.3*	The operations level responder shall receive additional training to meet applicable governmental occupational health and safety regulations.	1-1, 1-2, 1-4	1	2–11
5.1.2	**Goal**			
5.1.2.1	The goal of the competencies at this level shall be to provide operations level responders with the knowledge and skills to perform the core competencies in 5.1.2.2 safely.	1-2, 1-3	1	2–11
5.1.2.2	When responding to hazardous materials/WMD incidents, operations level responders shall be able to perform the following tasks:	1-2	1	2–11
(1)	Analyze a hazardous materials/WMD incident to determine the scope of the problem and potential outcomes by completing the following tasks:	1-2, 1-3, 1-4	1	2–11
(a)	Survey a hazardous materials/WMD incident to identify the containers and materials involved, determine whether hazardous materials/WMD have been released, and evaluate the surrounding conditions.	2-1, 2-2	2	16–61
(b)	Collect hazard and response information from MSDS; CHEMTREC/CANUTEC/SETIQ; local, state, and federal authorities; and shipper/manufacturer contacts.	3-1, 3-2, 3-3, 3-4, 3-8	3	78–108
(c)	Predict the likely behavior of a hazardous material/WMD and its container.	5-6	5	151–165
(d)	Estimate the potential harm at a hazardous materials/WMD incident.	4-1	4	114–124
(2)	Plan an initial response to a hazardous materials/WMD incident within the capabilities and competencies of available personnel and personal protective equipment by completing the following tasks:	4-4	4	114–124
(a)	Describe the response objectives for the hazardous materials/WMD incident.	5-1	5	138–151
(b)	Describe the response options available for each objective.	5-1	5	138–151
(c)	Determine whether the personal protective equipment provided is appropriate for implementing each option.	4-4	4	124–133
(d)	Describe emergency decontamination procedures.	5-7	5	165–177
(e)	Develop a plan of action, including safety considerations.	5-1	5	138–151
(3)	Implement the planned response for a hazardous materials/WMD incident to favorably change the outcomes consistent with the emergency response plan and/or standard operating procedures by completing the following tasks:	5-1	5	138–151
(a)	Establish and enforce scene control procedures, including control zones, emergency decontamination, and communications.	5-1, 5-2, 5-3, 5-4	5	138–151
(b)	Where criminal or terrorist acts are suspected, establish means of evidence preservation.	7-3	7	226
(c)	Initiate an incident command system (ICS) for hazardous materials/WMD incidents.	5-1	5	138–151
(d)	Perform tasks assigned as identified in the incident action plan.	5-1	5	138–151
(e)	Demonstrate emergency decontamination.	5-7	5	165–177

Reprinted with permission from NFPA 472, Professional Competence of Responders to Hazardous Materials Incidents, copyright © 2008, National Fire Protection Association. This is not the complete and official position of the NFPA, which is represented solely by the Standard in its entirety.

	NFPA 472, Chapter 5, Operations	Learning Objective(s)	Chapter(s)	Page(s)
(4)	Evaluate the progress of the actions taken at a hazardous materials/WMD incident to ensure that the response objectives are being met safely, effectively, and efficiently by completing the following tasks:	5-1	5	138–151
(a)	Evaluate the status of the actions taken in accomplishing the response objectives.	5-1	5	138–151
(b)	Communicate the status of the planned response.	5-1	5	138–151
5.2	**Core Competencies—Analyzing the Incident.**			
5.2.1*	**Surveying Hazardous Materials/WMD Incidents.** Given scenarios involving hazardous materials/WMD incidents, the operations level responder shall survey the incident to identify the containers and materials involved, determine whether hazardous materials/WMD have been released, and evaluate the surrounding conditions and shall meet the requirements of 5.2.1.1 through 5.2.1.6.			
5.2.1.1*	Given three examples each of liquid, gas, and solid hazardous material or WMD, including various hazard classes, operations level personnel shall identify the general shapes of containers in which the hazardous materials/WMD are typically found.	1-2, 2-1, 2-4, 2-5, 2-6, 3-6	1, 2, 3	2–11, 17–71, 78–108
5.2.1.1.1	Given examples of the following tank cars, the operations level responder shall identify each tank car by type, as follows:	2-1, 2-6, 2-7	2	54–57
(1)	Cryogenic liquid tank cars	2-1, 2-6, 2-7	2	54–57
(2)	Nonpressure tank cars (general service or low pressure cars)	2-1, 2-6, 2-7	2	54–57
(3)	Pressure tank cars	2-1, 2-6, 2-7	2	54–57
5.2.1.1.2	Given examples of the following intermodal tanks, the operations level responder shall identify each intermodal tank by type, as follows:	2-1, 2-3, 2-5, 2-7	2	52–53
(1)	Nonpressure intermodal tanks	2-1, 2-3, 2-5, 2-7	2	52–53
(2)	Pressure intermodal tanks	2-1, 2-3, 2-5, 2-7	2	52–53
(3)	Specialized intermodal tanks, including the following:	2-1, 2-3, 2-5, 2-7	2	52–53
(a)	Cryogenic intermodal tanks	2-1, 2-3, 2-5, 2-7	2	52–53
(b)	Tube modules	2-1, 2-3, 2-5, 2-7	2	52–53
5.2.1.1.3	Given examples of the following cargo tanks, the operations level responder shall identify each cargo tank by type, as follows:	2-1, 2-3, 2-5, 2-7	2	40–52
(1)	Compressed gas tube trailers	2-1, 2-3, 2-5, 2-7	2	51
(2)	Corrosive liquid tanks	2-1, 2-3, 2-5, 2-7	2	45
(3)	Cryogenic liquid tanks	2-1, 2-3, 2-5, 2-7	2	49–50

Reprinted with permission from NFPA 472, Professional Competence of Responders to Hazardous Materials Incidents, copyright © 2008, National Fire Protection Association. This is not the complete and official position of the NFPA, which is represented solely by the Standard in its entirety.

NFPA 472, Chapter 5, Operations		Learning Objective(s)	Chapter(s)	Page(s)
(4)	Dry bulk cargo tanks	2-1, 2-3, 2-5, 2-7	2	51
(5)	High pressure tanks	2-1, 2-3, 2-5, 2-7	2	51
(6)	Low pressure chemical tanks	2-1, 2-3, 2-5, 2-7	2	44–45
(7)	Nonpressure liquid tanks	2-1, 2-3, 2-5, 2-7	2	42–43
5.2.1.1.4	Given examples of the following storage tanks, the operations level responder shall identify each tank by type, as follows:	2-1, 2-3, 2-5, 2-7	2	57–61
(1)	Cryogenic liquid tank	2-1, 2-3, 2-5, 2-7	2	60–61
(2)	Nonpressure tank	2-1, 2-3, 2-5, 2-7	2	57–60
(3)	Pressure tank	2-1, 2-3, 2-5, 2-7	2	57–60
5.2.1.1.5	Given examples of the following nonbulk packaging, the operations level responder shall identify each package by type, as follows:	2-1, 2-3, 2-5, 2-7	2	35–39
(1)	Bags	2-1, 2-3, 2-5, 2-7	2	36
(2)	Carboys	2-1, 2-3, 2-5, 2-7	2	35
(3)	Cylinders	2-1, 2-3, 2-5, 2-7	2	37
(4)	Drums	2-1, 2-3, 2-5, 2-7	2	36
(5)	Dewar flask (cryogenic liquids)	2-1, 2-3, 2-5, 2-7	2	37
5.2.1.1.6*	Given examples of the following radioactive material packages, the operations level responder shall identify the characteristics of each container or package by type, as follows:	2-1, 2-3, 2-5, 2-7	2	39–40
(1)	Excepted	2-1, 2-3, 2-5, 2-7	2	39–40
(2)	Industrial	2-1, 2-3, 2-5, 2-7	2	39–40
(3)	Type A	2-1, 2-3, 2-5, 2-7	2	39–40
(4)	Type B	2-1, 2-3, 2-5, 2-7	2	39–40
(5)	Type C	2-1, 2-3, 2-5, 2-7	2	39–40
5.2.1.2	Given examples of containers, the operations level responder shall identify the markings that differentiate one container from another.	2-1, 2-3, 2-5, 2-7	2	33
5.2.1.2.1	Given examples of the following marked transport vehicles and their corresponding shipping papers, the operations level responder shall identify the following vehicle or tank identification marking:	2-1, 2-3, 2-5, 2-7	2	17–30

Reprinted with permission from NFPA 472, Professional Competence of Responders to Hazardous Materials Incidents, copyright © 2008, National Fire Protection Association. This is not the complete and official position of the NFPA, which is represented solely by the Standard in its entirety.

NFPA 472, Chapter 5, Operations		Learning Objective(s)	Chapter(s)	Page(s)
(1)	Highway transport vehicles, including cargo tanks	2-1, 2-3, 2-5, 2-7	2	17–30
(2)	Intermodal equipment, including tank containers	2-1, 2-3, 2-5, 2-7	2	17–30
(3)	Rail transport vehicles, including tank cars	2-1, 2-3, 2-5, 2-7	2	17–30
5.2.1.2.2	Given examples of facility containers, the operations level responder shall identify the markings indicating container size, product contained, and/or site identification numbers.	2-1, 2-3, 2-5, 2-7	2	17–31
5.2.1.3	Given examples of hazardous materials incidents, the operations level responder shall identify the name(s) of the hazardous material(s) in 5.2.1.3.1 through 5.2.1.3.3.	2-1, 2-3, 2-5, 2-7	2	17–34
5.2.1.3.1	The operations level responder shall identify the following information on a pipeline marker:	2-1, 2-3, 2-5, 2-7	2	32
(1)	Emergency telephone number	2-1, 2-3, 2-5, 2-7	2	32
(2)	Owner	2-1, 2-3, 2-5, 2-7	2	32
(3)	Product	2-1, 2-3, 2-5, 2-7	2	32
5.2.1.3.2	Given a pesticide label, the operations level responder shall identify each of the following pieces of information, then match the piece of information to its significance in surveying hazardous materials incidents:	2-1, 2-3, 2-5, 2-7	2	33–34
(1)	Active ingredient	2-1, 2-3, 2-5, 2-7	2	33–34
(2)	Hazard statement	2-1, 2-3, 2-5, 2-7	2	33–34
(3)	Name of pesticide	2-1, 2-3, 2-5, 2-7	2	33–34
(4)	Pest control product (PCP) number (in Canada)	2-1, 2-3, 2-5, 2-7	2	33–34
(5)	Precautionary statement	2-1, 2-3, 2-5, 2-7	2	33–34
(6)	Signal word	2-1, 2-3, 2-5, 2-7	2	33–34
5.2.1.3.3	Given a label for a radioactive material, the operations level responder shall identify the type or category of label, contents, activity, transport index, and criticality safety index as applicable.	2-1, 2-3, 2-5, 2-7	2	25–26, 34
5.2.1.4*	The operations level responder shall identify and list the surrounding conditions that should be noted when a hazardous materials/WMD incident is surveyed.	3-5, 7-2, 7-3, 7-7	3, 7	78–108, 210–233
5.2.1.5	The operations level responder shall give examples of ways to verify information obtained from the survey of a hazardous materials/WMD incident.	3-1, 3-2, 3-3, 3-4, 3-5	3	78–108
5.2.1.6*	The operations level responder shall identify at least three additional hazards that could be associated with an incident involving terrorist or criminal activities.	7-2, 7-3, 7-4	7	210–233

Reprinted with permission from NFPA 472, Professional Competence of Responders to Hazardous Materials Incidents, copyright © 2008, National Fire Protection Association. This is not the complete and official position of the NFPA, which is represented solely by the Standard in its entirety.

Reprinted with permission from NFPA 472, Professional Competence of Responders to Hazardous Materials Incidents, copyright © 2008, National Fire Protection Association. This is not the complete and official position of the NFPA, which is represented solely by the Standard in its entirety.

NFPA 472, Chapter 5, Operations		Learning Objective(s)	Chapter(s)	Page(s)
(8)*	Describe the properties and characteristics of the following:	2-8	2	68–71
(a)	Alpha radiation	2-8	2	68–71
(b)	Beta radiation	2-8	2	68–71
(c)	Gamma radiation	2-8	2	68–71
(d)	Neutron radiation	2-8	2	68–71
5.2.3*	**Predicting the Likely Behavior of a Material and Its Container.** Given scenarios involving hazardous materials/WMD incidents, each with a single hazardous material/WMD, the operations level responder shall predict the likely behavior of the material or agent and its container and shall meet the following requirements:			
(1)	Interpret the hazard and response information obtained from the current edition of the DOT *Emergency Response Guidebook*, MSDS, CHEMTREC/CANUTEC/SETIQ, governmental authorities, and shipper and manufacturer contacts, as follows:	3-5, 3-6, 3-7	3	78–94
(a)	Match the following chemical and physical properties with their significance and impact on the behavior of the container and its contents:	2-8, 5-6	2, 5	68, 151–165
i.	Boiling point	2-8, 5-6	2, 5	62, 151–165
ii.	Chemical reactivity	2-8, 5-6	2, 5	66, 151–165
iii.	Corrosivity (pH)	2-8, 5-6	2, 5	66, 151–165
iv.	Flammable (explosive) range [lower explosive limit (LEL) and upper explosive limit (UEL)]	2-8, 5-6	2, 5	67, 151–165
v.	Flash point	2-8, 5-6	2, 5	67, 151–165
vi.	Ignition (autoignition) temperature	2-8, 5-6	2, 5	67, 151–165
vii.	Particle size	2-8, 5-6	2, 5	69, 151–165
viii.	Persistence	2-8, 5-6	2, 5	64, 151–165
ix.	Physical state (solid, liquid, gas)	2-8, 5-6	2, 5	62, 151–165
x.	Radiation (ionizing and non-ionizing)	2-8, 5-6	2, 5	68, 151–165
xi.	Specific gravity	2-8, 5-6	2, 5	64, 151–165
xii.	Toxic products of combustion	2-8, 5-6	2, 5	71, 151–165
xiii.	Vapor density	2-8, 5-6	2, 5	64, 151–165
xiv.	Vapor pressure	2-8, 5-6	2, 5	63, 151–165
xv.	Water solubility	2-8, 5-6	2, 5	65, 151–165
(b)	Identify the differences between the following terms:	2-8, 5-6	2, 5	61–71
i.	Contamination and secondary contamination	5-7	5	165–177
ii.	Exposure and contamination	5-7	5	165–177
iii.	Exposure and hazard	5-7	5	165–177
iv.	Infectious and contagious	5-7	5	165–177
v.	Acute effects and chronic effects	5-7	5	165–177
vi.	Acute exposures and chronic exposures	4-1	4	114–116
(2)*	Identify three types of stress that can cause a container system to release its contents.	5-2, 5-6	5	151–165
(3)*	Identify five ways in which containers can breach.	5-6	5	151–165

Reprinted with permission from NFPA 472, Professional Competence of Responders to Hazardous Materials Incidents, copyright © 2008, National Fire Protection Association. This is not the complete and official position of the NFPA, which is represented solely by the Standard in its entirety.

	NFPA 472, Chapter 5, Operations	Learning Objective(s)	Chapter(s)	Page(s)
(4)*	Identify four ways in which containers can release their contents.	5-6	5	151–165
(5)*	Identify at least four dispersion patterns that can be created upon release of a hazardous material.	5-2	5	151–165
(6)*	Identify the time frames for estimating the duration that hazardous materials/WMD will present an exposure risk.	5-2, 5-3, 5-4, 5-5	5	151–165
(7)*	Identify the health and physical hazards that could cause harm.	4-1	4	114–122
(8)*	Identify the health hazards associated with the following terms:	4-1	4	114–122
(a)	Alpha, beta, gamma, and neutron radiation	4-1	4	114–122
(b)	Asphyxiant	4-1	4	114–122
(c)*	Carcinogen	4-1	4	114–122
(d)	Convulsant	4-1	4	114–122
(e)	Corrosive	4-1	4	114–122
(f)	Highly toxic	4-1	4	114–122
(g)	Irritant	4-1	4	114–122
(h)	Sensitizer, allergen	4-1	4	114–122
(i)	Target organ effects	4-1	4	114–122
(j)	Toxic	4-1	4	114–122
(9)*	Given the following, identify the corresponding UN/DOT hazard class and division:	7-7	7	226–233
(a)	Blood agents	7-7	7	226–233
(b)	Biological agents and biological toxins	7-7	7	226–233
(c)	Choking agents	7-7	7	226–233
(d)	Irritants (riot control agents)	7-7	7	226–233
(e)	Nerve agents	7-7	7	226–233
(f)	Radiological materials	7-7	7	226–233
(g)	Vesicants (blister agents)	7-7	7	226–233
5.2.4*	**Estimating Potential Harm.** Given scenarios involving hazardous materials/WMD incidents, the operations level responder shall estimate the potential harm within the endangered area at each incident and shall meet the following requirements:			
(1)*	Identify a resource for determining the size of an endangered area of a hazardous materials/WMD incident.	3-5	3	78–108
(2)	Given the dimensions of the endangered area and the surrounding conditions at a hazardous materials/WMD incident, estimate the number and type of exposures within that endangered area.	3-5	3	78–108
(3)	Identify resources available for determining the concentrations of a released hazardous material/WMD within an endangered area.	3-5	3	78–108
(4)*	Given the concentrations of the released material, identify the factors for determining the extent of physical, health, and safety hazards within the endangered area of a hazardous materials/WMD incident.	3-5	3	78–108
(5)	Describe the impact that time, distance, and shielding have on exposure to radioactive materials specific to the expected dose rate.	4-1	4	114–122

Reprinted with permission from NFPA 472, Professional Competence of Responders to Hazardous Materials Incidents, copyright © 2008, National Fire Protection Association. This is not the complete and official position of the NFPA, which is represented solely by the Standard in its entirety.

Reprinted with permission from NFPA 472, Professional Competence of Responders to Hazardous Materials Incidents, copyright © 2008, National Fire Protection Association. This is not the complete and official position of the NFPA, which is represented solely by the Standard in its entirety.

	NFPA 472, Chapter 5, Operations	Learning Objective(s)	Chapter(s)	Page(s)
(2)	Identify the personal protective clothing required for a given option and the following:	4-4, 4-5, 4-6, 4-7	4	124–133
(a)	Identify skin contact hazards encountered at hazardous materials/WMD incidents.	4-4, 4-5, 4-6, 4-7	4	124–133
(b)	Identify the purpose, advantages, and limitations of the following types of protective clothing at hazardous materials/WMD incidents:	4-4, 4-5, 4-6, 4-7	4	124–133
i.	Chemical-protective clothing: liquid splash–protective clothing and vapor-protective clothing	4-4, 4-5, 4-6, 4-7	4	124–133
ii.	High temperature–protective clothing: proximity suit and entry suits	4-4, 4-5, 4-6, 4-7	4	124–133
iii.	Structural fire-fighting protective clothing	4-4, 4-5, 4-6, 4-7	4	124–133
5.3.4*	**Identifying Decontamination Issues.** Given scenarios involving hazardous materials/WMD incidents, operations level responders shall identify when emergency decontamination is needed and shall meet the following requirements:			
(1)	Identify ways that people, personal protective equipment, apparatus, tools, and equipment become contaminated.	5-7	5	165–177
(2)	Describe how the potential for secondary contamination determines the need for decontamination.	5-7	5	165–177
(3)	Explain the importance and limitations of decontamination procedures at hazardous materials incidents.	5-7	5	165–177
(4)	Identify the purpose of emergency decontamination procedures at hazardous materials incidents.	5-7	5	165–177
(5)	Identify the factors that should be considered in emergency decontamination.	5-7	5	165–177
(6)	Identify the advantages and limitations of emergency decontamination procedures.	5-7	5	165–177
5.4	**Core Competencies—Implementing the Planned Response.**			
5.4.1	**Establishing and Enforcing Scene Control Procedures.** Given two scenarios involving hazardous materials/WMD incidents, the operations level responder shall identify how to establish and enforce scene control, including control zones and emergency decontamination, and communications between responders and to the public and shall meet the following requirements:			
(1)	Identify the procedures for establishing scene control through control zones.	5-4	5	138–165
(2)	Identify the criteria for determining the locations of the control zones at hazardous materials/WMD incidents.	5-4	5	138–165
(3)	Identify the basic techniques for the following protective actions at hazardous materials/WMD incidents:	5-3, 5-4, 5-5	5	138–165
(a)	Evacuation	5-2, 5-3, 5-4	5	138–151
(b)	Sheltering-in-place	5-2, 5-3, 5-4	5	138–151
(4)*	Demonstrate the ability to perform emergency decontamination.	5-7	5	165–177
(5)*	Identify the items to be considered in a safety briefing prior to allowing personnel to work at the following:	5-7	5	138–165

Reprinted with permission from NFPA 472, Professional Competence of Responders to Hazardous Materials Incidents, copyright © 2008, National Fire Protection Association. This is not the complete and official position of the NFPA, which is represented solely by the Standard in its entirety.

	NFPA 472, Chapter 5, Operations	Learning Objective(s)	Chapter(s)	Page(s)
(a)	Hazardous material incidents	5-1	5	138–165
(b)*	Hazardous materials/WMD incidents involving criminal activities	5-1, 7-3	5, 7	138–165, 210–233
(6)	Identify the procedures for ensuring coordinated communication between responders and to the public.	5-1, 5-3, 5-5, 7-3	5, 7	138–165, 210–233
5.4.2*	**Preserving Evidence.** Given two scenarios involving hazardous materials/WMD incidents, the operations level responder shall describe the process to preserve evidence as listed in the emergency response plan and/or standard operating procedures.			
5.4.3*	**Initiating the Incident Command System.** Given scenarios involving hazardous materials/WMD incidents, the operations level responder shall initiate the incident command system specified in the emergency response plan and/or standard operating procedures and shall meet the following requirements:			
(1)	Identify the role of the operations level responder during hazardous materials/WMD incidents as specified in the emergency response plan and/or standard operating procedures.	5-1	5	138–165
(2)	Identify the levels of hazardous materials/WMD incidents as defined in the emergency response plan.	5-5	5	177
(3)	Identify the purpose, need, benefits, and elements of the incident command system for hazardous materials/WMD incidents.	5-1, 5-5	5	177
(4)	Identify the duties and responsibilities of the following functions within the incident management system:	5-1	5	138–165
(a)	Incident safety officer	5-1	5	138–165
(b)	Hazardous materials branch or group	5-1	5	138–165
(5)	Identify the considerations for determining the location of the incident command post for a hazardous materials/WMD incident.	5-1	5	138–165
(6)	Identify the procedures for requesting additional resources at a hazardous materials/WMD incident.	5-1	5	138–165
(7)	Describe the role and response objectives of other agencies that respond to hazardous materials/WMD incidents.	5-1	5	138–165
5.4.4	**Using Personal Protective Equipment.** The operations level responder shall describe considerations for the use of personal protective equipment provided by the AHJ, and shall meet the following requirements:			
(1)	Identify the importance of the buddy system.	5-1	5	138–165
(2)	Identify the importance of the backup personnel.	5-1	5	138–165
(3)	Identify the safety precautions to be observed when approaching and working at hazardous materials/WMD incidents.	4-2	4	130–132
(4)	Identify the signs and symptoms of heat and cold stress and procedures for their control.	4-7	4	130–132
(5)	Identify the capabilities and limitations of personnel working in the personal protective equipment provided by the AHJ.	4-4, 4-7	4	130–132
(6)	Identify the procedures for cleaning, disinfecting, and inspecting personal protective equipment provided by the AHJ.	4-7	4	126

Reprinted with permission from NFPA 472, Professional Competence of Responders to Hazardous Materials Incidents, copyright © 2008, National Fire Protection Association. This is not the complete and official position of the NFPA, which is represented solely by the Standard in its entirety.

Operations Level, Chapter 6

Reprinted with permission from NFPA 472, Professional Competence of Responders to Hazardous Materials Incidents, copyright © 2008, National Fire Protection Association. This is not the complete and official position of the NFPA, which is represented solely by the Standard in its entirety.

NFPA 472, Chapter 6 Product Control	Learning Objective(s)	Chapter(s)	Page(s)
6.6.1.1.4* The operations level responder assigned to perform product control at hazardous materials/WMD incidents shall receive the additional training necessary to meet specific needs of the jurisdiction.			
6.6.1.2 **Goal.**			
6.6.1.2.1 The goal of the competencies in this section shall be to provide the operations level responder assigned to product control at hazardous materials/WMD incidents with the knowledge and skills to perform the tasks in 6.6.1.2.2 safely and effectively.			
6.6.1.2.2 When responding to hazardous materials/WMD incidents, the operations level responder assigned to perform product control shall be able to perform the following tasks:			
(1) Plan an initial response within the capabilities and competencies of available personnel, personal protective equipment, and control equipment and in accordance with the emergency response plan or standard operating procedures by completing the following tasks:	5-1, 6-2	5, 6	143–148, 151–165, 184–192
(a) Describe the control options available to the operations level responder.	5-1, 6-2	5, 6	143–148, 151–165
(b) Describe the control options available for flammable liquid and flammable gas incidents.	5-1, 5-6, 6-2	5, 6	143–148, 151–165
(2) Implement the planned response to a hazardous materials/WMD incident.	5-1, 6-2	5, 6	143–148, 151–165
6.6.2 **Competencies—Analyzing the Incident. (Reserved)**			
6.6.3 **Competencies—Planning the Response.**			
6.6.3.1 **Identifying Control Options.** Given examples of hazardous materials/WMD incidents, the operations level responder assigned to perform product control shall identify the options for each response objective and shall meet the following requirements as prescribed by the AHJ:			
(1) Identify the options to accomplish a given response objective.	5-1, 6-2, 7-3	5, 6, 7	143–148, 151–165, 184–192, 224–226
(2) Identify the purpose for and the procedures, equipment, and safety precautions associated with each of the following control techniques:	6-1	6	184–192
(a) Absorption	6-1	6	184–192
(b) Adsorption	6-1	6	184–192
(c) Damming	6-1	6	184–192
(d) Diking	6-1	6	184–192
(e) Dilution	6-1	6	184–192
(f) Diversion	6-1	6	184–192
(g) Remote valve shutoff	6-1	6	184–192
(h) Retention	6-1	6	184–192
(i) Vapor dispersion	6-1	6	184–192
(j) Vapor suppression	6-1	6	184–192

*Reprinted with permission from NFPA 472, Professional Competence of Responders to Hazardous Materials Incidents, copyright © 2008, National Fire Protection Association. This is not the complete and official position of the NFPA, which is represented solely by the Standard in its entirety.

NFPA 472, Chapter 6 Product Control	Learning Objective(s)	Chapter(s)	Page(s)
6.6.3.2 **Selecting Personal Protective Equipment.** The operations level responder assigned to perform product control shall select the personal protective equipment required to support product control at hazardous materials/WMD incidents based on local procedures *(see Section 6.2)*.	4-4, 4-5, 4-6, 4-7	4	124–132
6.6.4 **Competencies—Implementing the Planned Response.**			
6.6.4.1 **Performing Control Options.** Given an incident action plan for a hazardous materials/WMD incident, within the capabilities and equipment provided by the AHJ, the operations level responder assigned to perform product control shall demonstrate control functions set out in the plan and shall meet the following requirements as prescribed by the AHJ:			
(1) Given the required tools and equipment, demonstrate how to perform the following control activities:	5-1, 6-1	5, 6	143–148, 151–165, 184–192
(a) Absorption	5-1, 6-1	5, 6	143–148, 151–165, 184–192
(b) Adsorption	5-1, 6-1	5, 6	143–148, 151–165, 184–192
(c) Damming	5-1, 6-1	5, 6	143–148, 151–165, 184–192
(d) Diking	5-1, 6-1	5, 6	143–148, 151–165, 184–192
(e) Dilution	5-1, 6-1	5, 6	143–148, 151–165, 184–192
(f) Diversion	5-1, 6-1	5, 6	143–148, 151–165, 184–192
(g) Retention	5-1, 6-1	5, 6	143–148, 151–165, 184–192
(h) Remote valve shutoff	5-1, 6-1	5, 6	143–148, 151–165, 184–192
(i) Vapor dispersion	5-1, 6-1	5, 6	143–148, 151–165, 184–192
(j) Vapor suppression	5-1, 6-1	5, 6	143–148, 151–165, 184–192

*Reprinted with permission from NFPA 472, Professional Competence of Responders to Hazardous Materials Incidents, copyright © 2008, National Fire Protection Association. This is not the complete and official position of the NFPA, which is represented solely by the Standard in its entirety.

NFPA 472, Chapter 6 Product Control	Learning Objective(s)	Chapter(s)	Page(s)
(2) Identify the location and describe the use of emergency remote shutoff devices on MC/DOT-306/406, MC/DOT-307/407, and MC-331 cargo tanks containing flammable liquids or gases.	2-6	2	43
(3) Describe the use of emergency remote shutoff devices at fixed facilities.	2-5	2	57–60
6.6.4.2 The operations level responder assigned to perform product control shall describe local procedures for going through the technical decontamination process.			
6.6.5 Competencies—Evaluating Progress. (Reserved)			
6.6.6 Competencies—Terminating the Incident. (Reserved)			

*Reprinted with permission from NFPA 472, Professional Competence of Responders to Hazardous Materials Incidents, copyright © 2008, National Fire Protection Association. This is not the complete and official position of the NFPA, which is represented solely by the Standard in its entirety.

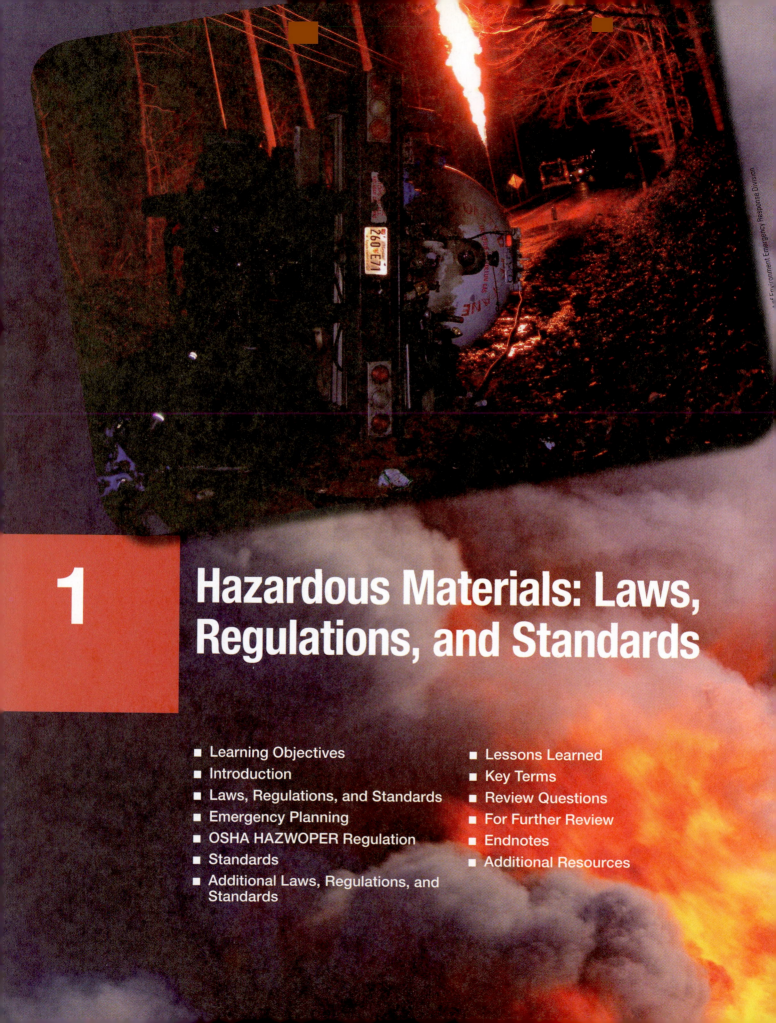

Courtesy of Environment Emergency Response Division

1

Hazardous Materials: Laws, Regulations, and Standards

We arrived on scene to find the patient lying beside the hot dusty road, baking in the summer sun. He was a male in his mid-thirties, wearing torn blue jeans, old tennis shoes, and no shirt. A local farmer was on the scene. He was driving home when he noticed the man beside the road. It was obvious the man was in distress, so the farmer called 9-1-1 from a cellular phone.

When we got off the fire engine, it was apparent that the man was struggling to breathe. He was sunburned, sweaty, and unable to respond to any of our questions. He was shaking slightly but it just did not look like the typical epileptic seizure. We put the man on high-flow oxygen and began to assess his vital signs. It was the size of his pupils that tipped us off. In addition to the pinpoint pupils, the patient was incontinent and producing volumes of frothy sputum. We stripped off his clothes, decontaminated him with water, and alerted the incoming ambulance by radio that we might have an organophosphate poisoning. Fortunately, we had recently had some hazardous materials training on pesticide exposures. Had we not been armed with that information, we may not have picked up on the clues so quickly.

The transport ambulance arrived what seemed like hours later, and by that time we were assisting his respirations with a bag valve mask and suctioning his airway. An IV was started, and atropine was administered in large doses to counteract the effects of a suspected pesticide. Unfortunately, we were never able to identify the substance he was exposed to or even know the duration of the exposure.

Sometimes, despite the greatest efforts and the best technology available, people do not survive. He fought hard to breathe and even harder to live, but in the end, neither we nor the hospital staff were able to save his life. He died an hour or so after we first saw him, probably never knowing why he became so sick.

I think about him from time to time, wondering if the call could have gone smoother or we could have done something different, but I always arrive at the same conclusion: You do your best, learn from any mistakes or situations you find yourself in, and just hope you have a better outcome the next time around.

We could have suffered a secondary exposure because we were not properly protected, but we did not. We could have had an inhalation exposure from his contaminated clothing, but we did not. We could all have ended up as patients, but we did not. It was only luck that brought us through without harm. Would I do it differently the next time? Absolutely. Our first responsibility as firefighters is to not make the problem worse when we show up! You are no good to anyone if you become part of the problem or end up being a victim yourself. Use common sense, wear the right PPE for the right situation, and, above all, think before you act. Do not count on luck to bring you back to the station.

*—Street Story by Rob Schnepp, Assistant Chief,
Alameda County, CA, Fire Department*

LEARNING OBJECTIVES

After completing this chapter, the reader should be able to:

1-1. Explain the local emergency response plan and standard operating guidelines.

1-2. Explain the student's role within these documents at the awareness level.

1-3. Explain the notification process to request assistance.

1-4. Explain the role of the local emergency planning committee.

1-5. Explain the role of the SARA Title III regulation and emergency response.

1-6. Explain other regulations that have an effect on fire department activities.

INTRODUCTION

Hazardous materials response, as shown in **Figure 1-1,** is typically one of the specialty fields within the fire service. When the fire department is called to a hazardous materials release, the actual handling of chemical releases is usually done by a well-trained team whose function it is to handle such emergencies, but in almost all cases firefighters are required to assist in the effort.

Firefighters and EMS providers are bombarded with exposures to hazardous materials each and every day in their regular fire suppression and EMS duties. A large number of firefighters and EMS providers are exposed to bloodborne pathogen materials each day, sometimes suffering fatal effects from this exposure. By the nature of their jobs they are exposed to numerous toxic and cancer-causing agents, many times in non-emergency situations. When confronted with emergency situations, the exposure and risks increase as well. This chapter covers some hazardous materials risks and some fundamental response profiles that will help keep firefighters safe. As the fire service furthers its efforts to improve firefighter health and safety, the response to chemical accidents is one area in which extra precautions are needed.

NOTE

Many times at hazardous materials incidents the actions that the first-in firefighters take in the first five minutes determine how the next five hours/days/months will go.

FIGURE 1-1   A hazardous material team member surveys a chemical agent lab using air monitors.

NOTE

This book uses the term *personal protective equipment* (PPE) to describe a variety of protective clothing. In the firefighting portion of this text the use of PPE indicates firefighting protective clothing. In the hazardous materials section, PPE can also mean chemical protective equipment. To differentiate between the two, the terms **firefighters protective clothing (FFPC)** and *hazardous materials protective clothing* will be used.

SAFETY

In a fire situation, the injuries can be acute, for instance, if a building collapses during firefighting operations. A chemical exposure can also kill a firefighter immediately with a deadly dose, or it can cause serious illness or injury that prolongs death for ten to twenty years.

NOTE

There are two options for hazardous materials emergency response training. One can be trained to an Occupational Safety & Health Administration (OSHA) level or to meet the National Fire Protection Association levels. The requirements vary from state to state, but the minimum level is the OSHA training. The term *certification* is commonly used, and typically only the employer can certify the training. Neither OSHA nor the NFPA certifiy responders; their fire department certifies that they have met the training objectives outlined by OSHA or the NFPA. The NFPA requirements are much more stringent but better reflect the current emergency responders' actual duties. If trained to meet the NFPA training objectives, a responder would meet the OSHA requirements.

When firefighters receive their basic training, it is typically based on the NFPA training objectives for Firefighters. The NFPA Standard 1001, also known as the Standard for Firefighter Professional Qualifications, has been revised and the 2008 edition will be published in July 2007. As part of the revision to be certified as a Firefighter I, the student will also have to be certified at the Operations level according to NFPA 472, which is the Competence of Responders to Hazardous Materials/Weapons of Mass Destruction Incidents Standard. All of this information is important for firefighters' survival since no matter which state the training is provided in, firefighters are required to have some form of hazardous materials training. The training can be based on the OSHA regulations or the NFPA standards. The information in this text exceeds the requirements for OSHA training and covers both awareness and operations levels, and in some cases exceeds the Operations level of the NFPA requirements. See the discussion on NFPA standards and OSHA's HAZWOPER for more information on Awareness and Operations level training.

As society has become more environmentally aware, so has the fire service. Hazardous materials response is still a very young field, and most communities have had fire department–based hazardous materials response teams for only nineteen years or less. The Jacksonville, Florida, team is the oldest hazardous materials response team in the country, and it is only twenty-six years old. This field is constantly changing and dynamic, and the response to hazardous materials emergencies has undergone three distinct changes in response methodology. Tactics used to handle incidents thirteen years ago certainly were appropriate then but might bring criticism if used today.

Technology is rapidly changing as well. **Air monitoring devices** that were used in the past strictly by hazardous materials response teams are now common on engine, truck, medic, and squad companies. On the hazardous materials response team side, technology has also increased to a level that requires a high degree of training and expertise. Response to chemical accidents is a rapidly changing field and brings new experiences every day. From building fires, tank truck fires, **clandestine drug labs,** shipboard incidents, pressurized cylinder leaks, and overturned tank trucks to the emerging incidents of terrorism, hazardous materials response teams face new challenges on every response. This chapter provides firefighters with a practical framework that can help keep them and their crews alive in situations that involve hazardous materials.

The next few sections describe in detail the many aspects of hazardous materials and appropriate actions when working those incidents. To establish the groundwork for these sections, it is necessary to provide a definition for hazardous materials. The text that follows provides these standard definitions. Out of these definitions, the following may be considered the best definition: "A hazardous material is any substance that jumps out of its container when something goes wrong, and hurts and harms the things it touches." Ludwig Benner Jr., who worked for the National Transportation Safety Board, provided this early definition. It is a perfect definition for this field.

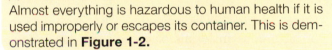

CAUTION

Almost everything is hazardous to human health if it is used improperly or escapes its container. This is demonstrated in **Figure 1-2.**

Even the most toxic material is not hazardous as long as it is used correctly and stays in its intended container. The other principle that is important here is that even if the material escapes its container, it must have the ability to cause harm to humans, property, or the environment. Every day hazardous materials are packaged to be transported and are intended to stay in that package. When firefighters become involved in an incident that involves a hazardous material, an increased level of concern results, but as long as the material remains packaged, and the firefighters do not open the package, the risk to them is very minimal. The material may be toxic, flammable, corrosive, or reactive, and these properties remain no matter what happens—but to reach out and harm people it must escape or be forced from the container. The following are some standard definitions:

Hazardous Materials: A substance (either matter—solid, liquid, or gas—or energy) that when released is capable of creating harm to people, the environment, and property, including weapons of mass destruction (WMD) as defined in 18 U.S.

FIGURE 1-2 The material shown here is an example of one that ignites when it escapes its container and comes in contact with the air. A material that is air reactive is known as *pyrophoric*.

Code, Section 2332a, as well as other criminal use of hazardous materials, such as illicit labs, environmental crimes, or industrial sabotage. Some specific agency definitions follow.

DOT hazardous material: Any substance or material in any form or quantity that poses an unreasonable risk to the safety, health, and property when transported in commerce. The DOT also provides more exacting definitions with each hazard class.

EPA hazardous substances (40 CFR300.5): A chemical released into the environment that could be potentially harmful to the public's health or welfare.

OSHA hazardous chemicals (29 CFR1910.1200): Those chemicals that would be a risk to employees if exposed in the workplace.

LAWS, REGULATIONS, AND STANDARDS

It is important for the first responder to have a basic understanding of the legislative history of hazardous materials. It was in this area that fire departments

first became regulated as to how they would respond and how they would be trained. Many environmental and safety regulations affect how firefighters respond to emergencies. This section provides an overview of the major federal laws, regulations, and standards. Readers should consult their local environmental and OSHA offices to learn about state and local laws and regulations.

Development Process

It is important to understand the differences among laws, regulations, and standards. **Laws** result from legislation passed by Congress and signed by the president. **Regulations,** on the other hand, are developed by government agencies, like OSHA or the EPA, and have the weight of law but are not passed by Congress nor signed by the president. In some cases laws require the development of regulations and provide a framework for the government agency to follow. **Standards** are developed and reviewed by a nongovernmental consensus committee, such as the NFPA. These do not have the weight of law but could be applied by a regulating agency or in court. As citizens and as members of an industry, it is important for firefighters to participate in the development and review of laws, regulations, and standards.

NOTE

As time goes on, standards are being applied with the weight of law, typically by OSHA.

EMERGENCY PLANNING

The first law that regulated how fire departments respond to emergencies was the **Superfund Amendments and Reauthorization Act,** commonly referred to as **SARA.** This law was passed in 1986 for the protection of emergency responders and the community. Its intent was to inform emergency responders of chemical hazards within their community. The main component of the law is called **Emergency Planning and Community Right to Know Act (EPCRA).** It is divided into two sections: planning for emergencies and providing a mechanism to get chemical storage information to emergency responders.

State and Local Emergency Response Committees

The planning portion established a requirement to provide a **State Emergency Response Committee (SERC)** in each state. It is the responsibility of the

SERC to ensure that the state has the resources necessary to respond safely and effectively to chemical releases. The SERC also had to provide the framework to implement **Local Emergency Planning Committees** known as **LEPCs.** The LEPC is a group composed of the representatives of the community, emergency responders, industry, hospitals, media, and other government agencies. Most LEPCs are set up on a county basis,[1] although larger communities may have their own. The LEPC is responsible for the development of a hazardous materials emergency plan and its annual revision.

Local Emergency Response Plans

The emergency plan is an important component to a successful response. The emergency plan should outline emergency contacts and procedures. Target facilities such as those that store extremely hazardous substances are to be included, along with specific information and response tactics. A decision tree for evacuation and in-place sheltering is usually included. If the decision is made to evacuate citizens, the plan provides the mechanism to shelter and take care of these evacuees. It is important for personnel to have an understanding of this plan and know how to access its resources during an emergency.

The LEPC is also responsible for ensuring that local resources are adequate to handle a chemical release in the community. Ensuring that local responders receive the proper amount of training and that the emergency plan is evaluated annually by conducting an exercise (drill) is also part of their responsibility. It is important for the emergency services to be an integral player in the LEPC. Decisions regarding a fire department's response to chemical incidents may be planned at the LEPC level. Many of the facilities using hazardous materials will have representatives at these meetings, and before an emergency incident is a better time to meet these representatives. Funding for training and planning is also available through some LEPCs, since most federal HAZMAT grants are provided through the LEPC.

Chemical Inventory Reporting

The other section of EPCRA, usually referred to as SARA Title III, is the chemical reporting portion of the act. It requires some facilities to report chemical information to the state, LEPC, and local fire department. Failure to report this information on an annual basis can result in fines of $25,000 a day.

To qualify as a reporting facility, the facility has two methods of reaching a reporting threshold. Most facilities meet the reporting threshold of storing more than 10,000 pounds of a chemical. The 10,000-pound number is for a single chemical and is not the total of all chemicals on site. A facility that has more than 1,300 gallons of acetone is required to report because the amount represents more than 10,000 pounds of acetone. A facility that stores drums of ammonia and water is only required to report the ammonia, not the total weight of the drum. They do have to account for all of the ammonia at the facility, regardless of the location or type of container.

Another method of meeting the threshold of reporting is to store one of 366 chemicals that the EPA considers an **extremely hazardous substance (EHS).** The EHSs have separate reporting requirements and lower reporting thresholds. Some must be reported at 100 pounds. If these chemicals are released, then the facility has a whole host of responsibilities to comply with, including immediate notification of the LEPC, usually by dialing 9-1-1 or some other listed emergency contact number as determined by the LEPC. The common EHSs are chlorine, sulfur dioxide, anhydrous ammonia, and sulfuric acid. Because EHS materials can present an extreme threat to the community, it is important for emergency responders to become familiar with the facilities that use them and to use caution when responding to potential releases involving these materials.

Facilities that are required to report as SARA facilities are required to submit either a Tier 1 or Tier 2 Chemical Inventory Report. The Tier 2 report, **Figure 1-3,** is the most common and lists the chemical name, storage amount in a range, storage location and information, and emergency contact information, including twenty-four-hour phone numbers. The facility is also required to submit a list of chemicals or **Material Safety Data Sheets (MSDS).** A site plan may also be required that outlines the storage areas. EHS facilities are also required to submit a copy of their emergency plan. Depending on the state or locality, reporting requirements may be more stringent, although all have to meet the minimum listed. Note, however, that the information contained on a Tier 2 form regarding the chemicals that are located on site is for the prior year. Facilities report from January 1 to March 1 for the prior year.

The basis for this reporting is to inform the emergency responders of the hazard of responding to a SARA facility. It also allows the emergency responder to make informed decisions as to site entry, tactical decisions, and potential evacuation. MSDS are a valuable tool in obtaining chemical information, and SARA[2] requires their availability to emergency responders.

Tier Two	Facility Identification		Owner/Operator Name		Page ___ of ___ pages

Tier Two

Emergency and Hazardous Chemical Inventory

Specific Information by Chemical

Facility Identification

Name _____
Street _____
City _____ State _____ Zip _____

SIC Code _____ Dun & Brad Number _____

For Official Use Only — ID# _____ Date received _____

Owner/Operator Name

Name _____ Phone _____
Mail Address _____

Emergency Contact

Name _____ Title _____
Phone _____ 24 hr phone _____

Name _____ Title _____
Phone _____ 24 hr Phone _____

Important: Read all instructions before completing form   Reporting Period From January 1 to December 31, 20_____   ☐ Check if information is identical to the information submitted last year.

Chemical Description	Physical and Health Hazards	Inventory	Container Type / Temperature / Pressure	Storage Codes and Locations (Non-confidential) Storage Locations	Optional

CAS _____ Trade Secret ____
Chem. Name _____

Check all that apply — Pure Mix Solid Liquid Gas EHS

EHS Name _____

Fire
Sudden release of pressure
Reactivity
Immediate (acute)
Delayed (chronic)

Max Daily Amount (code)
Avg. Daily Amount (code)
No. Days on site (days)

☐

CAS _____ Trade Secret ____
Chem. Name _____

Check all that apply — Pure Mix Solid Liquid Gas EHS

EHS Name _____

Fire
Sudden release of pressure
Reactivity
Immediate (acute)
Delayed (chronic)

Max Daily Amount (code)
Avg. Daily Amount (code)
No. Days on site (days)

☐

CAS _____ Trade Secret ____
Chem. Name _____

Check all that apply — Pure Mix Solid Liquid Gas EHS

EHS Name _____

Fire
Sudden release of pressure
Reactivity
Immediate (acute)
Delayed (chronic)

Max Daily Amount (code)
Avg. Daily Amount (code)
No. Days on site (days)

☐

Certification (Read and sign after completing all sections)
I certify under penalty of law that I have personally examined and am familiar with the information in pages one through ___ and that based on my inquiry of those individuals responsible for obtaining information, I believe that the submitted information is true, accurate, and complete.

Name and official title of owner/operator or owner/operator authorized representative   Signature   Date signed

Optional Attachments
☐ I have attached a site plan
☐ I have attached a list of site coordinate abbreviations
☐ I have attached a descriptions of dikes and other safeguard measures

(A)

FIGURE 1-3 This form is an example of what facilities are required to submit to the fire department and the Local Emergency Planning Committee on an annual basis.

OSHA HAZWOPER REGULATION

Another part of SARA was the requirement for OSHA to develop a regulation covering activities that involve hazardous materials. The regulation, known as **Hazardous Waste Operations and Emergency Response,** became final on March 6, 1989. It is referred to as **HAZWOPER** or by its identification number[3] of 29 CFR 1910.120. There was considerable discussion within the fire service when this regulation was established as to whether or not it applied to all of the fire service. The primary unknown was whether this regulation applied to volunteer firefighters. In some states Hazwoper only affected the private sector and not career firefighters, while in other states only the public sector was subject to this regulation. In many cases volunteer firefighters were determined to be employees of the local government since they received workers' compensation benefits or received safety equipment from the employer, and so this regulation applied. To end this argument and confusion, the EPA adopted the same regulation and issued it on the EPA's behalf. The EPA regulation is also called HAZWOPER and is referenced by 40 CFR 311. It is interesting to note that OSHA now puts a provision in some of its regulations that specifically mandates the coverage of volunteer employees.

Tier Two	Facility Identification	Owner/Operator Name	

Tier Two

Emergency and Hazardous Chemical Inventory

Specific Information by Chemical

Facility Identification

Name _____
Street _____
City _____ State _____ Zip _____

SIC Code _____,_____   Dun & Brad Number _____

For Official Use Only

ID# _____
Date received _____

Owner/Operator Name

Name _____ Phone _____
Mail Address _____

Emergency Contact

Name _____ Title _____
Phone _____ 24 hr phone _____

Name _____ Title _____
Phone _____ 24 hr Phone _____

Page ____ of ____ pages

Important: Read all instructions before completing form | Reporting Period From January 1 to December 31, 20____ | ☐ Check if information is identical to the information submitted last year.

Confidential Location Information Sheet

Container Type | Temperature | Pressure

Storage Codes and Locations (Confidential) Storage Locations

Optional

CAS # _____

Chemical Name _____

☐

CAS # _____

Chemical Name _____

☐

CAS # _____

Chemical Name _____

☐

Certification (Read and sign after completing all sections)
I certify under penalty of law that I have personally examined and am familiar with the information in pages one through ___ and that based on my inquiry of those individuals responsible for obtaining information, I believe that the submitted information is true, accurate, and complete.

Name and official title of owner/operator or owner/operator authorized representative | Signature | Date signed

Optional Attachments
☐ I have attached a site plan
☐ I have attached a list of site coordinate abbreviations
☐ I have attached a descriptions of dikes and other safeguard measures

(B)

FIGURE 1-3 (Continued)

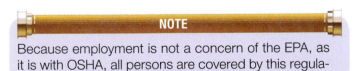

NOTE

Because employment is not a concern of the EPA, as it is with OSHA, all persons are covered by this regulation regardless of their employment status.

This regulation has had far-reaching effects for the fire service. It has required that certain training be provided, requires the development of standard operating procedures, and mandates certain requirements when handling chemical releases. It only allows persons trained in the handling of chemical releases to respond to chemical incidents. Only certain levels allow operation in and around chemical releases. Even just to be present at a chemical accident requires training, and as firefighters' activity levels increase, so does their need for further training.

Paragraph q

The majority of this OSHA regulation covers employers' responsibilities at hazardous waste sites. The last section, **paragraph q**, covers emergency response and applies to the fire service. It established five levels of training, as well as a requirement for annual refresher training. It requires that an incident command system be used and that there be an incident commander at chemical releases. At larger incidents a safety officer must be appointed. As firefighters progress up the training levels the requirements become more detailed.

The two-in/two-out rule originated from this OSHA regulation.

The five training levels established by OSHA mirror the four levels established by the NFPA. Their basic responsibilities are as follows:

Awareness Level. Persons trained at the awareness level respond to possible hazardous materials/Weapons of Mass Destruction (WMD) incidents, identify the potential for a hazardous materials release, call for trained assistance, and stand by isolating the area and denying entry to other persons. Persons trained at the Awareness level cannot take any action beyond this. This level is intended for police officers, public works employees, and other government employees.

Operations Level. The next level of training above awareness level that provides the foundation that allows for the responder to perform defensive activities at a hazardous materials/WMD incident. Acting defensively means that the responder does not enter a hazardous area, but can set up dikes, dams, and other confinement measures. Training at the operations level allows the firefighter to assist technicians in setting up the various activities that are required at an incident. Operations training can be expanded to include specialized activities such as decontamination, so that persons trained at this level can assist with this activity. This level is intended for the fire and EMS service.

Incident Commander Level. This level is designed for a person who has received operations level training and has received training in incident command procedures. This person will be in charge of the incident. To be the incident commander does not mean that this person has the highest level of chemical response training but is the senior response official. The IC must rely on the expertise of the other responders, such as the hazardous materials team, facility officials, or other technical specialists, to make strategic and tactical decisions.

Technician Level. This is the level at which offensive activities can be completed in the hazard area. Other than some specific restrictions outlined in HAZWOPER, there are no general restrictions as to the activities technicians can perform, as long as the activities fit within the scope of their training. This is the level at which leaks can be stopped and mitigation of the incident can be completed. Hazardous materials technicians are expected to mitigate or stop the incident from progressing.

Specialist Level. This level is identified only in HAZWOPER and is intended to be a level that has received a higher level of training above a technician, or may be someone who specializes in a specific chemical or area of expertise. The training concentrates on chemistry and the identification of unknown materials. In some instances, the specialist supervises the technicians. The NFPA removed this level in 1992.

Medical Monitoring

Other sections of this regulation include a requirement for medical monitoring of certain employees. An individual's responsibilities mandate whether the employer must provide that individual with an annual physical. A physical is required if an individual meets any of these requirements:

- Was exposed to a chemical above the permissible exposure limit.
- Wears a respirator or is covered by the OSHA respiratory regulation (29 CFR 1910.134).
- Was injured due to a chemical exposure.
- Is a member of a hazardous materials team.

The physician determines the extent of the exam and can establish that the exams be given every two years if that time period is determined to be acceptable. For the fire service the only persons required to be given a physical are the members of a hazardous materials team, although OSHA 1910.134 (Respiratory Protection) has a requirement that a medical survey questionnaire be answered by every firefighter. Depending on the answers to the questions, a physical may or may not be required. If a firefighter is exposed to a hazardous material above the OSHA-specified level, then the department is required to provide a physical exam to the firefighter. The physician determines the extent of the exam and any other future visit or tests.

OSHA requires that the medical records be kept by the employer for a period of thirty years past the last date of employment. This applies to the HAZWOPER regulation as well as other OSHA regulations. In simple terms any medical record generated by the employer, or for the employer, must be kept and must be available for employees if they request it.

STANDARDS

The group that establishes most of the standards for the fire service is the National Fire Protection Association. The NFPA establishes a variety of committees that develop standards, which are then made available for review and comment by the public. After the review process, each standard is voted on by the NFPA committee and, if passed, then goes to the next public meeting where it is voted on by the group in attendance. Once passed it then is sent to the Standards Council, where it is voted on again. If it passes there, it becomes a standard. As with emergency medical services in which an acceptable standard of care applies, typically based on a national or regional standard, the NFPA establishes a standard of care. A person cannot be held criminally liable for violating an NFPA standard, but may be held civilly liable.

> **NOTE**
>
> OSHA has used what is known as the general duty clause, which means that all employers have a general duty or an obligation to provide a workplace free from hazards.

One of the areas in which the NFPA standards have been applied as having the weight of a regulation is in the hazardous materials arena. OSHA has used this clause to cite employers for violating an NFPA standard.

NFPA 472

Prior to the 2008 edition, the NFPA had three hazardous materials response standards: NFPA 471, 472, and 473. In the 2008 edition, the NFPA eliminated the 471 Standard that covered the response to hazardous materials emergencies and moved much of its information into 472. The NFPA Hazardous Materials committee felt that training standards adequately covered the response objectives, and that only two standard documents were required. NFPA 472, Professional Competence of Responders to Hazardous Materials/Weapons of Mass Destruction Incidents is the listing of objectives required to meet the training levels established by the NFPA. OSHA established only basic criteria to meet in order for the employer to certify their employees as to their ability to respond to hazardous materials incidents. This NFPA document expands the requirements in order for the employer to certify their employees. The levels mirror those of the HAZWOPER regulation, but in 1992 the NFPA removed the specialist category. In the 1997 edition they added the following competencies: private sector specialist employee, hazardous materials branch officer, hazardous materials branch safety officer, tank car specialist, cargo tank specialist, and inter-modal specialist.

In the 2002 edition the NFPA added objectives related to terrorism response. These changes were part of the 1997 edition but were listed as tentative interim amendments[4] (TIAs). The 2002 version adopts them as part of the standard, and they are covered by this text. The 2008 edition adds several law enforcement competencies for the collection of evidence in hazardous situations. There are several criminal scenarios that involve the potential use of hazardous materials or weapons of mass destruction and present significant risk to investigators. The revision to this standard provides training objectives for response personnel who may be collecting evidence when hazardous situations exist. Also in the 2008 revision is the addition of **mission-specific competencies,** which used to be reserved for hazardous materials technicians. These competencies reflect the realities of real-world incidents. Many firefighters have access to and the ability to use air monitoring devices, which used to be a technician-level skill. The revision to the standard provides that, if properly trained and equipped, an operations-level responder can use some air monitoring devices.

NFPA 473

The third standard is NFPA 473, Competencies for EMS Personnel Responding to Hazardous Materials/Weapons of Mass Destruction Incidents. This standard adds some additional competencies above NFPA 472 with regard to EMS issues. It provides EMS Level I and Level II training levels. The competencies in this edition have been updated and use common terminology known to the EMS response community. Instead of Level I and II, the standard now relies on BLS and ALS providers. To meet the competencies of the standard, all EMS personnel will meet the core competencies of Hazardous Materials Operations and some of the Mission Specific Competencies based on their on-scene responsibilities. *Emergency Medical Response to Hazardous Materials Incidents,* authored by Richard Stilp and Armando Bevelacqua, offers more information on this topic.

Standard of Care

As mentioned before, emergency responders have to abide by a standard of care. In years past it was deemed acceptable to wash spilled gasoline down a storm drain. To do that now is illegal, and personnel could face state and federal charges for violating the Clean Water Act. This standard of care is composed of the laws, regulations, standards, local protocols, and experience detailed earlier. Violations of this standard of care are based on three theories: **liability**, **negligence,** and **gross negligence.**

When operating in and around hazardous materials, liability becomes a very real issue. Other than EMS responses like those shown in **Figure 1-4,** firefighters at a chemical release are likely to find themselves personally liable for any wrongdoing. Liability is being responsible for personal actions. In the example of the

FIGURE 1-4 Just as EMS responders have to follow a standard of care so that the patient is provided an appropriate level of care, HAZMAT response has a similar standard of care. *(Courtesy of Cambria County, Pennsylvania, Emergency Services)*

individual who made the decision to wash gasoline down a storm drain, that individual could end up in jail or have to pay a substantial fine.

> **NOTE**
>
> Environmental and safety regulations are designed to place blame on the person who made the decision, not on the organization itself, unless the organization has policies that violate these regulations.

Negligence is not following the standard of care or an accepted practice. Gross negligence is the willful disregard for the standard of care. In other words, a conscious decision was made not to follow the standard of care. To avoid becoming liable, firefighters should receive the training that is required for their positions and act only to that level of training. If firefighters are not sure what the next step should be, they should consult with their local HAZMAT team or other local or state environmental protection agency.

ADDITIONAL LAWS, REGULATIONS, AND STANDARDS

The items discussed next are ones that firefighters should be aware of, because they are commonly encountered or applied in chemical releases.

Hazard Communication

The hazard communication regulation issued by OSHA (29 CFR 1910.1200) requires that employers provide an MSDS for all chemicals located at a fa-

cility at quantities above "household quantities." The term household quantities allows for small amounts, such as those that would be purchased for home use, such that MSDS are not required. An example would be a gallon of household ammonia for use in an office. This would not require an MSDS, because this is a normal household quantity. If, however, a case of household ammonia were brought in, an MSDS would be required because this is not a normal household quantity.

The employer must provide a list of these MSDS materials and when they arrived on site. The employer must provide training on these MSDS materials and the hazard communication program. One of the important components of this program is the requirement for the facility to make these available for emergency responders. Fire departments[5] themselves are responsible for following this regulation.

Superfund Act

The Comprehensive Environmental Response, Compensation and Liability Act (CERCLA) is commonly referred to as the Superfund law. This law was established for the cleanup of toxic waste sites around the country. It was the first law to set the groundwork for the regulating of the fire service with regard to response to chemical emergencies. When responding to a Superfund site, some additional concerns and requirements must be followed.

The site has an existing emergency response plan, of which the local fire department should have a copy. The site should have its access limited, hazard zones established, and, depending on the nature of the work involved, may have a decontamination corridor created. Prior to work beginning on a Superfund site, the local fire department should meet with the site supervisor to learn the hazards of the site and any protection measures that may be in place. Because Superfund sites vary from old landfills to mercury dumping sites, it is not possible to provide here a breakdown of the potential site hazards and precautions that the site may hold. Prior to becoming a Superfund site, the area may be listed as a National Priority List (NPL) site. Firefighters should contact their state EPA or federal regional EPA office to learn of any Superfund or NPL sites in their area.

Clean Air Act

The Clean Air Act Amendments (CAAA), passed in 1990, have some language that affects the fire service. It requires that certain facilities file additional planning documents and that the LEPC and the local fire service be involved in training and exercises. As of June 1999, facilities were required to submit emergency plans, many of which must be coordinated with the local fire department. Other requirements, such as exercises and worst-case scenarios, may also involve the fire department.

Respiratory Protection

OSHA's respiratory protection regulation (29 CFR 1910.134) also impacts the fire service, because it adds additional requirements that were previously exempted. The biggest changes are the inclusion of the two-in/two-out rule, which was in place for chemical incidents but now also includes structural fire situations. The fire service is now required to fit test all firefighters and provide them with a medical survey or a physical exam.

Record keeping has also changed. Specific records must be kept by the fire department regarding the employees and the respiratory protection program. Records include the listing of personnel training, daily equipment checks, periodic maintenance, routine service, review of the respiratory protection program, and other items.

Firefighter Safety

NFPA Standard 1500, Fire Department Occupational Safety and Health Program, is sometimes referred to when discussing hazardous materials issues. This standard is a "broad-based" program for providing a safe workplace for firefighters. It has been applied by OSHA using the general duty clause, specifically regarding the two-in/two-out requirement.

NFPA Chemical Protective Clothing

NFPA Standards 1991 and 1992 are standards for chemical protective clothing ensembles, including encapsulated suits, splash-protective suits, and support garments. They establish design and use requirements.

NFPA 1994, Standard on Protective Ensembles for Chemical/Biological Terrorism Incidents, has three levels of protective equipment that can be used in the event of a chemical or biological attack.

LESSONS LEARNED

The maze of laws, regulations, and standards can be confusing and sometimes overwhelming. Most laws and regulations are not easy to read and it can take years to comprehend their true intent. They are subject to interpretation and change frequently. Emergency responders must keep abreast of the ones that affect their everyday jobs. In larger departments, staff may be assigned to monitor health, safety, and environmental laws and regulations. In smaller departments, this may not be missed, but the fire service is an industry the same as the chemical plant or gas station in any community. The fire service has to follow the same health, safety, and environmental laws and regulations. For everyone's health and safety, firefighters should keep abreast of these items, so they can help make communities safer places to live and work.

KEY TERMS

Air Monitoring Devices Used to determine oxygen, explosive, or toxic levels of gases in air.

Awareness Level The basic level of training for emergency response to an incident involving hazardous materials/weapons of mass destruction (WMDs), the basis of which is the firefighters' ability to recognize a hazardous situation, protect themselves and call for trained assistance, and secure the scene.

Clandestine Drug Labs Illegal labs set up to manufacture street drugs.

Emergency Planning and Community Right to Know Act (EPCRA) The portion of SARA that specifically outlines how industries report their chemical inventory to the community.

Extremely Hazardous Substances (EHS) A list of 366 substances that the EPA has determined present an extreme risk to the community if released.

Firefighters Protective Clothing (FFPC) Protective clothing that firefighters use for firefighting, rescue, and other first-response activities; also known as turnout or bunker gear.

Gross Negligence Occurs when an individual disregards training and continues to act in a manner without regard for others.

Hazardous Waste Operations and Emergency Response (HAZWOPER) The OSHA regulation that covers safety and health issues at hazardous waste sites, as well as response to hazardous materials incidents.

Incident Commander Level A training level that encompasses the operations level with the addition of incident command training. Intended to be the person who may command a chemical incident.

Laws Legislation that is passed by the House and Senate, and signed by the president.

Liability The possibility of being held responsible for individual actions.

Local Emergency Planning Committee (LEPC) A group composed of members of the community, industry and emergency responders to plan for a chemical incident, and to ensure that local resources are adequate to handle an incident.

Material Safety Data Sheet (MSDS) Information sheet for employees that provides specific information about a chemical, with attention to health effects, handling, and emergency procedures.

Mission-Specific Competencies The knowledge, skills, and abilities to perform the mission of the organization. Examples are provided in NFPA 472.

Negligence Acting in an irresponsible manner or different from the way in which someone was trained; that is, differing from the standard of care.

Operations Level The next level of training above awareness that provides the foundation which allows for the responder to perform defensive activities at a hazardous materials incident.

Paragraph q The paragraph within HAZWOPER that outlines the regulations that govern emergency response to hazardous materials incidents.

Regulations Developed and issued by a governmental agency and have the weight of law.

Specialist Level A level of training that provides for a specific type of training, such as railcar specialist; someone who has a higher level of training than a technician.

Standards Usually developed by consensus groups establishing a recommended practice or standard to follow.

State Emergency Response Committee (SERC) A group that ensures that the state has adequate training and resources to respond to a chemical incident.

Superfund Amendments and Reauthorization Act (SARA) A law that regulates a number of environmental issues, but is primarily for chemical inventory reporting by industry to the local community.

Technician Level A high level of training that allows specific offensive activities to take place, to stop or handle a hazardous materials incident.

REVIEW QUESTIONS

1. How does the response to hazardous materials differ from a structural fire response?
2. How can hazardous materials be easily defined?
3. Who is to receive the chemical reporting required under SARA Title III?
4. Who is responsible for ensuring the proper response to a hazardous materials emergency?
5. Which regulation covers all emergency responders who respond to chemical accidents?
6. What is the difference between negligence and gross negligence?
7. Which regulation requires employers to maintain Material Safety Data Sheets?
8. What was the largest change to the 2008 Edition of NFPA 472?
9. Which training regulations are most firefighters required to follow?
10. To perform offensive activities at a hazardous materials release, what level of training is required?

FOR FURTHER REVIEW

For additional review of the content covered in this chapter, including activities, games, and study materials to prepare for the certification exam, please refer to the following resources:

Firefighter's Handbook Online Companion
Click on our Web site at **http://www.delmarfire.cengage.com** for FREE access to games, quizzes, tips for studying for the certification exam, safety information, links to additional resources and more!

Firefighter's Handbook Study Guide
Order#: 978-1-4180-7322-0
An essential tool for review and exam preparation, this Study Guide combines various types of questions and exercises to evaluate your knowledge of the important concepts presented in each chapter of *Firefighter's Handbook*.

ENDNOTES

1. Some LEPCs are established on a city basis; some states may only have one. The format varies from state to state.

2. The OSHA hazard communication regulation (29 CFR 1910.1200) also makes this requirement.

3. Federal regulations are published in the Federal Register, which comes out daily. Annually, they are published in a text called the Code of Federal Regulations (CFR). Each federal agency is identified by a two-digit number (29 for OSHA, 40 for EPA, 49 for DOT), and each regulation is given a number to identify it.

4. A TIA is tentative because it has been established outside of the normal NFPA process, in between revisions. When the standard comes up for revision it will be subject to the normal review process.

5. The applicability varies from state to state, depending on OSHA's jurisdiction. Some states may have more stringent regulations.

ADDITIONAL RESOURCES

Bevelacqua, Armando S., *Hazardous Materials Chemistry,* 2nd ed. Thomson Delmar Learning, Clifton Park, NY, 2006.

Bevelacqua, Armando S. and Richard Stilp, *Hazardous Materials Field Guide,* 2nd ed. Thomson Delmar Learning, Clifton Park, NY, 2007.

Bevelacqua, Armando S. and Richard Stilp, *Terrorism Handbook for Operational Responders,* 3rd ed. Delmar, Cengage Learning, Clifton Park, NY.

Hawley, Chris, *Hazardous Materials Air Monitoring and Detection Devices,* 2nd ed. Thomson Delmar Learning, Clifton Park, NY, 2006.

Hawley, Chris, *Hazardous Materials Incidents,* 3rd ed. Thomson Delmar Learning, Clifton Park, NY, 2007.

Henry, Timothy V. *Decontamination for Hazardous Materials Emergencies.* Delmar Publishers, Albany, NY, 1998.

Laughlin, Jerry, and David Trebisacci, eds., *Hazardous Materials Response Handbook,* 4th ed. National Fire Protection Association, Quincy, MA, 2002.

Lesak, David, *Hazardous Materials Strategies and Tactics.* Prentice-Hall, 1998.

Noll, Gregory, Michael Hildebrand, and James Yvorra, *Hazardous Materials Managing the Incident,* 3rd ed. Red Hat Publishing, Chester, MD, 2005.

Schnepp, Rob and Paul Gantt, *Hazardous Materials: Regulations, Response & Site Operations.* Delmar Publishers, Albany, NY, 1998.

Stilp, Richard and Armando Bevelacqua, *Emergency Medical Response to Hazardous Materials Incidents.* Delmar Publishers, Albany, NY, 1996.

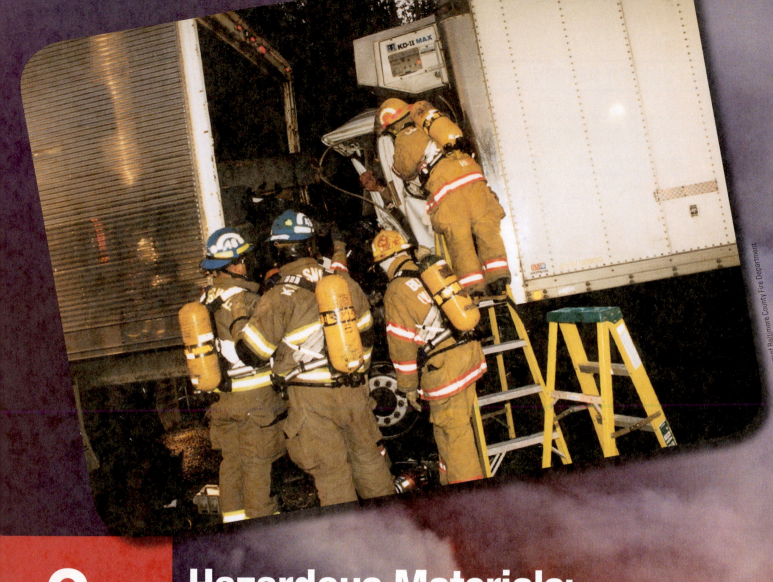

Courtesy of Baltimore County Fire Department.

2

Hazardous Materials: Recognition and Identification

While the news media tends to focus on the large and spectacular, the reality is that every day the fire service responds to literally hundreds of hazardous materials–related incidents. These incidents usually do not get much attention because they are small in scope and are handled with a minimal amount of fire department resources. Common examples include natural gas leaks and flammable and combustible liquid spills.

We have an old saying in the hazardous materials community that the initial ten minutes of an incident will dictate the tone for the first hour of an incident. Clearly, the actions—or inactions—of the first responders will set the tone for an incident. I think back to two incidents in my career that reinforce this point.

The first incident involved an engine company being sent to investigate a call regarding an unknown liquid spilled along a highway. Upon arrival, the engine company officer found a very viscous red liquid spilled along a long distance of a road, apparently from the rear of a tractor-trailer. Unsure of the identity and the potential hazard of the liquid, the officer requested a hazardous materials unit to respond to the scene and provide assistance. Using their monitoring and detection equipment, responders were eventually able to determine that the liquid posed no hazard to the community. In fact, the unknown liquid was eventually identified as strawberry syrup! Although the officer took some ribbing from his peers, he clearly made the proper decisions to ensure the safety of both his personnel and the community.

In the second incident, a police officer was called to provide assistance to a public works road crew that had discovered what appeared to be a 5-pound portable fire extinguisher wrapped in duct tape with a fuse on the top. Although he believed it to be a hoax, the officer still requested the response of the fire department bomb squad. After conducting a thorough risk assessment and requesting the additional assistance of a hazardous materials response team, the bomb squad was able to determine that the perceived hoax was, in fact, an actual explosive device. When the device was disrupted (i.e., blown up) by the bomb squad, it made a lasting impression on the police officer!

In both of these instances, recognition and identification by the first responders set the tone for the incident. If you do not know what to do, isolate the area, deny entry, and call for help.

—*Street Story by Gregory G. Noll, Emergency Planning and Response Consultant, Hildebrand and Noll Associates, Lancaster, Pennsylvania*

LEARNING OBJECTIVES

After completing this chapter, the reader should be able to:

2-1. Explain the various tools for recognition and identification of hazardous materials.

2-2. Identify the hazards associated with the nine hazard classes as defined by DOT.

2-3. Identify the standard occupancies where hazardous materials may be used or stored.

2-4. Identify the standard container shapes and sizes and common products.

2-5. Identify both facility- and transportation-related markings and warning signs.

2-6. Identify the standard transportation types for highway and rail.

2-7. Explain the use of transportation containers in identifying possible contents.

2-8. Explain the importance of understanding chemical and physical properties of hazardous materials.

INTRODUCTION

This chapter is of primary importance to the emergency responder.

The inability to recognize the potential for chemicals to be present and the inability to identify the chemical hazard can place firefighters in severe danger. Firefighting is inherently dangerous, but the response to a hazardous materials release creates an additional risk. Not only can there be immediate effects from some materials, but multiple exposures can have far-reaching effects. As depicted in the opening photo, firefighters are using proper PPE but taking a considerable risk to rescue a live victim. More information on this rescue and risk is provided in the case study in Chapter 5. Although fires have killed hundreds before, these types of

fires are rare. Hazardous materials incidents have killed thousands and injured countless more. In 1984, a release of methyl isocyanate in Bhopal, India, killed several thousand and injured tens of thousands more. This incident was the basis for the Emergency Planning and Community Right to Know Act (EPCRA) because several facilities in the United States use this material.

Four basic clues to recognition and identification are (1) location and occupancy; (2) placards, labels, and markings; (3) container types; and (4) the senses. A mere suspicion in any of these areas should be enough to place a first responder on guard for the possibilities of a chemical release and its associated hazards, **Figure 2-1.**

SAFETY

It is through recognition and identification (R&I) that firefighters can impact their ability to stay alive.

NOTE

In the hazardous materials section, NFPA 1001 will require that the student receive hazardous materials training at the Operations level. The information in this text covers both the Awareness level and the Operations level. In some cases it exceeds the Operations level. All of this information is important for firefighters' survival. See the discussion on NFPA standards and OSHA's HAZWOPER for more information on Awareness and Operations level training.

FIGURE 2-1 The four basic clues to recognition and identification are location and occupancy; placards, labels, and markings; container types; and senses.

LOCATION AND OCCUPANCY

<table>
<tr><td>CAUTION</td></tr>
</table>

The size of the community does not impact the potential for hazardous materials; every community has hazardous materials.

Most communities have a gasoline station or a hardware store. The average home has a large amount of hazardous materials that can cause enormous prob-lems during a response. In rural communities, farms present unique risks due to the storage of pesticides and fertilizers, **Figure 2-2.** All of these locations and occupancies provide the potential for the storage of hazardous materials. In general, the more industrialized a community is, the more hazardous materials the community will contain. Communities adjacent to industrialized areas or along major transportation corridors (interstate highways, rail, water) may also have the same hazards because these materials can travel through the community, **Figure 2-3.** Buildings that typically store hazardous materials include hardware stores, hospitals, auto part supply stores, dry cleaners, manufacturing facilities, print shops, doctors' offices, photo labs, agricultural supply stores, semiconductor manufacturing facilities, electronics manufacturing facilities, light to heavy industrial facilities, marine terminals, rail yards, airport terminals and fueling areas, pool chemical stores, paint stores, hotels, swimming pools, and food manufacturing facilities.

PLACARDS, LABELS, AND MARKINGS

This section examines the first concrete evidence of the presence of hazardous materials. A number of systems are used to mark hazardous materials containers, buildings, and transportation vehicles. The systems result from laws, regulations, and standards, and in some cases from a combination of the three. As an example the **Building Officials Conference Association (BOCA)** code, which has been adopted as a regulation in local communities, requires the use of the NFPA 704M marking system, **Figure 2-4,** for certain occupancies.

FIGURE 2-2 Agricultural supply stores have a large quantity of hazardous materials, including pesticides, herbicides, and fertilizers. In many cases they also have fuels, including propane.

FIGURE 2-3   If a community has a road, the potential for a hazardous materials incident exists. One of the most common chemical releases is a gasoline spill.

FIGURE 2-4   This NFPA 704M symbol is used to warn of potential chemical dangers in the building. It warns of fire, health, reactivity, and special hazards.

Placards

The most commonly seen item for identifying the location of hazardous materials is the placard. The Department of Transportation (DOT) regulates the movement of hazardous materials ("dangerous goods" in Canada) by air, rail, water, roadway, and pipeline by means of 49 CFR 170-180. After meeting certain guidelines a shipper must placard a vehicle to warn of the storage of chemicals on the vehicle, an example of which is shown in **Figure 2-5. Table 2-1** and **Table 2-2** provide further explanations of the placarding system.

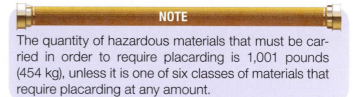

NOTE

The quantity of hazardous materials that must be carried in order to require placarding is 1,001 pounds (454 kg), unless it is one of six classes of materials that require placarding at any amount.

The DOT has established a system of nine hazard classes that uses more than twenty-seven placards to identify a shipment. The idea behind these hazard classes is to provide a general grouping to a shipment and to provide some basic information regarding the potential hazards. The placards, like the one shown in **Figure 2-6,** are 10¾ inches by 10¾ inches and are to be placed on four sides of the vehicle. Labels are 3⁹⁄₁₀ inches by 3⁹⁄₁₀ inches and are affixed near the shipping name on the container. Labels, for the most part, are smaller versions of placards and are designed to provide warnings about the package contents. There are labels for some materials that do not have or require placards.

The system is designed so that materials designated by the DOT as potentially harmful to the

FIGURE 2-5 The DOT requires some shippers of hazardous materials to provide placards to warn responders of chemicals that may be on the truck.

TABLE 2-1 Materials That Require Placarding at Any Amount (DOT Table 1)	
Hazard Class or Division	**Placard Type**
1.1	Explosives 1.1
1.2	Explosives 1.2
1.3	Explosives 1.3
2.3	Poison gas
4.3	Dangerous when wet
5.2 (Organic peroxide, Type B, liquid or solid, temperature controlled)	Organic peroxide
6.1 (Inhalation hazard, Zone A or B)	Poison inhalation hazard
7 (Radioactive label III only)	Radioactive

TABLE 2-2 Materials That Require Placarding at 1,001 Pounds (454 Kg) (DOT Table 2)	
Class or Division	**Placard Type**
1.4	Explosives 1.4
1.5	Explosives 1.5
1.6	Explosives 1.6
2.1	Flammable gas
2.2	Nonflammable gas
3	Flammable
Combustible Liquid	**Combustible**
4.1	Flammable solid
4.2	Spontaneously combustible
5.1	Oxidizer
5.2 (Other than organic peroxide, Type B, liquid or solid, temperature controlled)	Organic peroxide
6.1 (Other than Inhalation hazard, Zone A or B)	Poison
6.1 (PG III)	PG III
6.2	None
8	Corrosive
9	Class 9
ORM-D	None

FIGURE 2-6 "Corrosive" placard. Not to scale.

environment, humans, and animals are easily identified. Such a material has to present an unreasonable risk to the safety, health, or property upon contact. The DOT has two placarding tables, which are called Table 1 and Table 2. The materials that are on Table 1 (as shown in **Table 2-1**) are those that are most hazardous and require the use of a placard no matter what the quantity being shipped. The materials in DOT Table 2 (as shown in **Table 2-2**) are those that require placarding at 1,001 pounds. The criteria establishing this 1,001-pound (454 kg) rule are not clearly defined, but responders should be aware that a spill of 999 pounds (453 kg) of a material can be as hazardous as 1,001 pounds (454 kg). The shipper uses the hazardous materials tables (49 CFR 172.504) to determine which labels and placards are required. The tables may also list a packing group for the material, which indicates the danger associated with the material being transported.

- Packaging Group I: Greatest danger
- Packaging Group II: Medium danger
- Packaging Group III: Minor danger

TABLE 2-3 Packing Groups: Flammability

Flammability—Class 3 Packing Group (PG)	Flash Point	Boiling Point
I		≤ 95°F (35°C)
II	≤ 73°F (23°C)	> 95°F (35°C)
III	≥ 73°F (23°C) and ≤ 141°F (60.5°C)	> 95°F (35°C)

TABLE 2-4 Packing Groups: Poisonous Gas, Division 2.3

Hazard Zone	Inhalation Toxicity—Lethal Concentration (LC_{50})*
Hazard Zone A	LC_{50} less than or equal to 200 ppm
Hazard Zone B	LC_{50} greater than 200 ppm and less than or equal to 1,000 ppm
Hazard Zone C	LC_{50} greater than 1,000 ppm and less than or equal to 3,000 ppm
Hazard Zone D	LC_{50} greater than 3,000 ppm and less than or equal to 5,000 ppm

*LC_{50} is the lethal concentration (gases) to 50 percent of an exposed population.

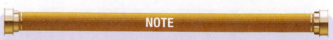

NOTE

Placards are useful because they take advantage of four ways to communicate the hazard class they represent.

NOTE

The DOT uses several terms to provide a measure of toxicity. The DOT establishes ranges for the shipment of toxic materials. The determination of how toxic a material is can be based upon exposure limits established for many chemicals. LC_{50} and LD_{50} are terms used to describe a concentration or amount of a material that causes lethal effects to 50 percent of an exposed population. LC is lethal concentration, used to measure toxicity of gases, and LD is a lethal dose of solids and liquids. To determine an LC_{50} or LD_{50}, researchers expose a population of test animals to a given amount of a substance. At the level that 50 percent of the test population dies, this establishes the LC_{50} or LD_{50}.

When discussing toxicity, two common units of measure for the amount of material required to cause toxic effects are parts per million (ppm) and milligrams per cubic meter cubed (mg/M³). The use of ppm usually is in reference to a gas and represents a part of a million units, as one marble is one part per million of a million marbles. The use of mg/M³ usually refers to an amount of a solid or liquid material. It is a comparison of the amount of chemical in milligrams (mg) as compared to a cubic meter of air, meaning that amount of the material that could cause toxic effects within that cubic meter. Similarly, the use of mg/L is a reference to the amount (mg) of a given toxic chemical within the space of a liter. More information related to toxicity is provided in Chapter 4.

TABLE 2-5 Packing Group Division 6.1

For packing materials that are toxic via a route other than inhalation, the following are used:

Packing Group	Oral Toxicity—Lethal Dose (LD_{50})* (mg/kg)	Dermal Toxicity—Lethal Dose (LD_{50}) (mg/kg)	Inhalation via Dusts and Mists (LC_{50}) (mg/L)
I	≤ 5	≤ 40	≤ 0.5
II	> 5 and ≤ 200	> 40 and ≤ 200	> 0.5 and ≤ 2
III	Solids: > 50 and ≤ 200; Liquids: > 50 and ≤ 500	> 200 and ≤ 1000	> 2 and ≤ 10

*(LD_{50}) Lethal dose (solids and liquids) to 50 percent of the exposed population.

TABLE 2-6 Division 6.1, Toxic Materials Poisonous by Inhalation

Packing Group	Vapor Concentration and Toxicity
I (Hazard Zone A)	$V \geq 500\ LC_{50}$ and $LC_{50}\ 200\ mL/M^3$

FIGURE 2-7 There are three ways to signify the four-digit ID number for bulk shipments. The most common is for the four-digit DOT identification number to be placed in the middle of the placard.

Packaging groups are only assigned to classes 1, 3 through 6, 8, and 9 as shown in **Tables 2-3** through **Table 2-6.** These are determined based on flash points, boiling points, and toxicity.

Placards have distinct colors, they have a picture at the top of the triangle depicting a representation of the hazard, they state the hazard class in the middle of the placard, and they display the hazard class and division number in the bottom triangle.

As an example the placard shown in Figure 2-6 is black and white, which represents the corrosive class. The top of the placard has a picture of a hand and a steel bar being eaten away by the corrosive material being poured on it, the middle of the placard states corrosive, and at the bottom the class number 8 shows.

The DOT also requires the addition of a four-digit number, known as the United Nations/North America (UN/NA) identification number, either on a placard or on an adjacent orange rectangle. This identifies a bulk shipment of over 119 gallons (450 liters) and provides an identity to the material. A tank truck carrying gasoline, which would be considered a bulk shipment, would display a Flammable Liquid placard[1] with the number 1203 either in the middle of the placard or on an orange rectangle adjacent to the placard, as shown in **Figure 2-7.** This provides an additional bit of information to the responder, because without the UN/NA number, the only information provided would be that a flammable liquid was on board.

The nine hazard classes and subdivisions[2] are discussed next.

Class 1, Explosives (Figure 2-8)

- *Division 1.1.* Mass explosion hazard, such as black powder, dynamite, ammonium perchlorate, detonators for blasting, and RDX explosives.
- *Division 1.2.* Projectile hazard, such as aerial flares, detonating cord, detonators for ammunition, and power device cartridges.
- *Division 1.3.* Fire hazard or minor blast hazard. Examples include liquid-fuel rocket motors and propellant explosives.
- *Division 1.4.* Minor explosion hazard, which includes line throwing rockets, practice ammunition, detonation cord, and signal cartridges.
- *Division 1.5.* Very insensitive explosives, which do have mass explosion potential but during normal shipping would not present a risk. Ammonium nitrate and fuel oil (ANFO) mixtures are an example of this division.
- *Division 1.6.* Also very insensitive explosives that do not have mass explosion potential. Materials that present an unlikely chance of ignition are part of this grouping.

The placards for divisions 1.1, 1.2, or 1.3 will have the division number and compatibility group located just above the class number for any quantity, as shown in Figure 2-8. For divisions 1.4, 1.5, and 1.6, the placard will have a compatibility group number when a vehicle is transporting 1,001 pounds (454 kg) or more. The DOT provides compatibility charts that dictate that certain high-hazard shipments cannot be mixed with other hazardous materials. The compatibility charts are designed to prevent the mixing of certain materials in the event of an accident. Some materials when mixed can create substantial issues such as explosions, fires, or release of toxic gases.

SAFETY

Incidents involving explosives can be very dangerous, especially when involved in fire.

Making a tactical decision to attack a fire involving explosives can endanger the responders, especially if the fire has reached the cargo area of the vehicle. The recommendations in the DOT **Emergency Response Guidebook** (**ERG**—Guide page 112) should be followed, paying particular attention to the isolation and evacuation distances. When explosives are involved in traffic accidents not involving fire, the actual threat is minimized depending on the circumstances. As long as the explosives were transported legally and as they were intended to be transported, the responders should face little danger. Some cities require an escort and that the explosives be transported at nonpeak hours.

Spilled explosive materials may present a health hazard if inhaled or absorbed through the skin. When explosives are involved in a fire situation, the best course of action would be the immediate withdrawal from the area as well as isolating and denying entry. The area should be evacuated following recommendations from the current version of the DOT ERG. Many chemicals used in explosives are powerful oxidizers that if ignited can be very aggressive and will burn with considerable energy, and in some cases are explosive in nature.

FIGURE 2-8 "Explosive" placards and labels.

ANFO EXPLOSION

On November 29, 1988, an engine company from the Kansas City, Missouri, Fire Department was dispatched to a reported pickup truck fire. While en route to the incident the engine company was told to use caution because explosives were reportedly involved. When they arrived they found that they had two separate fires, one in the pickup truck and the other in a trailer. The first engine began to extinguish the fire in the pickup truck and requested a second engine for assistance as well as the district battalion chief. They attempted to contact the second engine to warn them of the explosives on fire on top of the hill. The second engine arrived and began to attack the fire on top of the hill. They requested the assistance of the first engine and also requested a squad for water. From the radio communications with the battalion chief the crews thought that the explosives had already detonated.

The battalion chief was a quarter of a mile away where he had stopped to talk with the security guards when the explosives detonated. The explosion moved the chief's car 50 feet (15 m) and blew in the windshield. The blast was heard for 60 miles (87 Km) and damaged homes within 15 miles (24 Km). The chief requested additional assistance and staged the responding companies. There was a report that there were more explosives on the hill that had not yet detonated. Luckily the chief did not let any other responders into the scene because shortly after pulling back, a second explosion went off, reportedly larger than the first one.

The next morning a team of investigators went to the site to begin the investigation. They discovered that six firefighters were killed in the blast and the explosion was very devastating. Only one engine was recognizable; the other was reduced to the frame rail. The first blast was from 17,000 pounds (7,711 kg) of an ammonium nitrate, fuel oil, and aluminum mixture. It also had 3,500 pounds (1,588 kg) of ANFO. The second explosion involved 30,000 pounds (13,608 kg) of the mixture. The use of ANFO is common throughout the United States for blasting purposes. Ammonium nitrate is commonly used as a fertilizer in the agricultural business and is used on residential lawns. The Oklahoma City Alfred P. Murrah Federal Building explosive was devised of ammonium nitrate and nitromethane, very similar to ANFO. The World Trade Center bombing in 1993 used an explosive made up primarily of urea nitrate and three cylinders of hydrogen. Both explosives are comparable to each other, and when they are used improperly or for criminal purposes the results can be devastating.

FIGURE 2-9 "Gas" placards and labels.

Class 2, Gases (Figure 2-9)

- *Division 2.1.* Flammable gases that are ignitable at 14.7 psi (1 bar) in a mixture of 13 percent or less in air, or have a flammable range with air of at least 12 percent regardless of the lower explosive limit (LEL). Propane and isobutylene are examples of this division.

- *Division 2.2.* Nonflammable, nonpoisonous, compressed gas, including liquefied gas, pressurized **cryogenic gas,** and compressed gas in solution. Carbon dioxide, liquid argon, and nitrogen are examples.

- *Division 2.3.* Poisonous gases that are known to be toxic to humans and would pose a threat during transportation. Chlorine and liquid cyanogen are common examples of this division. Gases assigned to this division are also assigned a letter code identifying the material's toxicity levels. These levels are discussed further in the section on toxicology. The hazard zones associated with this division are:

  Hazard Zone A: LC_{50} less than or equal to 200 ppm

  Hazard Zone B: LC_{50} greater than 200 ppm and less than or equal to 1,000 ppm

  Hazard Zone C: LC_{50} greater than 1,000 ppm and less than or equal to 3,000 ppm

  Hazard Zone D: LC_{50} greater than 3,000 ppm and less than or equal to 5,000 ppm

The hazard zones are a quick way to determine how toxic a material is. Hazard Zone A materials are more toxic than Hazard Zone B materials, and so forth. The shipping papers will identify these by the addition of Poison Inhalation Hazard (PIH) Zone A, B, C, or D.

NOTE

Remember that many of these gases have multiple hazards. Consider that based on information that can be obtained from a **Material Safety Data Sheet (MSDS)** and the DOT Emergency Response Guide (ERG), some of these products such as 2.2 can also be flammable and cause unforeseen problems.

Class 3, Flammable Liquids (Figure 2-10)

- Flammable liquids have a flash point of less than 141°F (60.5°C). Gasoline, acetone, and methyl alcohol are examples.

- Combustible liquids are those with flash points above 100°F (37.7°C) and below 200°F (93°C). The DOT allows liquids with a flash point of 100°F (37.7°C) to be shipped as a combustible liquid. Examples include diesel fuel, kerosene, and various oils.

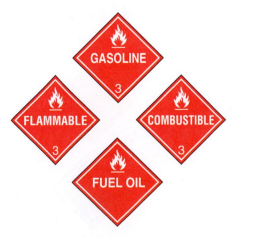 

FIGURE 2-10 "Flammable" and "Combustible" placards and label.

CAUTION

The difference between a flammable liquid and a combustible liquid is based on the material's flash point. The flash point is the temperature of the liquid at which there could be a flash fire if an ignition source is present. The fire service usually refers to a flammable liquid as one that has a flash point of 100°F (37.7°C) or lower. The DOT classifies any liquid with a flash point of 141°F (60.5°C) or lower as being a flammable liquid. Any liquid with a flash point greater than 141°F (60.5°C) is considered combustible by the DOT. It can be confusing when using the terms flammable and combustible to describe a liquid material. Most references, with the exception of the DOT, use 100°F (37.7°C) as the criteria for flammable and combustible.

Class 4, Flammable Solids (Figure 2-11)

- *Division 4.1.* Includes wetted explosives, self-reactive materials, and readily combustible solids. Examples include magnesium ribbons, picric acid, explosives wetted with water or alcohol, or plasticized explosives.

- *Division 4.2.* Composed of spontaneously combustible materials including pyrophoric materials or self-heating materials. An example is zirconium powder.

- *Division 4.3.* Dangerous-when-wet materials are those that when in contact with water can ignite or give off flammable or toxic gas. Calcium carbide when mixed with water makes acetylene gas, which is very flammable as well as unstable in this

form. Sodium is another example of a material that when wet can ignite explosively. Lithium and magnesium are not as explosive as sodium but will react with water. Magnesium if on fire will react violently if water is used in an attempt to extinguish the fire.

Class 5, Oxidizers and Organic Peroxides (Figure 2-12)

- *Division 5.1.* The class assigned to materials that readily release oxygen; by yielding oxygen, an **oxidizer** can easily cause or enhance the combustion of other materials. Oxidizers can dramatically increase the rate of burning when combustible material is ignited. Ammonium nitrate and calcium hypochlorite are examples.

- *Division 5.2.* The organic peroxides, which have the ability to explode or polymerize, which if contained is an explosive reaction. These are further subdivided into seven types:

  Type A: Can explode upon packaging. These are DOT forbidden, which means they cannot be transported and must instead be produced on site.

  Type B: Can thermally explode, considered a very slow explosion.

  Type C: Neither detonates nor **deflagrates** rapidly, and will not thermally explode.

  Type D: Only detonates partially or deflagrates slowly, and has medium or no effect when heated and confined.

  Type E: Shows low or no effect when heated and confined.

  Type F: Shows low or no effect when heated and confined, and has low or no explosive power.

  Type G: Is thermally stable and is desensitized.

FIGURE 2-11 "Flammable solids" placards and labels (Class 4).

FIGURE 2-12 "Oxidizers and organic peroxides" placards and labels (Class 5).

SAFETY

A chemical that polymerizes is one that presents an extreme danger to firefighters. Polymerization is a runaway chain reaction that once started cannot be stopped. If the material is contained, it will rupture the container in a violent manner, sending pieces of the container for a considerable distance. Polymerization can be best described as a chain. It starts off as one link, and once the reaction is started, many other links develop and attach themselves to the other links. Each link develops a great deal of heat and for each 41°F (5°C) of heat that develops, the reaction rate doubles. Thus, they quickly duplicate themselves, eventually rupturing the container. The reaction can result in a fire or explosion, since many of the chemicals that can polymerize are also flammable liquids. A container of spray foam insulation that is used for sealing spaces in walls around doors and windows is a good example. When the foam hits the air, it reacts and rapidly expands to a much greater volume. This reaction is uncontrollable and will continue to expand for quite some time until the polymerization is completed. A small container that holds a few ounces does not present much risk, but a railcar full of a chemical that may polymerize presents an extreme risk to the community.

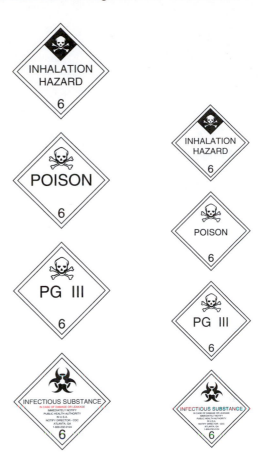

FIGURE 2-13 "Poisonous materials" placards and labels (Class 6).

Class 6, Poisonous Materials (Figure 2-13)

- *Division 6.1.* Materials that are so toxic to humans that they would present a risk during transportation. Examples include arsenic and aniline.
- *Division 6.2.* Composed of microorganisms or their toxins, which can cause disease to humans or animals. Anthrax, rabies, tetanus, and botulism are examples.

Hazard zones are associated with Class 6 materials:

- *Hazard Zone A:* LC_{50} less than or equal to 200 ppm
- *Hazard Zone B:* LC_{50} greater than 200 ppm and less than or equal to 1,000 ppm.

Class 7, Radioactive Materials (Figure 2-14)

- Those materials determined to have radioactive activity at certain levels.
- Although there is only one placard, the labels shown in Figure 2-14 are further subdivided into Radioactive I, II, and III, with level III being the highest hazard. The designation I, II, or III is dependent on two criteria: the transport index and the radiation level coming from the package. The transport index, which is comparable to the United Nations and IAEA criticality safety index is the degree of control the shipper is to use and is based

on a calculation of the radiation threat the package presents. Responders should understand that a package labeled radioactive may be emitting radiation. These emissions of radiation are legal within certain guidelines:

- *Radioactive I label*—less than 0.005 mSv/hr (0.5 mR/hr)
- *Radioactive II label*—more than Radioactive I and less than 0.5 mSv/hr (50 mR/hr)
- *Radioactive III label*—more than Radioactive II and less than 2 mSv/hr (200 mR/hr)
- *Fissile material label*—for radioactive material that can be used as a fuel. Common examples include uranium-233, uranium-235, uranium-239, and uranium-241.

Packages emitting more than 2 mSv/hr require special handling and transportation and are subject to additional regulations.

Class 8, Corrosives (Figure 2-15)

- Includes both acids and bases, and is described by the DOT as a material capable of causing visible destruction in skin or corroding steel or aluminum. Examples include sulfuric acid and sodium hydroxide.

FIGURE 2-14 "Radioactive materials" placards and labels (Class 7).

FIGURE 2-15 "Corrosives" placards and labels (Class 8).

Class 9, Miscellaneous Hazardous Materials (Figure 2-16)

- A general grouping that is composed of mostly hazardous waste. Dry ice, molten sulfur, liquid asphalt, and polymeric beads are examples that would use a Class 9 placard.

- This is known as a catch-all category. If a substance does not fit into any other category and presents a risk during transportation then it becomes Class 9.

Other Placards and Labels

- The Dangerous placard is used when the shipper is shipping a mixed load of hazardous materials. If the shipper ships 2,000 pounds (907 kg) of cor-

FIGURE 2-16 "Miscellaneous hazardous materials" placards and labels (Class 9).

rosives and 2,000 pounds (907 kg) of a flammable liquid, then instead of displaying two placards the shipper can display a Dangerous placard, as shown in **Figure 2-17.** If any of the items exceeds 2,205 pounds (1,000 kg) and is picked up at one location, then in addition to the Dangerous placard the shipper is to display the placard for the material that exceeds the 2,205 pounds (1,000 kg).

- Specific-name placards or placards with "Inhalation Hazard," as shown in **Figure 2-18,** are sometimes used in place of the Poison Gas placard.

- The PG III placard and label has replaced the Stow Away from Foodstuffs placard and indicates that a poisonous material is being transported, but it is not poisonous enough to meet the rules to be placarded as a poison. Chloroform is an example that would use a placard like the one shown in **Figure 2-19.**

- "Other Regulated Material—Class D" (ORM-D) is a classification that is left over from a previous DOT regulation. It is a subdivision that includes ammunition and consumer commodities, such as cases of hair spray. The package will have the printing "ORM-D" on the outside of the package. The previous regulation used to have ORM-A through

FIGURE 2-17 "Dangerous" placard.

FIGURE 2-18 "Inhalation" placard.

FIGURE 2-19 Another form of a poison placard and label.

TABLE 2-7 Examples of Materials of Trade (MOT)	
Class or Division	**Example**
Flammable gases (Division 2.1)	Acetylene, propane
Non-flammable gases (Division 2.2)	Oxygen, nitrogen
Flammable or combustible liquids (Class 3)	Paint, paint thinners, gasoline
Flammable solids (Division 4.1)	Charcoal
Dangerous-when-wet materials (Division 4.2)	Some fumigants
Oxidizers (Division 5.1)	Bleaching compounds
Organic peroxides (Division 5.2)	Benzoyl peroxides
Poisons (Division 6.1)	Pesticides
Some infectious substances (Division 6.2)	Diagnostic specimens
Corrosive materials (Class 8)	Muratic acid, drain cleaners, battery acid
Miscellaneous hazardous materials (Class 9)	Asbestos, self-inflating life boats
Consumer commodities (ORM-D)	Hair spray, spray paints

ORM-E, but these are now grouped together in Class 9, Miscellaneous Hazardous Materials.

> **NOTE**
>
> The DOT has changed some terminology for Other Regulated Materials, commonly referred to as ORMs. Typically, these were shipments of consumer goods, which are now known as Materials of Trade (MOT). Some common examples are shown in **Table 2-7.** The DOT has changed some of the rules when transporting smaller quantities of hazardous materials when used in a business. With the exception of tanks containing diluted mixtures of class 9 materials, a person cannot transport more than 440 pounds (200 kg) of a material. If the material is a high-hazard material (packing group I), the maximum amount of material allowed in one package is one pound (0.5 kg) for solids and one pint (0.5 liters).

- "Marine Pollutant" is displayed on shipments that, if the material were released into a waterway, would damage the marine life, **Figure 2-20.**
- Elevated temperature material will have a "HOT" label either to the side of or on the placard, as shown in **Figure 2-21,** if it meets one of the following criteria:
  - Is a liquid above 212°F (100°C).
  - Is a liquid that is intentionally heated and has a flash point above 100°F (37.7°C).
  - Is a solid at 464°F (240°C) or above.

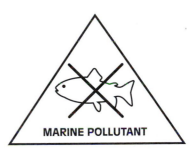

FIGURE 2-20 "Marine Pollutant" marking.

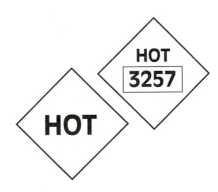

FIGURE 2-21 "Hot" placard.

- "Infectious Substances" or "BioHazard" are labels like those shown in **Figure 2-22** that are sometimes used on the outside of trucks. It is not required by the DOT but may be required by other agencies such as Health and Human Services or a state agency. The DOT now calls medical waste "regulated medical

FIGURE 2-22   "BioHazard" labels.

waste" or RMW. An infectious substance is a material known or reasonably expected to contain a pathogen. A pathogen is a microorganism (including bacteria, viruses, ricksettia, parasites, and fungi) or other agent, such as a proteinaceous infectious particle (prion), that can cause disease in humans or animals. They have also split this category into two subdivisions. Category A is an infectious substance in a form capable of causing permanent disability or life-threatening or fatal disease in otherwise healthy humans or animals when an exposure occurs. A category B material presents a lesser hazard, but the infectious substance is not in a form generally capable of causing permanent disability or life-threatening or fatal disease in otherwise healthy humans or animals when exposed.

■ The Fumigated marker, like the one shown in **Figure 2-23,** is used when a trailer or railcar has been fumigated or is being fumigated with a poisonous material. This placard is commonly found near ports where containers are frequently fumigated after arriving from a foreign port.

A white square background as shown in **Figure 2-24** is used in the following situations:

■ On the highway for controlled Radioactive III shipments.
■ On rail for:
   ■ Explosives 1.1 or 1.2
   ■ Division 2.3 Hazard Zone A (poison gas) materials
   ■ Division 6.1 Packing Group I Hazard Zone A (poison)
   ■ Division 2.1 (flammable gas) in a DOT 113 tank car
   ■ Division 1.1 or 1.2, which is chemical ammunition that also meets the definition of a material that is poisonous by inhalation.

FIGURE 2-23   "Fumigation" marking.

FIGURE 2-24   White-squared "Flammable Gas" placard.

Problems with the Placarding System

The placarding system relies on a human to determine the extent of the load, determine the appropriate hazard classes, and interpret difficult regulations to determine if a placard is required. The placard then must be affixed to all four sides of the vehicle before shipment. Placards are only required for shipments that exceed 1,001 pounds (454 kg), except for materials listed in Table 2-1, which require placarding at any amount. It is suspected that 10 to 20 percent of the trucks traveling the highway are not placarded at all or not placarded correctly.

CAUTION

Given the restrictions of many cities, bridges, and tunnels where hazardous materials are not allowed, many trucks are probably not carrying the proper designations.

A placard can come off during transport and legally may not have to be immediately replaced. The fact that an incident involves a truck or train should alert the first responder to the potential for hazardous materials, and when a placard is involved extra precautions should be taken.

Labels

A tractor trailer was involved in an accident and had jackknifed in Baltimore County, Maryland. The hazardous materials team was called to assist with the fuel spill. Upon arrival the hazardous materials team inquired as to the contents of the truck and was told by the first responders that the truck was carrying rocking chairs. While working to control the fuel leak the hazardous materials team noticed a greenish blue liquid coming from the front of the trailer, where it had been damaged in the accident. They located the driver and asked him about the contents of the trailer, which did not display a placard. The driver stated he was carrying rocking chairs, which is what the shipping papers showed the load to be.

The crews opened the back of the trailer to inspect the cargo and found rocking chairs. Upon closer inspection of the front of the truck, several drums were noticed in the very front of the trailer. When the driver was confronted by the hazardous materials team and a state trooper, the driver confessed he had picked up a load of unknown waste from his first stop. He did not know the contents of the drums and it took several days to determine the exact contents of the drums.

STREETSMART TIP
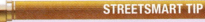

Beware; it is unlikely that someone trying to evade the law will mark the shipment appropriately.

Labeling and Marking Specifics

Package markings must include the shipping name of the material, the UN/NA identification number, and the shipping and receiving companies' names and addresses. Packages that contain more than a **Reportable Quantity (RQ)** of a material must also be marked with an RQ near the shipping name. Packages that are listed as ORM-D materials should be marked as such. Some packages with liquids in them must use orientation arrows. Materials that pose inhalation hazards must affix an Inhalation Hazard label next to the shipping name, as shown in **Figure 2-25**. Hazardous wastes will be marked "Waste" or will use the EPA labeling system to identify these packages.

Labels are identical to placards, other than their size.

NOTE

Materials that have more than one hazard may be required to display a primary hazard label and a subsidiary label.

FIGURE 2-25 The DOT adds the "Poison—Inhalation Hazard" label to those materials that present severe toxic hazards.

FIGURE 2-26 Primary and subsidiary placards.

FIGURE 2-27 NFPA 704 system marking.

The primary label will have the class and division number in the bottom triangle, while the subsidiary label will not have the number at all, as shown in **Figure 2-26.** As an example, the material acrylonitrile, inhibited, is required to be labeled "Flammable" with a subsidiary label of "Poison."

OTHER IDENTIFICATION SYSTEMS

There are several other identification systems that are used in private industry to mark facilities and containers. Military shipments and pipelines are also marked to provide a warning as to the potential for hazardous materials. Much like the transportation system the warnings are a clue to the potential presence of hazardous materials that could cause harm to the responders.

NFPA 704 System

One of the other more common systems used to identify the presence of hazardous materials is the NFPA 704 system. This system is designed for buildings, not transportation, and alerts the first responders to the potential hazards in and around a facility. The system is much like the placarding system and relies on a triangular sign that is divided into four areas, as shown in **Figure 2-27.** The four areas are divided by color as well and use a ranking system to identify severity. The four areas and colors are:

- Health hazard—blue
- Fire hazard—red
- Reactivity hazard—yellow
- Special hazards—white

The system uses a ranking of 0–4 with 0 presenting no risk and a ranking of 4 indicating severe risk. The specific listings are discussed next.

Health

This listing is based on a limited exposure to the materials using standard firefighting protective clothing as the protective clothing for the exposure.

4—Severe health hazard

3—Serious health hazard

2—Moderate hazard

1—Slight hazard

0—No hazard

Flammability

This listing pertains to the ability of the material to burn or be ignited.

4—Flammable gases, volatile liquids, pyrophoric materials

3—Ignites at room temperature

2—Ignites when slightly heated

1—Needs to be preheated to burn

0—Will not burn

Reactivity

This listing is based on the material's ability to react, especially when shocked or placed under pressure.

4—Can detonate or explode at normal conditions

3—Can detonate or explode if strong initiating source is used

2—Violent chemical change if temperature and pressure are elevated

1—Unstable if heated

0—Normally stable

Special Hazards

This listing is used to indicate water reactivity and oxidizers, which are included in the NFPA 704 system. In some cases other symbols may be used such as the tri-foil for radiation hazards, "ALK" for alkalis, and "CORR" for corrosives. In the presence of the slashed W there is also an accompanying ranking structure for water reactivity in addition to the hazards listed in the other triangles:

4—Not used with the slashed W and a reactivity ranking of 4

3—Can react explosively with water

2—May react with water or form explosive mixtures with water

1—May react vigorously with water

0—Slashed W is not used with a reactivity ranking of 0

Some potential problems exist with the NFPA 704 system, because it groups all of the chemical hazards listed in a building into one sign. If the sign is placed on a tank that contains one material, the system does a good job of warning about the contents of the tank, but does not provide the name of the product. For a facility that has hundreds—if not thousands—of materials, the system will only warn of the worst-case scenario. As an example, dramatically different tactics are used to handle a flammable gas incident versus a flammable liquids incident, but both can be classified as fire hazard 4. The system is best used to alert the first responder to the presence of hazardous materials and to warn of the worst-case scenario.

Hazardous Materials Information System

Commonly referred to as HMIS, the Hazardous Materials Information System was designed to provide a mechanism to comply with the federal hazard communication regulation, which requires that all containers be marked with the appropriate hazard warnings and the ingredients be provided on the label, **Figure 2-28.** Many products that come into the workplace are missing adequate warning labels. The HMIS is not a uniform system. It can be developed by the facility or by the manufacturer of the labels, so one system may vary from another. Most systems are similar to the NFPA 704 system and use blue, red, and yellow colors with a numbering system that provides an indication of hazard. The colors may be used in a triangle format or, in most cases, as stacked bars. The numbers are usually 0–4, the same as the NFPA system, but in rare cases may differ from the NFPA system. The facility manager or other representative should have the key to the symbols, or a chart should be provided somewhere in the facility indicating what the symbols and the warning levels are. The chart is usually stored with the MSDS. In most cases, a central location should be chosen for the MSDS and other hazard communication information.

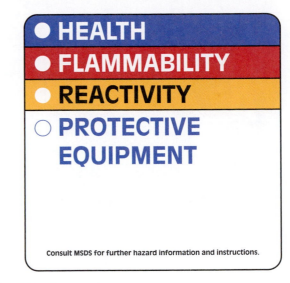

FIGURE 2-28 HMIS label.

In some systems, a picture is provided of the level of PPE required for the substance. Each HMIS is different, and responders should not assume any particular hazard level until the warning levels can be determined.

Military Warning System

The military uses the DOT placarding system when possible, but in some cases may use its own system. Within the DOT's Emergency Response Guidebook, an emergency contact number is given when responding to an incident involving a military shipment.

In most cases, for extremely hazardous materials, arms, explosives, or secret shipments, firefighters can assume prior to their arrival that the military is already aware of the incident and probably already responding. The higher the hazard the more likely there will be an escort for the shipment. There may be shipments in which the driver of the truck is not allowed to leave the cab of the truck and may provide warnings to stay away from the truck.

SAFETY

For high-security shipments the driver is armed as are the personnel in the escort vehicle. If an incident occurs involving one of these vehicles, firefighters must obey the commands of the escorting personnel and determine if they have made the appropriate notifications.

If the driver and escort crew are killed or seriously injured in the accident, it would be advisable to notify the military about the incident, although with satellite tracking help is probably already on the way. The phone number to contact the military is in the DOT

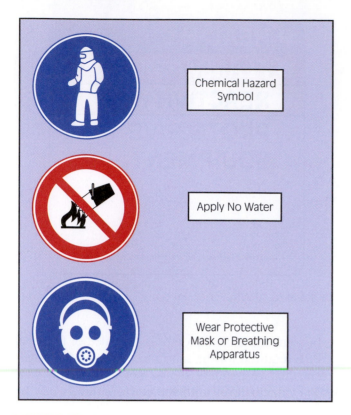

FIGURE 2-29 Military placards.

ERG, along with **Chemtrec's** and other emergency contact numbers.

Other incidents involving fuels, food, or military equipment may require notification of the military.

The military typically uses its own marking system at its facilities to mark the buildings. The military uses a series of symbols and a numerical ranking system as shown in **Figure 2-29.**

Pipeline Markings

Any place an underground pipeline crosses a mode of transportation, the pipeline owner is required to place a sign like the one shown in **Figure 2-30** that displays the wording "Warning, Caution, or Danger," the words "Petroleum Pipeline," or the hazardous contents of the pipe, as well as the owner's name and telephone number. The pipeline contents may be general, as in "petroleum products," because the same pipeline may be used to ship fuel oil, gasoline, motor oil, or other products. Dedicated pipelines that carry only one product will be marked with the specific product that they carry. The pipeline should be buried a minimum of 3 feet and should be adequately marked.

Many of the larger pipelines, such as the Colonial pipeline that originates in Texas and ends in New York, are 26 inches in diameter along the main pipeline. In the event of a release involving the pipeline, the line

FIGURE 2-30 The owner of a pipeline is required to provide a sign with the words "Warning," "Caution," or "Danger"; the words "Petroleum Pipeline" or the hazardous contents of the pipe; and the owner's name and telephone number.

will be shut down immediately. Even with this immediate shutdown, the potential exists to lose several hundred thousand gallons of hazardous materials because the distance between the shutoffs is substantial.

An incident involving a pipeline can be a serious event—firefighters should not underestimate the need for considerable local, state, and federal resources. Within the fire department alone, considerable resources may be required such as command staff, logistic support, communications, and tactical units. A fuel oil pipeline rupture in Reston, Virginia, resulted in the loss of more than 400,000 gallons (1,514,164 liters) of fuel, requiring considerable resources from several states to control the spill. The resources included emergency response organizations; local, state, and federal assistance; and a considerable number of private cleanup companies.

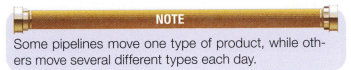

NOTE

Some pipelines move one type of product, while others move several different types each day.

The product in the pipelines varies from liquefied gases and petroleum products to slurried material. Pipeline companies are required to conduct in-service training and tours for the emergency responders in the communities their pipelines transverse. When firefighters have an incident on or near any pipeline, it is advisable to notify the pipeline owner of the incident, even if they are pretty sure the pipeline was not damaged. A train derailment in California caused a pipeline to shift, and the pipeline did not release any product until several days after the original derailment occurred. Most pipeline operators would like to have the opportunity to check the line as opposed to having a catastrophic release several days later because the line was not checked.

Container Markings

Most containers such as drums are marked with the contents of the drum, **Figure 2-31,** while cylinders have the name of the product stenciled on the side of the cylinder. In bulk shipments the bulk container will have the name of the product stenciled on the side. Trucks that are dedicated haulers will also stencil or mark the product name on two sides of the vehicle.

Pesticide Container Markings

Due to their toxicity, pesticides are regulated by the EPA as to how they are to be marked. The label on a pesticide container, such as the one shown in **Figure 2-32,** will have the manufacturer's name for the pesticide, which is not usually the chemical name for the product. The product name will usually not offer any clue as to the chemical makeup of the product, and in some cases may be difficult to research. In Chapter 3, there is a discussion of chlordane and the multiple names it's referred to by. The label will also contain a signal word such as "Danger," "Warning,"

FIGURE 2-31 The label describes the contents of the drum.

FIGURE 2-32 "Pesticide" labels.

or "Caution." The assignment of the signal word is the clue to how dangerous the material actually is. If the label indicates "danger" then extreme care should be used as the material is highly toxic. The designations of "warning" and "caution" present lesser hazards, but still should be considered very hazardous.

In the United States the EPA issues an EPA registration number and in Canada the label will have a pest control number. The label will also include a precautionary statement and a hazard statement, examples of which include "Keep from Waterways," and "Keep Away from Children." The active ingredients will be listed by name and percentage; in most cases the active ingredients are usually a small percentage of the product. Inert ingredients are also listed but not specifically named. It is the active ingredient that presents the major hazard. For liquid pesticides, the "inert" ingredient is usually water, or petroleum solvents, such as xylene, plus emulsifying agents.

Radiation Source Labeling

In 2007, the International Atomic Energy Association (IAEA), a group that sets nuclear and radiation standards worldwide, issued a new radiation warning label. The old label using the trefoil commonly referred to as a "propeller" isn't really associated with radiation or nuclear material. The label shows energy waves coming from a trefoil with skull and cross bones, with a human running away. The label shown in **Figure 2-33** more accurately represents the hazards associated with radiation. It is not mandated for use by DOT, and the DOT will still require the DOT placards and labels, but the new label will be more common in the future. Devices or containers that store radiation sources will be using this type of label. On devices that use a radiation source and are relatively large, the label will be placed near the access panel to the radiation source. Obviously, responders should not try to access the source and should request radiation specialists to assist them.

CONTAINERS

Hazardous materials come in a variety of containers of many shapes and sizes, from 1-ounce bottles and larger bags to tanks and ships carrying hundreds of thousands of gallons. A survey of the materials in the average home will reveal a wide variety of storage containers. Compressed gas cylinders hold propane; steel containers hold flammable and combustible liquids; bottles, jars, and small drums hold various products. Plastic-lined cardboard boxes, and bags of various types are also used for chemical storage.

The type of material and the end use for the product determine the packaging. Packaging used to store household or consumer commodities is usually different than the industrial version. In some cases the industrial version may be full-strength undiluted product, whereas the household version is only a small percentage of that strength mixed with a less hazardous substance such as water. The type of container usually provides a good clue as to the contents of the package.

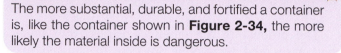
STREETSMART TIP

The more substantial, durable, and fortified a container is, like the container shown in **Figure 2-34,** the more likely the material inside is dangerous.

On the other hand, materials transported in fiberboard drums usually have no significance with regard

FIGURE 2-34 The type of container can provide some clues as to the contents of the container. Because this drum is reinforced, it has a high likelihood of containing an extremely hazardous material.

FIGURE 2-33 "IAEA Radiation" label.

to human health, although they may pose a risk to the environment.

When looking at the recognition and identification process, **first responders** should be alert for anything unusual when arriving at an incident. When on an EMS call to a residential home, it would be unusual to find a 55-gallon (208 liter) drum in a bedroom along with glassware associated with a lab environment. These types of recognition and identification clues should alert the first responders that additional assistance may be required. Arriving at an auto repair garage and finding 55-gallon (208 liter) drums and compressed gas cylinders should not be unexpected, however.

General

Containers come in a variety of sizes and shapes and the general category of containers is not exceptional. Most of the general containers are designed for household use but will be carried in large quantities when moved in transportation. When moving to bags and into drums and cylinders the move is made from household to industrial use. All of these types can be used in the home, but a super sack which can hold thousands of pounds of materials is not usually considered household.

Cardboard Boxes

With the popularity of shopping clubs and discount warehouses, more and more homeowners are buying materials by the case, when in the past they bought in much smaller quantities. Cardboard boxes are used to ship and contain hazardous materials. They can hold glass, metal, or plastic bottles. In some cases they may have a plastic lining, such as the box shown in **Figure 2-35,** which holds sulfuric acid. Many household pesticides, insecticides, and fertilizers are contained in cardboard boxes. With the exception of these products and sulfuric acid, most products contained in boxes are usually not extremely toxic to humans, but may present an environmental threat. Materials in transport to suppliers may be transported in larger cardboard boxes and then broken down at the retail level. Responders should note any labels on these packages, but the absence of any labels does not indicate that hazardous materials are not present.

Bottles

From 1-ounce bottles to 1-gallon (3.8 liter) bottles, the variety of containers is endless and the types of products contained in them too numerous to mention. In recent times manufacturers have begun to take precautions when packaging their materials for transport and use, especially when glass bottles are used. Now-

FIGURE 2-35 Typically, chemicals that can cause harm are not packaged in cardboard. This sulfuric acid is one example of a material that can cause harm, but unfortunately is packaged in cardboard.

FIGURE 2-36 To protect the glass bottle, which has a corrosive in it, a carboy is used in case the bottle is dropped.

adays, when chemicals are shipped in glass bottles, the bottles are usually packed in cardboard boxes and insulated from potential damage. One-gallon glass containers are usually shipped in what is known as carboys, like the one shown in **Figure 2-36.** Carboys provide a protective cover to protect against potential damage during transportation. If the container is dropped, the bottle should survive the fall. Carboys are usually seen in laboratories and in smaller chemical production facilities.

Ensuring the material's compatibility with the container it will be stored in is important, but the one area that usually results in a release is the use of an improper cap. The chemical must not only be compatible with the glass, it must also be compatible with the material the lid is composed of. A variety of materials are used in the manufacturing of lids. Many new glass containers, like the one shown in **Figure 2-37,** are coated with plastic to avoid the bottle being broken when dropped. Even if the bottle is cracked, the contents are supposed to remain sealed within the plastic coating.

Bags

Bags are also commonly used as containers for chemicals. Bags can be as simple as paper bags or plastic-lined paper bags to fiber bags, plastic bags, and the reinforced super sacks or tote bags. It might be a surprise to open the back of a trailer and find four super sacks like those shown in **Figure 2-38** carrying a material that is classified as a poison. Bags carry anything from food items to poisonous pesticides, and the method of transportation varies widely.

Drums

When discussing hazardous materials, drums are the containers with which most responders are familiar. They vary from 1-gallon (3.8 L) sizes up to a 95-gallon (359.6 liter) overpack drum. The construction varies from fiberboard to stainless steel. The typical

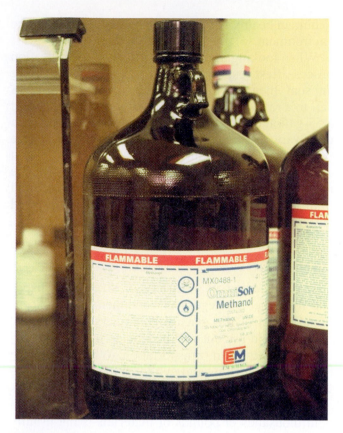

FIGURE 2-37 This glass jar is coated with a plastic coating that will not allow the liquid to spill out if the glass is broken or dropped.

drum holds 55 gallons (208 liters) and weighs 400 to 1,000 pounds (181–454 kg). It is possible to get an idea of what a drum may contain by the construction of the drum. **Table 2-8** provides an indication of potential drum contents, but this is not an absolute listing; contents can and do vary from drum to drum.

FIGURE 2-38 It can be quite surprising to open the back of a tractor trailer and find these super sacks. They can hold solid materials, some of which can be toxic.

TABLE 2-8 Drum Contents

Type of Drum	Possible Contents (In Order of Likelihood)
Fiberboard (cardboard), unlined	Dry, granular material such as floor sweep, sawdust, fertilizers, plastic pellets, grain, etc.
Fiberboard, plastic lined	Wetted material, slurries, foodstuffs, material that may affect cardboard or could permeate the cardboard
Plastic (poly)	Corrosives such as hydrochloric acid and sodium hydroxide, some combustibles, foodstuffs such as pig intestines
Steel	Flammable materials such as methyl alcohol, combustible materials such as fuel oil, motor oil, mild corrosives, and liquid materials used in food production
Stainless steel	More hazardous corrosives such as oleum (concentrated sulfuric acid)
Aluminum	Pesticides or materials that react with steel and cannot be shipped in a poly drum (e.g., concentrated hydrogen peroxide or beer)

FIGURE 2-39 Cylinders present additional risks to responders because not only can the contents be hazardous, but if the cylinder is involved in a fire it may explode.

Cylinders

Cylinders, like those shown in **Figure 2-39,** come in 1-pound (0.45 kg) sizes up to several thousand pounds and carry a variety of chemical products. The product and its chemical and physical properties will determine in what type of container the product is stored.

> **CAUTION**
> Other than the hazard of the chemical itself, the big hazard of all cylinders is that they are pressurized.

The pressures range from a low of 200 psi to a high of 5,000 psi (14–345 bar). One of the most common cylinders firefighters run across is the propane tank, which ranges from 1 pound up to millions of gallons. In residential homes, firefighters will find everything from the 20-gallon (75.7 L) cylinder for barbecues to the 100- to 250-pound (45–113 kg) cylinders used as a fuel source for the home. In some areas it is not uncommon to find 1,000-pound (454 kg) cylinders and often they are buried underground.

Specialized cylinders that hold cryogenic gases (extremely cold) appear to be high pressure, but in reality are low pressure. The bulkiness of the cylinders is a result of the large amount of insulation required to keep the material cold. It is not uncommon to find cryogenic cylinders in convenience stores or fast food restaurants, which presents an unusual hazard. They typically are storing carbon dioxide (CO_2) for the soda-dispensing machines. If the system were to leak, a large amount of CO_2 would be released presenting an asphyxiation risk. Cylinders usually have **relief valves** or **frangible disks** in the event they are overpressurized or are involved in a fire. The United States is one of the few countries that mandates relief valves, and most cylinders used in other countries do not have this feature. Incidents aboard ships may involve these types of cylinders as the cylinders are being trans-shipped and are destined for delivery in another country. Most communities, regardless of their size, have cylinders of chlorine and sulfur dioxide used in water treatment. These cylinders come in 100- to 150-pound (45–68 kg) and 1-ton (907 kg) cylinders, which could create a major incident if they were ruptured or suffered a release.

NOTE

The area affected by a 100-pound (45 kg) chlorine cylinder release can be several miles, **Figure 2-40,** causing serious injuries if not fatal effects.

There was an attempt to allow the use of composite propane cylinders, both inside and outside of buildings. The draft version of NFPA 58, which is the Liquefied Petroleum Gas code, had language that would have allowed composite propane cylinders to be used in small portable heaters. The final version only allows the use of composite cylinders outside of buildings. When this standard comes up for revision in 2010 this issue will come up again. These cylinders, which are constructed much like composite breathing apparatus cylinders, can be used in transportation through a special DOT permit. Composite cylinders can carry flammable gases, oxygen, and many other gaseous materials. Composite cylinders are in common use in Europe and are starting to become more popular in the United States. Two major tests have been conducted on these cylinders to determine the outcome if the cylinders were involved in a fire situation. From the initial testing composite cylinders appear to react to fire situations better than steel cylinders. As is discussed later in this section, BLEVEs are a major life threat to responders and are common with steel cylinders. None of the composite cylinders BLEVE'd or had a violent rupture during the testing process.

Totes and Bulk Tanks

Both **totes** and **bulk tanks** are becoming more common, sized as they are between drums and tank trucks, and are used for a variety of purposes. These totes and tanks are also called intermediate bulk containers and have capacities between 119 gallons and 793 gallons (450–3,001 liters). Used in industrial and food applications, they hold flammable, combustible, toxic, and corrosive materials. They are constructed of steel, aluminum, stainless steel, lined materials, poly tanks, and other products, **Figure 2-41.** The usual capacity is approximately 300 gallons (1,136 liters). They are transported on flatbed tractor trailers or in box-type tractor trailers.

A common incident with totes can occur during offloading. Tanks are offloaded from the bottom through a swinging valve, such as the one shown in **Figure 2-42.** It is a common occurrence for this valve to swing out during transport and get knocked off during movement.

One unusual tote is made to transport calcium carbide, a material that when it gets wet forms acetylene gas, which is reactive and very flammable.

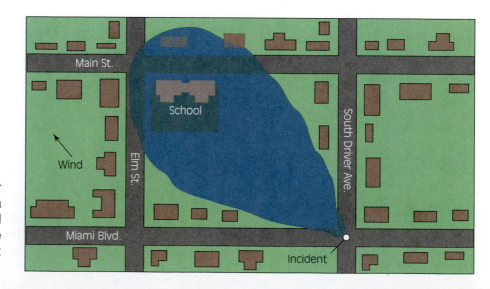

FIGURE 2-40 The type of vapor cloud commonly referred to as a plume varies with the terrain and buildings in the vicinity. The plume here represents one of the most common types.

FIGURE 2-41 A tote that holds approximately 300 gallons. This particular one is plastic with a steel frame for reinforcement.

FIGURE 2-42 The most common type of spill occurs when a valve is knocked off, releasing the contents.

Pipelines

Pipelines vary in size and pressure, but can be sized between ½ inch (1.27 cm) and more than 6 feet (1.9 meters). They are commonly buried underground. The most common products they carry are natural gas, propane, and assorted liquid petroleum products. The larger petroleum pipelines originate in Texas and Louisiana and then proceed up throughout the East Coast. The West Coast also has its share of large pipelines, with Alaska having a majority. Pipelines can originate from any bulk storage facility and can cross many states, and some type of pipeline system is found in every state. Because the amount in the pipelines varies it is important that first responders know the location of the pipelines and emergency contact names and phone numbers so if there is a suspected problem they can notify the pipeline owner immediately.

Radioactive Material Containers

The transportation of radioactive or nuclear materials is highly regulated by the DOT and the **Nuclear Regulatory Commission (NRC).** There are several types of containers that are used to transport these types of materials, all strengthened to keep the material inside the container. Radioactive material that is classified as low-level radioactive material, typically waste material, are required to be transported in what is called a strong, tight container. This container can be made of metal or wood but should be reinforced to prevent damage to an inner container holding the radioactive material. These materials would present little risk if they escaped the container and only emit radiation just above background levels.

Another type of package is the **excepted packaging** shown in **Figure 2-43,** which is designed to survive the normal conditions of transport. This packaging is designed to hold materials that have **low specific activity** (emitting low levels of radiation), also called **surface contaminated objects (SCO).** These are normally small-quantity shipments that contain natural or depleted uranium, or natural thorium. In some cases, the package does not have to be labeled since the material does not meet the DOT requirement for labeling and placarding. Industrial packaging is a container that is designed to be more substantial and has to pass some basic testing procedures, such as the drop test. Industrial packaging, shown in **Figure 2-44,** usually consists of metal boxes or drums. Industrial packaging is also used to transport low specific activity (LSA) or (SCO) materials. Commonly shipped materials are slightly contaminated tools, medical and research isotopes, and soil waste. For higher levels of radiation, a container known as a **Type A container,** shown in **Figure 2-45,** is used. The Type A container can be made of cardboard, wood, or most

FIGURE 2-43 Excepted packaging is used to transport low-risk radioactive materials; the packaging is exempted from DOT regulations. *(Courtesy of IAEA)*

FIGURE 2-44 Industrial packaging is used to transport low-risk radioactive materials, typically waste. *(Courtesy of Kirstie Hansen, IAEA)*

FIGURE 2-46 Type B containers, such as this container being loaded on a ship, are designed to hold high-risk radioactive materials. They are hardened transport cases capable of withstanding severe accidents. *(Courtesy of Pat Pavlicek, IAEA)*

FIGURE 2-45 Type A packaging is designed to hold low-risk radioactive material and offers moderate protection against accidents. *(Courtesy of John Mairs, IAEA)*

commonly, metal drums. The Type A containers are used to transport radioactive material (RAM) and are tested to withstand minor accidents. The level of radiation is still low because only materials that would not present a significant health risk can be placed in a Type A container. Materials transported in Type A packages are usually medical isotopes or industrial, agricultural, or research materials. The highest level of protection is offered by a **Type B container,** shown in **Figure 2-46,** and is used for the transportation of larger or more dangerous types of radioactive materials. The Type B containers must meet all types of testing procedures and must be able to withstand a severe accident. The Type B container is either a metal drum or a large, heavily shielded transport container. The Type B container may have 10 inches of lead shielding. **Spent nuclear fuel** is transported in Type B containers.

Highway Transportation Containers

The type of vehicle provides some important clues as to the possible contents of the vehicle. The most common truck is a tractor trailer or a box truck. There are four basic tank truck types that carry hazardous materials, with some additional specialized containers. Tractor trailers carry the whole variety of hazardous materials and portable containers. They can carry loose material that is not contained in any fashion other than by the truck itself. They can carry portable tanks that hold 500 gallons (1,893 liters) or bulk bags that weigh several tons.

CAUTION

When dealing with tractor trailers the rule is to expect the unexpected.

Nothing is routine. Until the driver has been interviewed, the shipping papers looked at, and the cargo actually examined, a firefighter cannot confirm or deny the presence of hazardous materials. Sometimes the signage on the trailer is an indication of the possible contents, and a trailer that has several placard holders is a likely candidate for hazardous materials transport. If a tractor trailer is refrigerated like the one shown in **Figure 2-47** and is carrying hazardous

materials, extra precautions must be taken because the materials may require the cold temperature to remain stable.

Leakage is often found in containers known as **intermodal containers** or, more commonly, **sea containers.** These types of containers are typically used on ships, then offloaded onto a tractor trailer or loaded directly onto a flatbed railcar. These containers come from all over the world and can contain any imaginable commodity.

NOTE

The types of containers that are shipped in these trailers vary from bags, boxes, and drums to bulk tanks and cylinders.

Although the driver is supposed to have the shipping papers for the contents, on occasion the paperwork is missing or is sealed in the back of the trailer. Determining the contents of a trailer can be very difficult and frustrating.

Tank trucks carry several hundred gallons up to a maximum of 10,000 gallons (37,854 liters). The DOT allows maximum loads by weight not by gallons, so the actual capacity varies state to state. The most common tank truck is the gasoline tank truck, which usually carries 5,000 to 10,000 gallons (18,927 to 37,854 liters). In September 1995 the DOT changed the regulations covering tank trucks, so two systems are used for identifying tank trucks, **Figure 2-48.** In the past the DOT wrote specifications as to how the manufacturer should build a tank truck. Today, they have established performance-based standards for the construction. The DOT allows trucks that were manufactured before the new regulation to remain on the road, as

FIGURE 2-47 Although in most cases refrigerated trailers are carrying food, there exists the possibility that they may have chemicals that require refrigeration to remain stable.

FIGURE 2-48 Many of the differences between a 306 and a 406 tank truck are internal. The biggest difference is that the dome covers on a 406 are less likely to open during a rollover, although the skin of the tank is thinner. *(Courtesy of Maryland Department of the Environment)*

long as they meet the applicable inspection requirements. The four basic types of tank trucks are:

- DOT-406/MC-306 gasoline tank truck
- DOT-407/MC-307 chemical hauler
- DOT-412/MC-312 corrosive tanker
- MC-331 pressurized tanker

The DOT numbers are the more recently manufactured tanks, and the MC (motor carrier) numbers identify those tanks manufactured prior to September 1995. There are some differences in construction of the two types—some of which favor the emergency responders, others favor the shipping company—but overall the newer tanks hold up better during accidents and rollovers. If unsure of the type of tank, all tank trucks have **specification (spec) plates,** which list all the pertinent information regarding that tank. The spec plate in many cases is located on the driver's side of the tank near the front of the tank. In some cases, shipping papers or MSDS are located in a paper holder as shown in **Figure 2-49** (tube).

DOT-406/MC-306

This is the most common tank truck on the road today, **Figure 2-50,** and for that reason, along with the large number of shipments, suffers the most accidents. Although the most common products carried on these trucks are gasoline and diesel fuel, almost any flammable or combustible liquid can be found on these types of trucks. This truck is known for its elliptical shape and is usually made of aluminum. Older style tanks used to be made of steel, which presented an explosive situation known as a BLEVE. The tanks generally have three or four separate compartments, but two to five compartments are not uncommon. Newer style tanks have considerable vapor recovery systems as well as rollover protection, and in some cases these features are combined. The valving and piping are contained on the bottom of the tank, as shown in **Figure 2-51,** and the number of pots is indicated by the number of outlets as well as the number of manhole assemblies located on top of the tank. The maximum pressure that this truck can hold is 4.2 psi. The compartments are separated by bulkheads, which usually fail during a rollover situation. In the event that the shipper wishes to carry mixed loads, they may add a double bulkhead to protect against the mixing of chemical products. During a

FIGURE 2-49 In most cases the shipping papers will be with the driver in the cab of the truck, but they may be in a special tube located on the trailer.

FIGURE 2-50 The DOT-406/MC-306 is the most common truck on the highway today and is referred to as a gasoline tanker. It can and frequently does carry other types of flammable and combustible liquids. *(Courtesy of Maryland Department of the Environment)*

violent rollover, even this double bulkhead is likely to fail. Although in most states the shipper is not allowed to ship a mixed load of flammable and combustible materials, widely differing loads can be encountered from compartment to compartment.

> **CAUTION**
>
> During a rollover, the bulkheads may shift, allowing all products to mix.

On initial examination it may appear that only one tank has ruptured and is leaking, but it is possible to lose the entire contents of the tank through a leak in one compartment, like that shown in **Figure 2-52.** In the majority of rollovers experience has shown that at least one bulkhead has separated in almost every accident, resulting in product mixing. In most cases this is not a big problem because it may only be the mixing of different grades of gasoline. Within the individual tanks themselves, baffles limit the movement of the product within the tank. The emergency shutoff valves, **Figure 2-53,** are located on the driver's side

FIGURE 2-51 Most 406/306 trucks have more than one tank (pots) and the number of tanks is indicated by the number of valves or the number of dome covers on top of the truck. *(Courtesy of Maryland Department of the Environment)*

FIGURE 2-53 On most trucks there is a minimum of one emergency shutoff, and with most there are two. The most common location is near the driver's door; the other is usually located near the valve area.

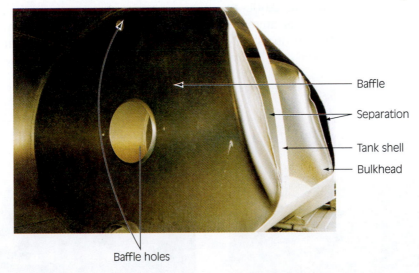

Baffle

Separation

Tank shell

Bulkhead

Baffle holes

FIGURE 2-52 When a truck rolls over on its side, the internal baffles and bulkheads may shift. The internal baffles reduce the amount of sloshing the liquid will do when the truck starts and stops. The baffles will allow product to move through the holes in the baffle wall. A bulkhead separates the compartments and does not allow the products to mix. Shown in the photograph is an MC-306 that rolled on its side. The side wall of the tank has been cut away, revealing a baffle and a bulkhead. Also shown is the separation that took place between the tank shell and the baffles/bulkhead. When the tank wall was cut away, strips of the tank shell were left where they should have connected to a baffle or bulkhead.

FIGURE 2-54 On the right is an insulated MC-307/DOT-407, and on the left is an uninsulated MC-307/DOT-407. Both trucks hold comparable products, but the insulated one holds products that are heated or may require heating to offload.

near the front of the tank and near the piping on the passenger side.

DOT-407/MC-307

The DOT-407/MC-307 tankers are the workhorses of the chemical industry, **Figure 2-54.** They carry a variety of materials including flammable, combustible, corrosive, poisonous, and food products. The two basic types of chemical tanks are insulated and uninsulated. The insulated tank can have a number of additional concerns that do not apply to an uninsulated tank. These tanks usually hold 2,000 to 7,000 gallons (7,571 to 26,498 liters), lower amounts than the 406/306 because most of the products they carry are heavier than petroleum products. The av-

erage amount found in these tanks is 5,000 gallons (18,927 liters).

The uninsulated tank is round and has stiffening rings around the tank. The offloading piping is located on the bottom or off the rear of the tank. These tanks are composed of only one compartment, and its loading piping and manhole are usually on the top in the middle. These tanks generally do not hold up as well during rollovers and accidents as the insulated version. The shell is made of stainless steel and can hold pressures up to 40 psi (2.8 bar).

The insulated tank, **Figure 2-55,** is a covered version of the uninsulated tank, although in some cases it has a slightly smaller inner tank. The inner tank is made of stainless steel with about 6 inches (15.24 cm) of insulation, and the outer shell is made of aluminum. The inner tank may also be lined with a fiberglass or other liner depending on the chemical that is carried. Due to the aluminum and insulation these tanks hold up remarkably well during rollovers.

SAFETY

One of the major problems with this insulated tank is that in the event of a leak, the location where the material leaks out of the outer shell is usually nowhere near the leak on the inner shell.

Within the insulation there can be heating and cooling lines depending on the product being carried, **Figure 2-56.** Products such as paint are shipped at 170°F (77°C) and need to be heated to that temperature for offloading. Some products need to remain at certain temperatures to remain stable, and first responders need to be aware of any special requirements.

In general, both types of tanks have rollover protection, similar piping, and relief valves that serve two

FIGURE 2-55 This insulated version is identical to the uninsulated tank but has an aluminum cover and insulation. Note the differences in these two trucks. The one on the right has safety railings around the manhole. Although not an absolute rule, the truck on the right would carry more dangerous products and would have other added safety features. Most of these items are not required but were added by the trucking company.

FIGURE 2-56 Products carried in an insulated 407/307 need to remain either heated or cooled. Some products may need to be heated for offloading. The heater coils that run around the tank heat the product up, allowing it to be offloaded.

FIGURE 2-57 The DOT-412/MC-312 is designed to carry corrosives and is similar in design to the uninsulated 307, although smaller. The inner tank may be lined with a variety of materials to prevent the corrosive from attacking the tank.

purposes: overpressurization and vacuum protection. The emergency shutoffs are located near the front of the tank on the driver's side and near the offloading piping.

DOT-412/MC-312

These tankers, **Figure 2-57,** carry a wide variety of corrosives, both acids and bases. These tankers are round and are smaller in diameter than the 306s and 307s due to the weight of the corrosives they carry. Most petroleum products weigh about 8 pounds per gallon (1 kg per liter), while some corrosives weigh up to 15 pounds per gallon (1.8 kg per liter). Because of the weight, the stiffening rings used are generally bulkier than the ones used on DOT-407 tanks. These tankers are constructed of a single tank that carries up to 7,000 gallons (26,498 liters), with most tanks holding 5,000 or fewer gallons (18,927 liters). The tanks are made of stainless steel and are usually lined with a rubber or ceramic coating to protect against corrosion. The piping can be on top of the tank located in the middle, but is usually located on the end of the tanker. The piping is usually contained within a housing that includes the manhole and offload piping. This housing protects the piping in the event of a rollover, **Figure 2-58.** The area around the manhole is usually coated with a material that resists the chemical being carried and is usually a black, tar-like coating.

MC-331

MC-331 tanks look like bullets and are noted for their rounded ends and smooth exterior, as shown in **Figure 2-59.** They carry liquefied gases that are liquefied by pressure. One of the most common products

FIGURE 2-58 The black coating around the manhole indicates that a corrosive is being carried. It is used to protect the tank from spillage. It is not required nor will it be found on all corrosive tanks.

carried in this type of tank is propane. They also carry ammonia, butane, and other flammable and corrosive gases. These tanks carry up to 11,500 gallons (45,532 liters) and have a general pressure of 200 psi (14 bar), although it can be as high as 500 psi (34.4 bar). The relief valves are located on top of the tank at the rear of the trailer, and they sometimes malfunction during a rollover. The tanks are made of steel, are uninsulated, and are heavily fortified with heavy bolts used in the piping and manholes. The tanks are usually painted white or silver to reduce the potential heating by sunlight.

The tanks contain a liquid along with a certain amount of vapor. The most liquid the tank is supposed

FIGURE 2-59 MC-331 tanks carry liquefied compressed gases such as propane and ammonia. They are made of steel and are designed to carry a variety of products.

to have is 80 percent to allow for expansion when heated. The liquid in the tanks is at atmospheric temperature but on release can go below 0°F (–18°C) and could cause frostbite upon contact. The pressure in these tanks is of concern when responding to incidents involving these tanks.

FIREFIGHTER FACT

When a propane tank is emptied and, hence, the pressure reduced, the temperature of the propane drops below 0°F (–18°C). Temperature, pressure, and volume are interrelated. Think of an SCBA bottle. When it gets filled it becomes hot because the pressure and volume are increasing. When the SCBA bottle is used, it becomes cold because it is losing pressure and volume. Any time one of the parameters is changed, there is a corresponding change in the other properties. When a pressurized gas is pressurized to a point that it becomes a liquid, as is the case with propane, it allows for a lot of propane to be stored in a small container. When released, however, the temperature will drop because the pressure is decreasing in the container. This is known as autorefrigeration.

SAFETY

If the pressure increases at a rate higher than the relief valve can handle, the tank will explode. These explosions have been known to send pieces of the tank up to a mile, with the ends of the tank typically traveling the farthest, although any part is subject to becoming a projectile.

Boiling Liquid Expanding Vapor Explosion (BLEVE)

When tanks, trucks, tank cars, or other containers are involved in a fire situation there are a number of hazards. One very deadly hazard is known as a boiling liquid expanding vapor explosion (BLEVE), **Figure 2-60.** A large number of firefighters have been killed by propane tank BLEVEs, and when a BLEVE occurs it usually results in more than one firefighter being killed at a single incident. The type of container and the product within the container will dictate how severe a BLEVE may be. The basis of a BLEVE is the fact that the pressure inside the container increases and exceeds the maximum pressure the container was designed to handle. The contents are violently released, and if the material is flammable, an explosion or large fireball occurs. In the recent past there have been several incidents involving BLEVEs that resulted in emergency responder deaths and injuries, thus emphasizing the need to recognize and prevent this event before it occurs.

Another phenomenon that transpires with containers is known as violent tank rupture (VTR), which occurs with nonflammable materials. The concept is the same as a BLEVE, but there is not a characteristic fireball and explosion. With both a BLEVE and a VTR there is some form of heat increase inside the container, typically from a nearby fire. The fire heats the container, which heats the contents. The contents will boil, which creates expanding vapors, which in turn increase the pressure inside the container. In some containers the relief valve will activate, relieving the pressure. In some cases the pressure inside the tank is greater than the relief valve can handle and the

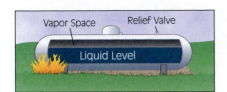

Fire impinging on a propane tank.

As heat increases inside of tank, the pressure also increases. The liquid will begin to boil.

As the pressure increases, the relief valve will open, releasing propane. The propane, being heavier than air, will sink.

The vapor will reach the fire and ignite.

The relief valve will ignite, also causing heat to be on the tank by that flame. The pressure will increase in the tank.

As the pressure in the tank is increasing, the tank may discolor and the pitch of the relief valve will get higher. Eventually the tank will rupture. This is known as a BLEVE.

FIGURE 2-60   Diagram of a BLEVE.

pressure continues to increase. One of two possibilities can occur with a BLEVE or a VTR. One possibility is that the relief valve will not be able to handle the increase in pressure and the tank will fail. The other possibility is that the fire or heat source that is creating the problem will weaken the container shell, and the resulting increasing pressure will vent at this weakened portion of the container. If the heat source is a fire and the container product is flammable, when the relief valve activates the raw product coming out of the container typically ignites. This may increase the temperature of the tank, increasing the pressure. It is never advisable to extinguish the fire coming from a relief valve, as that is a safety mechanism. It is possible to cool the top of the tank near the relief valve with an unstaffed hose stream. The difference between a BLEVE and a VTR occurs when the container fails. A BLEVE occurs with flammable liquids, such as propane. When the container fails, the vapors

of the flammable liquid ignite, creating the explosion. How the container ruptures and the amount of product released will determine how severe the explosion will be. With a VTR the container will fail. Since the product is nonflammable, the product will not ignite. The only event is a rupture of the container, spilling its contents. The container can still rocket, and the resulting release of pressure can be violent. A VTR can occur with a container of water or other "nonhazardous" material. As an example a 55-gallon (208 liter) steel drum of water, when heated, can travel several hundred feet depending on where the release point is on the container. In most cases the bungs (screw-top caps) will release, flying a considerable distance, and the container will remain mostly intact.

The failure of the container when impinged from a fire usually occurs as the fire is impacting the tank in the vapor space. When heat is applied to a section of the tank that has no internal mechanism to provide cooling, failure of the steel can occur. When fire impinges on the liquid portion of the tank, the liquid will distribute the heat spread internally and the steel tank typically will not weaken. The problem is that responding firefighters do not usually know the liquid level, and one cannot easily predict when a tank may fail. Any fire impingement on a tank is a serious problem, and withdrawing from the scene may be the best course of action. This brings up another interesting note: A casual observation of BLEVEs indicates that most BLEVEs and firefighter deaths occur within the first few minutes of arrival. The clock starts ticking on the BLEVE time bomb from the first minute heat is applied to the tank. The clock does not start with the 9-1-1 call or the arrival of firefighters. The critical time for safety may have already passed before firefighters arrive on the scene. If firefighters arrive at an incident involving a propane tank on fire or being impinged by fire, they are in extreme danger. If the relief valve is not operating, the danger is even more pronounced. Operating in close proximity to a tank in this situation can be a fatal mistake. Firefighters should follow this risk/benefit analysis: Risk a lot to save a lot, and risk a little to save a little.

One recent event involved the death of two firefighters and injuries to seven other emergency responders. The location was a farm, at which a propane tank was on fire. The relief valve on the tank was operating and the vapors from the relief valve were on fire. About eight minutes after the firefighters had arrived, the tank exploded into four separate parts. The four parts went in four different directions. The two firefighters who died were 105 feet away and were struck by one piece of the tank, dying instantly.

In two other farm incidents, firefighters lost their lives. In 1993 a fire and BLEVE in Ste. Elisabeth de Warwick, Quebec, Canada, killed three firefighters. In 1997 in Burnside, Illinois, two firefighters lost their lives in a fire and BLEVE. In all three of these fatal BLEVE events, the relief valves were operating upon arrival of the fire department.

One other event worth noting is the 1984 PEMEX LPG Terminal fire in Mexico City, Mexico. An 8-inch (20 cm) pipeline broke while filling a tank at the terminal. The resulting vapor cloud, which was 650 feet by 572 feet by 8 feet (198 × 174 × 2.4 meters), ignited. This resulted in an explosion, which included a ground shock, and a major fire. The terminal had two large spheres, four small spheres, and forty-eight horizontal tanks. About fifteen minutes into the fire the first BLEVE occurred, and for the next ninety minutes tanks continued to BLEVE. Tanks and liquid propane rained down on the adjacent community. The death toll exceeded 500 people, and thousands were injured.

In 1983 five Buffalo, New York, firefighters lost their lives when a three-story building collapsed. Nine other firefighters were injured in the massive explosion. The force of the explosion caused the first arriving apparatus, including a ladder truck, to be blown across the street. A propane tank inside the building had been struck and was leaking. An unknown ignition source sparked the explosion minutes after fire crews arrived, which caused the building to collapse.

In January 2003 a propane truck (MC-331) went over a guardrail and fell to the ground 35 feet (10.6 meters) below. The tank ruptured and the leaking propane ignited, resulting in a 600-foot-high (183 meter) fireball. The resulting explosion moved the truck several hundred feet away from the first impact area. The driver of the truck was killed in the accident.

As can be noted in the stories above, BLEVEs and VTRs can result in injuries and fatalities. Any time containers are under stress, such as during a fire, they can fail. Many times they fail violently and with severe consequences. Some containers have relief valves, while others do not. Materials that are highly poisonous such as chlorine will not have a relief valve; they will have a fusible metal plug that vents the pressure of the tank. This fusible metal plug is not like a relief valve that shuts off when the pressure inside the tank is decreased. Once a fusible metal plug melts, the contents of the tank come out, no matter the pressure. When a tank is on fire or is being impinged by fire, the tank is being weakened and the contents are being heated, creating increased pressure. Some of the general rules of firefighting and propane tanks (or other containers under pressure) are outlined in JPR 2-1.

Fighting a Fire for Tanks and Other Containers Under Pressure

A tank or container can fail at any time and it is impossible to determine the exact moment a tank is going to fail. The dangers associated with a BLEVE are:

- The fireball can engulf responders and exposures.
- Metal parts of the tank can fly considerable distances.
- Liquid propane can be released into the surrounding area and be ignited.
- The shock wave, air blast, or flying metal parts created by a BLEVE can collapse buildings or move responders and equipment.

1. Firefighters should withdraw immediately in the case of rising sound from venting relief valves or discoloration of the tank.

JOB PERFORMANCE REQUIREMENT 2-1

Fighting a Fire for Tanks and Other Containers Under Pressure

A A tank or container can fail at any time, and it is impossible to determine the exact moment that failure will occur. Firefighters should withdraw immediately when increasing sound from venting relief valves or a discoloration of the tank is noticed. In this photo, a high-pressure tube trailer carrying compressed hydrogen is on fire. *(Courtesy of Maryland Department of the Environment Emergency Response Division)*

B Fire must be fought from a distance with unstaffed hose holders or monitor nozzles. In this photo, a liquid propane tank has overturned and the propane is being flared (burned) off before the truck can be righted. *(Courtesy of Maryland Department of the Environment Emergency Response Division)*

2. Fire must be fought from a distance with unstaffed or unmanned hose holders or monitor nozzles. The tank should be cooled with flooding quantities long after the fire is out. A minimum of 500 gpm (1,893 liters per minute) at the point of flame impingement is recommended by the NFPA, **JPR 2-1A.**

- If the water is vaporizing on contact, firefighters are not putting enough water on the tank. Water should be running off the tank if it is being cooled.
- Firefighters should not direct water at relief valves or safety devices, as icing may occur. Icing would block the venting material, which could cause an increase in pressure inside the tank.
- The tank may fail from any direction, and any tank that is being exposed to a fire can fail at any moment.

3. For massive fire, it is recommended to use unstaffed or unmanned hose holders or monitor nozzles. If this is impossible, firefighters should withdraw from the area and let the fire burn, **JPR 2-1B.**

Specialized Tank Trucks

These types of tank trucks are used to carry unique chemicals or chemicals that have to be transported in a certain fashion. When gases are transported, they are transported as liquefied gases, as was described with the MC-331 tank trucks. They can also be transported as refrigerated gases or as compressed gases as will be described in this section. Other trucks are dry bulk which can carry a variety of products from grain to explosives. Materials that required high temperatures are transported in special vehicles to keep them hot. The intermodal series of tanks are cousins

FIGURE 2-61 The MC-338 carries cynogenic liquefied gases. The tank resembles a rolling Thermos bottle.

to their full size highway tanks but carry the chemicals in a comparable fashion.

MC-338 Cryogenic Tank Trucks

MC-338 tankers are uniquely constructed like the one shown in **Figure 2-61.** They have a tank with an outer shell. The inner container is steel or nickel, with a substantial layer of insulation; the exterior is aluminum or mild steel. The space between the shells is placed under a vacuum to assist in the cooling process. The ends of the tank are flat, and the piping is contained usually at the end of the tanker in a double door box. Relief valves are located on top of the tank, to the rear of the tank. The best way to describe this tanker is to compare it to a Thermos bottle on wheels. Cryogenic materials are gases that have been refrigerated to a temperature that converts them to liquids. Unlike liquefied gases, which use pressure to reduce them to liquids, these are cooled to the point of becoming liquid. To remain liquids, the material must be kept cool. To be a cryogenic material the liquid has to be at least $-150°F$ ($-101°C$) and can be as cold as $-456°F$ ($-271°C$).

The most common products are nitrogen, carbon dioxide, oxygen (liquid oxygen or LOX), argon, and hydrogen. The material inside is kept liquid by vacuum, and when on the road the tank will have a maximum of 25 psi (1.7 bar). As the truck travels the sun will heat the material, and the pressure will increase. As the pressure increases the relief valve will open up, relieving any access pressure into the atmosphere.

> **NOTE**
>
> It is not uncommon to see a white vapor cloud like that shown in **Figure 2-62** coming from the relief valves while the truck is traveling on the highway or sitting alongside the road. This is a normal occurrence and is not cause for alarm.

FIGURE 2-62 When transporting liquids it is not uncommon to see vapors coming from the relief valves. The truck can only travel with the tank at 25 psi or less. As it heats, the pressure increases, triggering the relief valve. This is a normal situation, and the tank will vent until below the 25 psi. When offloading the pressure can be increased to assist with the offloading procedure.

When the truck makes a delivery, the pressure needs to be increased to push the liquid out of the tank. The driver opens the piping and allows the material to flow into a heat exchanger, which is located just in front of the rear wheels of the tank. Once in the exchanger, the material will heat up and expand, increasing the pressure in the tank and forcing the liquid into the receiving tank. During transportation and offloading, a large amount of ice will build up on the piping and valves.

Tube Trailers

Tube trailers, **Figure 2-63,** contain several pressurized vessels, constructed much like the MC-331 tank. The DOT now refers to these as Multiple-Element Gas Containers (MEGC). They are constructed of steel and have pressures ranging from 2,000 to 6,000 psi (138–414 bar). They hold pressurized gases such as air, helium, and oxygen. The piping and controls are usually located on the rear of the trailer, but could be in the front. The typical delivery mode is that the driver will drop a full trailer off at a facility and pick up an empty one for refilling. Although they are not subject to BLEVEs because they only contain a gas, if involved in a fire, tube trailers can experience a VTR and rocket in the same fashion as a BLEVE.

FIGURE 2-63 The tubes on this trailer contain pressurized gases. The pressure can vary from 2,000 to 6,000 psi.

Dry Bulk Tanks

Dry bulk tanks resemble large uninsulated MC-307s in shape with bottom hoppers to unload the product, as shown in **Figure 2-64.** The tanks hold dry products and sometimes a slurry, like concrete. The most common products are fertilizers, lime, flour, grain, and other food products. The potential hazard when dealing with these tankers is predominantly environmental but at times these tankers contain toxic materials. They are usually offloaded using air pressure either from the truck itself or at the facility.

Hot Materials Tanker

Hot materials tankers vary in that they can be modified MC-306s, 307s, 406s, 407s, or dry bulk containers, **Figure 2-65.** They may have a mechanism to keep the material hot or it may be loaded hot. It may require heating prior to offloading. Common products are tar, asphalt, and molten sulfur, fuel oil #7 and #8.

> **SAFETY**
>
> The major problem with these tankers is the heated material itself. Anyone coming into contact with the material could be seriously burned. The molten material could ignite the truck or other combustibles.

If the material is allowed to cool, it can cause problems for the responders or the shipping company. Tar trucks may be transported with a propane or fuel oil flame ignited to heat the product en route to the job site, although this is illegal in most states.

FIGURE 2-64 Dry bulk tanks carry a variety of products. Some examples include fertilizers, explosives, and concrete.

FIGURE 2-65   These trucks carry molten products and can be heating the product while driving. This practice is illegal but is found on occasion. The fuels used to heat the product are either diesel/kerosene or propane.

TABLE 2-9   IMO Containers

Container	Maximum Capacity (Gal)	Pressures (Psig)	Products
IM-101 or IMO Type 1	6,340 (24,000 liters)	25.4–100 (1.8–7 bar)	Non-flammable liquids, mild corrosives, foods, and other products
IM-102 or IMO Type 2	6,340 (24,000 liters)	14.5–25.4 (1–1.7 bar)	Flammable materials, corrosives, and other industrial materials
Spec 51 or IMO Type 5	6,418 (24,295 liters)	100–500 (7–34.4 bar)	Liquefied gases, much like an MC-331
Specialized tanks (Cryogenic tanks are also known as Type 7 tanks)	Varies	Varies	Includes cryogenic tanks and tube banks (small tube trailers) that carry the same products as their highway counterparts

Intermodal Tanks

Intermodal tanks are increasing in use and carry the same types of products as their highway and rail companions, **Table 2-9** and **Figures 2-66, 2-67,** and **2-68.** They are called intermodal (IM) because they can be used on ships, railways, or highways.

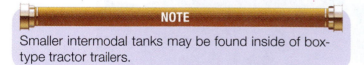

NOTE

Smaller intermodal tanks may be found inside of box-type tractor trailers.

Intermodals follow three basic types: nonpressurized, pressurized, and highly pressurized. They are built in two ways. In one, they sit inside a steel frame, called a box-type framework; in the other, the tank is part of the framework, called a beam-type intermodal. Like tank trucks they are assigned

FIGURE 2-66   This is an IMO-101 tank. Like the totes, these are bulk tanks capable of carrying a large quantity of product. These are normally placed on ships, then delivered locally by a truck, although trains can also be used.

FIGURE 2-67 This is an intermodal tank, commonly referred to as an IM or IMO. This is a bulk tank that carries an average of 3,000 to 5,000 gallons. This tank is an IMO-101, which is an atmospheric tank. The orange panel indicates that it has a United Nations (UN) hazard code of 60 and a UN number of 2572. The code of 60 indicates that the product is a toxic material. The UN number indicates that the product is phenylhydrazine. The name is also stenciled on the sides of the container, and there are toxic placards.

FIGURE 2-68 This is another IMO-101, and the photo shows that the orange panel indicates a hazard code of 66 and a UN number of 2810. The hazard code indicates that the product is a highly toxic material, and there are fifty-two different materials listed for UN 2810. One would have to look at the shipping papers to figure out what is being carried in this tank. The tank is marked "foodstuff only," which means that it is holding a product used in food. It is interesting to note that many of the chemicals listed for UN 2810 are chemical warfare agents such as sarin nerve agent and comparable materials. The most likely candidate, considering the "foodstuff only" label, would be a medical-type product.

specification numbers, IM-101, IM-102, Spec 51 (specification 51). When used internationally they are called IMOs. They are made to be dropped off at a facility and when empty picked up for refilling.

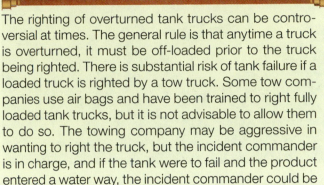

VIEWPOINT

The righting of overturned tank trucks can be controversial at times. The general rule is that anytime a truck is overturned, it must be off-loaded prior to the truck being righted. There is substantial risk of tank failure if a loaded truck is righted by a tow truck. Some tow companies use air bags and have been trained to right fully loaded tank trucks, but it is not advisable to allow them to do so. The towing company may be aggressive in wanting to right the truck, but the incident commander is in charge, and if the tank were to fail and the product entered a water way, the incident commander could be held responsible for violations of the Clean Water Act.

A tank truck is designed to function with its wheels down and when it is wheels up, the strength of the tank is in question. The stresses on a tank during a rollover cannot be seen and when rolled over there are more stresses placed on the tank. Gasoline tankers (MC-306/DOT-406) are especially susceptible to failure during a righting operation. The principal responsible party is required by law to provide response experts, either from within their company or an outside contractor, to mitigate this type of incident. In some cases the local public sector hazardous materials team would remain in command and monitor the scene until the scene is no longer an emergency. In some parts of the country, the local hazardous material team is aggressive in this mitigation and performs the offload. No matter who handles the incident the product needs to be offloaded from the truck.

There are several methods of removing the product from the tank, some easier than others. One method is to remove the product through existing valves and piping, which can sometimes be awkward when the truck is upside down. Another method is to attach a valve assembly to the existing valves and then offload the product. Comparable to this method are flexible bladders that attach to the valves and are connected to piping. Another method is to drill the tank to allow for the offload of the product. Each of the methods has benefits and pitfalls and may not apply in every case. First responders can speed up this process by calling for another compatible tank truck to transfer product into and calling for a vacuum truck or pump to conduct the transfer. These requests should be made in consultation with the responsible party and the hazardous materials team. In some states, box trailers that have any quantity of chemicals, including household commodities, are required to offload prior to righting, as shown in **Figure 2-69.**

FIGURE 2-69   Righting an overturned truck can be dangerous because of fuel and other motor fluids, and there is potential for the cargo to be further damaged. Any hazardous materials present could be released, endangering those in the vicinity and causing environmental damage. *(Courtesy of Maryland Department of the Environment Emergency Response Division)*

Rail Transportation

As with highway transportation there are only a few types of railcars, and they are similar to their highway counterparts. The piping and shape may be the same but that is the extent of the similarities.

> **NOTE**
>
> In rail transportation the quantities are greatly increased—up to 30,000 gallons (113,562 liters) for hazardous materials and up to 45,000 gallons (170,344 liters) for nonhazardous materials.

Rail incidents usually involve multiple railcars, whereas highway incidents usually involve one or two trucks. The incidents may occur in rural areas, away from water supplies and easy access. Rail incidents will involve multiple agencies, and the local community can expect a large contingent of assistance coming from the state and federal government, which in itself will be difficult to manage.

Railcars come in three basic types: nonpressurized, pressurized, and specialized cars. Although they are categorized in this fashion, the commodities they carry will determine the ultimate use of the car. As with highway transportation there are dedicated railcars that will be marked with the products they carry.

> **CAUTION**
>
> The term nonpressurized car is actually misapplied because it can have up to 100 psi (7 bar) in the tank.

A nonpressurized car carries chemicals, combustible and flammable liquids, corrosives, and slurries.

FIGURE 2-70   The indication that this is a nonpressurized railcar is given by the fact that all of the valves are on the outside of the tank and not contained within any protective housing.

The easiest way to determine if the car is nonpressurized, such as the one shown in **Figure 2-70,** is to observe whether the valves, piping, and other appliances are located outside of a protective housing. In some cases, the car will be bottom unloaded, with the ability to load the car through the top, or the car may be top loaded and unloaded. The cars will usually have a small dome cover, but the relief valves and other piping are located outside of this dome. Most of the piping is located on the catwalk on top of the car.

Some cars have unique paint schemes such as those shown in **Figure 2-71.** Nonpressurized cars can be insulated and may have heating and cooling lines around the tank. Prior to offloading, the tank may need to be heated to ease the offloading process. Some nonpressurized railcars have an expansion

FIGURE 2-71   This railcar is white with a red stripe, which is used to indicate hydrogen cyanide but on occasion is used to indicate dangerous materials. This car holds hydrogen fluoride anhydrous, which is a severe inhalation hazard and is corrosive. The term *anhydrous* indicates "no water" and means the hydrogen fluoride is pure.

FIGURE 2-72   The pressurized car is indicated by the valves being contained within the protective housing of the railcar.

dome. The piping valves and fittings sit on top of this dome, which was constructed to hold any potential expansion. These cars are not in regular service but may be seen in larger industrial facilities that have their own railroad service on site.

Pressurized tank cars, **Figure 2-72,** also carry a wide variety of products, including flammable gases like propane, and poisonous gases like chlorine and sulfur dioxide. The pressurized cars will have pressures in excess of 100 psi up to 600 psi (7–41.3 bar). Most pressure cars that carry flammable materials will be insulated with a spray-on insulation or may be thermally insulated. This insulation is 1 to 2 inches (2.5–5 cm) thick and helps reduce the chance of a BLEVE during a fire situation. To determine if the

FIGURE 2-73   This specialized railcar holds liquefied carbon dioxide, a cryogenic material.

railcar is a pressurized car the firefighter can look at the spec plate, which will have all of the valves, piping, and fittings located under the protective housing, on top of the railcar. A catwalk around the protective housing provides relatively easy access.

Specialized Railcars

Specialized railcars have the same characteristics as highway vehicles, and, in fact, in some cases highway box trailers are loaded onto flatbed railcars. When transported in this fashion they are referred to as trailers on flat cars (TOFC). Regular box trailers as well as refrigerated trailers can be found on flat cars. Much like highway trailers, there are freight boxcars that haul the same products as their highway trailers, with the exception of carrying much larger quantities. Examples of these cars are shown in **Figure 2-73.** The boxcars can carry boxes, cylinders, bulk tanks, totes, and super sacks. Refrigerated railcars are similar to highway and intermodal boxes and are referred to as reefers. They contain their own fuel source. Dry bulk closed railcars are common, and open hopper cars can also be found. Tube trailers for rail are also found, but they are rarely used; the most common is a highway tube trailer set on a flat car. Cryogenics are also carried in railcars and have the same low pressure (25 psi/1.7 bar) as in highway transportation, with the characteristic venting when the pressure increases.

Markings on Railcars

Railroads use the same placarding system as on highways, with the exceptions noted in the placard section. The placards are the same size, but additional information on the railcars is more extensive than their counterparts on the highway. The information on the car itself is printed larger as compared to highway

FIGURE 2-74   A railcar has a lot of information stenciled on the side. The FMLX identifies who owns the car. The X at the end indicates that the car is privately owned, that is, the railroad does not own the car. The 15020 is the car number and can be used to cross-reference the shipping papers. The photo on the left is the certification and testing data. The top line, DOT 111 A 100 W 1, provides most of the information. The tank is a DOT 111 specification, which means that it is a non-pressure car. The A indicates the type of couplers, and the 100 is the maximum pressure for the car. The W and the 1 indicate the type of welds and other tank construction information. The last line is also important, and it shows that the tank was lined on 3-94, which means that it has an internal lining. In a derailment the lining could separate from the tank and the chemical could react with the tank metal.

transport. By looking at the sides of a railcar, you can tell the specification type, maximum quantities, test pressures, relief valve settings, and other pertinent information. These types of markings are shown in **Figure 2-74**. In addition to a placard, the name of the hazardous materials will be stenciled on two sides of a dedicated railcar, as shown in **Figure 2-75.**

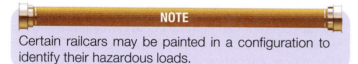

NOTE

Certain railcars may be painted in a configuration to identify their hazardous loads.

It was once common to paint a car carrying hydrogen cyanide white with a red stripe running around the middle; it was called a "candy-striper." The specification information listed on the car is coded to reveal a lot of information about the car. The tank's owner can be determined by the code on the top line, which consists of four letters followed by a series of numbers. If the fourth letter is an *X*, then the car is owned by someone other than the railroad. The numbers indicate the car number and can be cross-referenced to determine the tank's contents when contacting the railroad. An example of the coding on a railcar is as follows:

DOT 111 A 60 AL W 1

- *DOT* is the authorizing agency.
- *111* is the class or specification number.

FIGURE 2-75   Dedicated railcar stencil.

- The following markings in conjunction with the preceding markings will be found only on pressure railcars. The letters will be found between the authorizing agency and the pressure markings:
  - *A*—has top and bottom shelf couplers.
  - *S*—has features of an A car and head puncture resistance.
  - *J*—has features of A and S and jacketed thermal protection.
  - *T*—has features of A, S, and J plus spray-on insulation.

- *60* is the pressure rating of the tank.
- *AL* is the material that the tank is made up of.
  - No letter—carbon steel
  - AL—aluminum alloy
  - A—carbon steel
  - A-AL—aluminum alloy
  - AN—nickel
  - B—carbon steel, elastomer lined
  - C, D, or E—alloy steel
- *W* is the construction type of the tank.
  - W—fusion welded
  - F—forged welded
  - X—longitudinally welded
- *1* is a number that describes fittings, materials, and linings, which can be retained in rail-specific text.

Bulk Storage Tanks

Bulk storage tanks range in size from 250 gallons (946 liters) up to millions of gallons and store a variety of products, the most common being petroleum products. These tanks are seen in residential homes and rural areas, and are common at an industrial facility, such as the one shown in **Figure 2-76.** In residential homes the most common tank is a 250-gallon (946 liters) home heating oil tank, and some homes or small businesses may have gasoline or diesel fuel tanks that vary from 250 to 500 gallons (946 to 1,893 liters).

The two basic groupings of tanks are in-ground and aboveground. Since the passage of the EPA underground storage regulations, there has been considerable movement to remove **underground storage tanks (USTs)** and replace them with **aboveground storage tanks (ASTs).** If the facility owner elects to keep its tanks underground then additional requirements are placed on them for testing, containment, and leak prevention.

> **CAUTION**
>
> Leaks from these types of tanks can be overwhelming because the leak may not be detected for a period of time, giving it a chance to spread throughout the area, contaminating a large area.

A substantial release from a million-gallon tank of gasoline that escapes a facility would require an enormous response from the emergency responders and environmental contractors.

Underground tanks are usually constructed of fiberglass or are steel coated with an anticorrosion material. New tanks are usually double walled, that is, a tank within a tank, to prevent any spillage into the environment. Ethanol gasoline has created some problems because the ethanol may not be compatible with the fiberglass tanks and may cause a release. As new formulations for gasoline are developed, the true effect on the storage and piping systems may not be known for many years. The piping from the underground system comes up through the ground and its use will determine its route. The offload piping for an UST at a gasoline station comes up through the pump, which **Figure 2-77** demonstrates. The piping is manufactured so that in the event that a car knocks off the pump it will shut off the flow of gasoline and will snap the piping off at or near the ground level. The only spillage of gasoline would be from the

FIGURE 2-76 The sizes of tanks vary from a few hundred gallons up to several million gallons. A catastrophic failure of the tank may overwhelm the responders' ability to handle the incident.

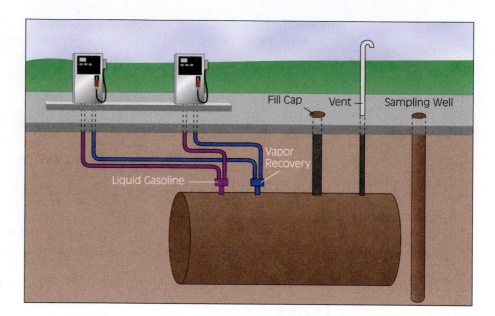

FIGURE 2-77 Piping system of a gas station.

amount in the piping aboveground and in the hose, if everything works properly.

The loading piping is located separately and all of the fill pipes are located in the same area for easy transfer from a tank truck. The fill pipes are usually color coded and marked so that the driver can differentiate between unleaded, unleaded super, and diesel fuel, although the color coding is not standard across the country and varies from company to company. At some location on the property there will be vent pipes for the tanks, generally away from the pumps and near the property line. There will also be other manhole covers approximately 6 to 8 inches (15.2–20 cm) in diameter that have a triangle on the top of the cover. These are inspection wells and typically surround the tank. The holes are drilled at various depths so that leaks can be easily detected by air monitoring of the well, or if water is in the well, by taking a sample. If a facility has had a leak, there may be a large number of these wells on and around the property. Most gas station tanks are 10,000 to 25,000 gallons (37,854–94,635 liters) in size and, if not properly monitored, can slowly release a substantial amount of product over a short period of time.

NOTE

It is not uncommon to find gasoline bubbling up in a basement miles away from the gas station and to find that the release occurred several years ago.

Another common problem arises when farms are redeveloped into housing developments. If unknown USTs are located on the property, they may or may not be discovered during the construction and can eventually leak, causing problems. The tank and envi-

ronmental industry refers to a **leaking underground storage tank** as a **LUST**.

SAFETY

A recent incident in Baltimore County, Maryland, demonstrates that regardless of how many protection systems may be in place, a release off the property is still possible. One afternoon a gas station was getting a tank filled with fuel, as per normal procedure, and the tank filling went as expected with no problems. The gas station had installed a state-of-the-art alarm system that would indicate a fault in the system and alert the owners to any potential releases. The system goes through a system check after each sale, if no pump is operated for a period of thirty seconds. If a pump is on, or the period between sales is less than thirty seconds, then the system check does not begin to work. During a busy afternoon, it could be estimated that the system check would not function for a long period of time, because at least one of the pumps is operating all the time.

In addition to the system check of each of the storage tank locations, some piping and the pumps all have electronic monitors that detect the presence of a liquid. When a liquid is detected, an alarm activates— the same type of alarm as if the system check fails the system. One of the potential problems with many self-service gas stations is that there is only one attendant, who for many reasons cannot leave the work area to either check on a problem outside or, depending on the alarm panel location, cannot check any alarms. In some gas stations the alarms are in other rooms and cannot be seen or heard by the only attendant. Another problem is that the liquid alarms will activate when it rains, tripping for rainwater.

The pumps that sit on top of the storage tanks have the ability to supply upward of 80 psi (5.5 bar) of pressure to all pump station nozzles, although this is governed down to an actual working pressure of 10 psi (0.7 bar) at all pumps. In this particular incident, a flange that comes out of the tank pump failed and separated from the tank pump. From the pump discharge there is a 2-inch- (5 cm) diameter opening, which at 10 to 80 psi (0.7–5.5 bar) could pump a considerable amount of fuel in a short time. One theory is that a customer may have set a nozzle down after not being able to pump fuel, or the amount of fuel being pumped was inadequate. This action would have resulted in the tank pump continuing to pump. A liquid alarm sounded for a number of areas at the gas station. The attendant notified the pump repair company as per protocol, but a response was delayed due to a number of extenuating circumstances.

Once the repair company arrived they found the tank empty; it was at this time the clerk informed them of the recent delivery. It was later found that an estimated 4,500 gallons (17,034 liters) were lost and were causing flammable vapor readings in a several-block area, although all the gasoline was later recovered. Within the gas delivery system, this station had the best protection in place, and had just retrofitted the station with a new piping system, but it shows that no matter how many protection systems are in place, it is still possible to have a release.

FIGURE 2-78 Due to increased environmental concerns, many tanks are being placed aboveground and are called aboveground storage tanks or ASTs.

Aboveground tanks are becoming more and more common, although they have been used for many years, **Figure 2-78.** They are of two basic construction types—upright and horizontal—and some are nonpressurized or atmospheric tanks and some are pressurized tanks. ASTs hold a wider variety of chemicals as opposed to their UST counterparts, but petroleum products are still a leading commodity stored in these type of tanks. They vary in size from the 250-gallon (946 liters) home heating oil tank to the several million gallon oil tank. In industrial and commercial applications, a containment area is required around the tank. The containment area must be able to hold the contents of the largest tank within the containment. Regular inspection of these areas is required to ensure that rainwater, snow, or ice does not cause a buildup of liquid in the containment area, which would reduce the containment's ability to hold its intended amount. Depending on the weather conditions, the containment area's gate valve may be left in an open position, which leaves the facility at risk for product to escape the facility through the open drain or gate valve.

Upright storage tanks come in three basic construction types: **ordinary tank, external floating**

FIGURE 2-79 This is a cone roof tank. It has a weak roof-to-shell seam so that in the event of an explosion the roof will come off, but the tank should remain intact.

roof tank, and **internal floating roof tank.** The ordinary tank, **Figure 2-79,** is constructed of steel and typically has a sloped or cone roof to shed rainwater, snow, and ice. The roof and tank shell seam is purposely made weak so that in the event of an explosion, only the top of the tank is relocated. One of the major problems with this type of tank is that when the tank is not full, there is room for vapors to accumulate.

CAUTION

Any time vapors are allowed to accumulate, the potential exists for a fire or explosion.

Some ordinary tanks are purposely constructed without a roof in place. These are typically used in safety vent situations and water treatment areas where a roof may cause additional problems. It is important to identify these tanks during pre-incident surveys so as not to misjudge the severity of an incident. In many chemical processes it is not uncommon to have a storage tank with piping into the tank just to catch overflow or the contents of a system in the event of overpressurization or failure. The materials can be hot and produce a large amount of vapors, which, if a roof were present, would allow a buildup of pressure, causing a catastrophic failure of the tank and roof.

External floating roof tanks are used to eliminate the buildup of vapors. They ride on top of the liquid, **Figure 2-80,** and since there is no space between the roof and the liquid, vapors cannot accumulate and create problems. External floaters can be seen from the top of the tanks, and a ladder is affixed to the roof to allow access. A common incident with these types of tanks results from a lightning strike, which can cause a fire in the roof/shell interface area. This type of fire is difficult to extinguish and can result in roof/tank failure if not controlled quickly. It is also possible that water used for firefighting could also sink the roof, causing further problems.

Internal floaters are constructed in the same manner as external floaters but have an additional roof over the top of the tank. An example is shown in **Figure 2-81.** The type of roof varies, but is usually a slightly coned roof with vent holes along the outer edge of the tank shell, or a geodesic type of roof that is usually made of fiberglass. The internal floater suffers from the same type of fire problem as the external, although it is a reduced risk. If a fire were to start

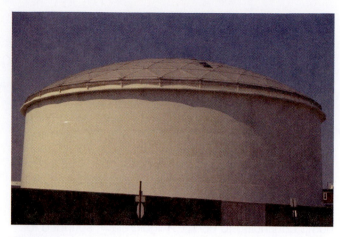

FIGURE 2-81   This is a covered floating roof tank, which is the same as an open floating roof tank but has a cover to keep out snow, rain, and debris. Another term for this type of tank is *geodesic domed tank.*

in the roof/shell interface, the roof makes it more difficult to extinguish.

Specialized Tanks

Specialized tanks are a combination of the tank types discussed earlier in this section and include pressurized tanks and cryogenic tanks. The larger pressurized vessels are divided into two categories: low-pressure and high-pressure tanks. The low-pressure tanks hold flammable liquids, corrosive liquids, and some gases, up to 15 psig (1 bar). Common high-pressure commodities are liquefied propane, liquefied natural gas, or other gaseous or liquefied petroleum gases. An acid like hydrochloric acid, which has a high vapor pressure, would not be uncommon in a tank like this.

These types of tanks may have an external cover, which appears to be a tank within a tank. The pressurized tanks are larger versions of the propane tanks discussed earlier and can have a capacity of up to 9 million gallons (34,068,706 liters) although less than 40,000 gallons (151,416 liters) is typical. Liquefied petroleum gases are not only stored in pressurized tanks such as the one shown in **Figure 2-82;** these types of gases are also stored in other locations such as caves carved out of mountains, although this is rare. Because these facilities are not required to report under the SARA Title III regulations, firefighters will have to contact their local gas suppliers to find out how they store their gas products.

Upright cryogenic storage tanks, **Figure 2-83,** are located in almost every community, especially if the community has a hospital or medical center. Most hospitals are supplied with liquid oxygen (LOX) through the cryogenic tank. Facilities that sell or distribute compressed gases are likely to have cryogenic tanks.

FIGURE 2-80   This is an open floating roof tank, in which the roof floats on top of the product. This reduces the release of vapors, as there is no vapor space, and reduces the fire potential.

FIGURE 2-82 The specialized tank such as the propane tank shown here has some of the same properties as its transportation equivalents.

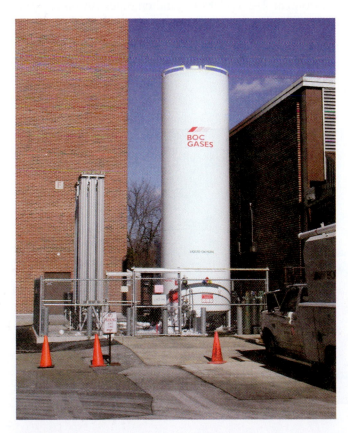

FIGURE 2-83 This tank holds cryogenic liquid oxygen and is typical of a cryogenic upright tank.

Fast-food restaurants and convenience stores are now using cryogenic tanks to supply carbon dioxide to the soda dispensing machines.

SENSES

Although the use of senses is listed as one of the four major recognition and identification categories, it is one that presents the most risk. Responders should never use smell, taste, or touch to assist in the hazardous materials identification process. If victims were unfortunately exposed to a material, responders should use the information gleaned from these persons, but only after they have been decontaminated. Responders can use hearing and vision, as well as other senses, to assist in the recognition and identification process. The ability to recognize potentially hazardous situations is key to survival, and there are typically visual clues available to responders. Each of the other recognition and identification categories relies on the use of the visual sense. The venting of a relief valve may only present an audible indication that there is a problem. As the pitch of the relief valve increases, this is an indication that the pressure is increasing inside of the container. Responders should never place themselves in harm's way to smell, taste, or touch hazardous materials; doing so may be the last act they perform.

Many chemicals are in a category known as desensitizers, which overwhelm the olfactory (sense of smell) system. Hydrogen sulfide, a toxic material, has a unique and offensive, rotten egg–like odor. After a few minutes of exposure, a person breathing in hydrogen sulfide will no longer have the ability to smell the toxic gas. The gas is still present, and could even be present in higher levels, but the exposed person no longer has the ability to smell, as their brain has told the olfactory system to ignore the smell. The odorant in natural gas, known as mercaptan, has the same effect.

> **CAUTION**
> The lack of an odor cannot be equated with a lack of toxicity. Many severely toxic materials are colorless and odorless.

CHEMICAL AND PHYSICAL PROPERTIES

Although not intended to be a full chemistry lesson, the material in this section is a key component of safety. The chemical and physical properties outlined here are appropriate for a firefighter's level of response. The identification and use of these key terms can determine the outcome of an incident and firefighter well-being. As a firefighter progresses up through the response levels the need for additional chemistry also increases. When in doubt, the firefighter should consult with a hazardous materials team or other resources such as Chemtrec or a local specialist. A lot of the terms to be discussed next can be applied throughout the entire firefighter text. The basis of a fire is a chemical reaction. The better that firefighters understand this chemical reaction, the better off they will be, which will also benefit the citizens of their communities.

States of Matter

The basic chemical and physical properties that are important to understand are the **states of matter**: **solid**, **liquid**, and **gas** as shown in **Figure 2-84.** The severity of an incident can be determined by knowing if the material is a solid chunk of stuff, a pool of liquid, or an invisible gas.

> **NOTE**
>
> The level of concern rises with each change of state; relatively speaking, a release of a solid material is much easier to handle than a liquid release, and it is nearly impossible to control a gas.

The control methodology for each state increases in difficulty from simple controls with a solid, to difficult with a gas. Evacuations have to be larger for releases involving gases, whereas a minimal evacuation would be required for most solid materials.

How the chemicals can hurt someone also varies with the state of material. Solids usually can only enter the body through contact or ingestion, although inhalation of dusts is possible. Liquids can be ingested, absorbed through the skin, and, if evaporating, inhaled. Gases on the other hand can be absorbed through the skin, and inhaled, and to some extent ingested.

Adjoining the states of matter are melting point, freezing point, **Figure 2-85,** boiling point, **Figure 2-86,** and condensation point. All of these are related because they are the points at which a material changes its state. The **melting point** is the temperature at which solid must be heated to transform the solid to the liquid state. Ice, for instance, has a melting point of 32°F (0°C). The **freezing point** is the temperature of a liquid when it is transformed into a solid. For water, the freezing point is 32°F (0°C). The actual temperatures vary by tenths of a degree, but are very close. The **boiling point** is reached when the liquid is heated to the point at which evaporization takes place, that is, the liquid is being changed into a gas. Another way of defining boiling point is the temperature of a liquid when the vapor pressure exceeds the atmospheric pressure and a gas is produced. Water boils at 212°F (100°C) and changes into a gaseous state. The important thing to remember about boiling points is the fact that when the liquid approaches this temperature, vapors are being produced that can cause larger problems.

Block   Balls   Dust/Dirt Pile

Cylinder

Solids Can Exist in a Variety of Forms   Liquids   Gases

FIGURE 2-84   States of matter.

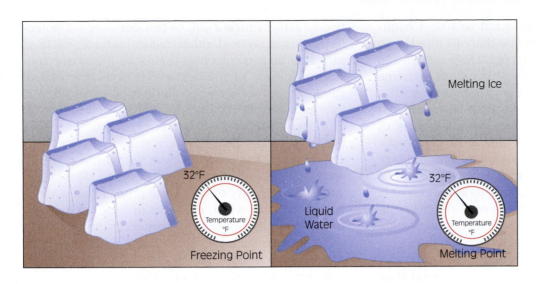

FIGURE 2-85 Melting and freezing points.

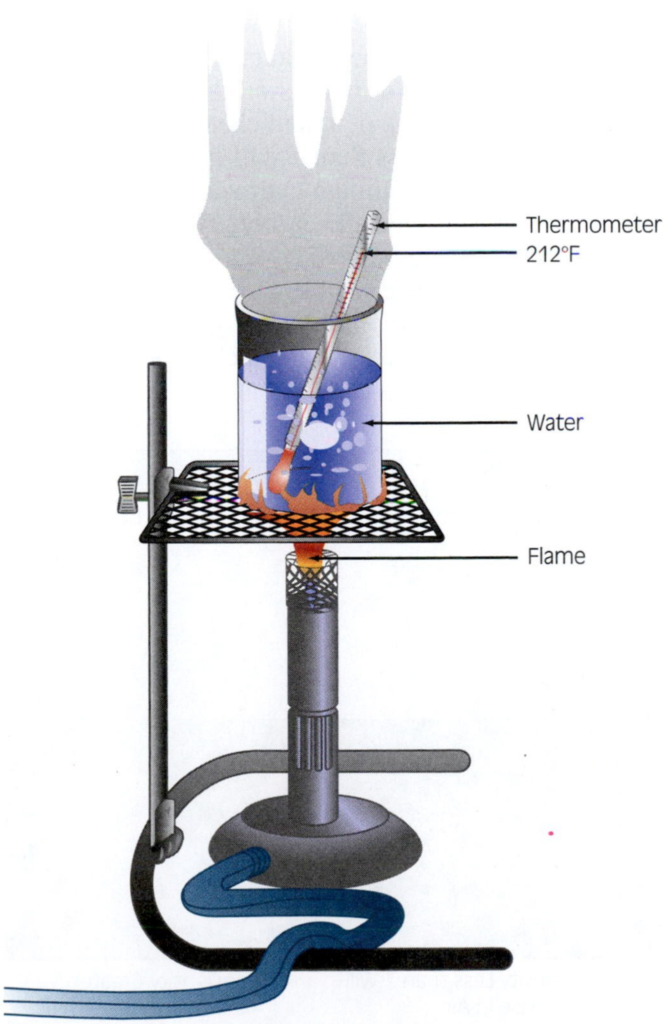

Vapor Pressure

Out of all of the chemical and physical properties, vapor pressure is one of the most important to a hazardous materials responder. If a product has a high **vapor pressure,** it can be very dangerous. A material with a low vapor pressure is typically not a major concern. Vapor pressure has to do with the amount of vapors released from a liquid or a solid, **Figure 2-87.** The true definition is the pressure that is exerted on a closed container by the vapors coming from the liquid or solid. Vapor pressure can be related with the ability of a material to evaporate, in that the material is not really disappearing, it is just moving to another state of matter. Chemicals like gasoline, acetone, and alcohol all have high vapor pressures, whereas diesel fuel, motor oil, and water all have low vapor pressures.

Vapor pressures are measured in three ways: millimeters of mercury (mm Hg), pounds per square inch (psi), and atmospheres (atm). Normal or average vapor pressures are 760 mm Hg, 14.7 psi, and 1 atm. Although these figures are used to describe normal vapor pressure, they best describe atmospheric pressure. The temperature which is used is 68°F (20°C), which is the standard temperature. If a temperature is not provided, then it can be assumed it is 68°F. Chemicals that have a true vapor hazard are those in excess of

A material with a vapor pressure is pushing against the sides of the container. If the vapor pressure is high and the material is in the wrong type of container, the container could fail.

FIGURE 2-86 Boiling point.

FIGURE 2-87 Vapor pressure.

40 mm Hg, and they are considered volatile. Chemicals with a vapor pressure of less than 40 mm Hg do not travel throughout the area, but if close enough can harm through inhalation; they still present extreme toxicity through skin absorption. Chemicals with a vapor pressure above 40 mm Hg can be considered inhalation hazards in addition to any other route of exposure they may possess. **Table 2-10** lists vapor pressures for some common products. The military uses the term **persistence,** which is an indication of the time that a material will remain as a liquid, and is directly related to the vapor pressure of the material. A material that is known to be persistent is one that has a low vapor pressure and will remain a liquid.

A unique chemical phenomenon called **sublimation** occurs when a chemical goes from a solid state of matter to a gas. The material never enters the liquid phase. The sublimation ability means that some solids have a vapor pressure and can move directly to the gaseous stage. Some solids such as dry ice (carbon dioxide) move quickly to the gaseous stage, while others, such as mothballs (naphthalene or paradichlorobenzene), move slowly.

Vapor Density

It is easy to confuse vapor pressure with vapor density, but they have two entirely different meanings. Vapor density determines whether the vapors will rise or fall, **Figure 2-88.** When deciding on potential evacuations and other tactical objectives (e.g., sampling and monitoring) this is an important consideration. One of the primary reasons that natural gas leaks do not ignite or flash back more often is the vapor density of natural gas. Air is given a value of 1, and all other gases are compared to air. Gases that have a vapor density of less than 1 will rise in air and will dissipate, whereas gases with a vapor density greater than 1 will stay low to the ground. Natural gas has a vapor density of 0.5, while propane has a vapor density of 1.56, which causes it to seek out any low spots, like gullies or sewers. Propane is more likely to ignite or flash back because it has a greater potential of finding an ignition source. Most gases and vapors will stay low to the ground. There are only eleven common gases that rise in air, which are shown in **Table 2-11.**

Specific Gravity

This chemical property is similar to vapor density in that it determines whether a material sinks or floats in water, **Figure 2-89.** Specific gravity is of prime concern when efforts are being taken to limit the spread of a spill by the use of booms or absorbent material. Water is given a value of 1, and chemicals that have a specific gravity of less than 1 will float on water. Fuels, oils, and other common hydrocarbons (chemicals composed of hydrogen and carbon, typically combustible and flammable liquids) have a specific gravity of less than 1 and, hence, will float on water.

Materials that have a specific gravity of greater than 1 will sink in water. It is much easier to recover

TABLE 2-10	**Common Products and Their Vapor Pressures**	
Name	**Vapor Pressure @ 68°F/20°C (mm Hg)**	**Boiling Point (°F)**
Water	25*	212 (100°C)
Acetone	180	134 (57°C)
Gasoline	300–400	399 (204°C)
Diesel fuel	2–5	304–574 (151–301°C)
Methyl alcohol	100	149 (65°C)
Ethion (pesticide)	0.0000015	304 (151°C) (decomposes)
Sarin nerve agent	2.1	316 (158°C)

*The vapor pressure of water has been reported in various texts as between 17 and 25 mmHg. This text uses 25 mmHg, as it is the highest reported vapor pressure for water.

Vapor Density Less than 1 Will Rise in Air   Vapor Density Greater than 1 Will Sink

FIGURE 2-88 Vapor density.

TABLE 2-11   Common Gases That Rise

Gas Name	Vapor Density	Gas Name	Vapor Density
Diborane (B_2H_6)	0.97	Ammonia (NH_3)	0.60
Methane (CH_4)	0.55	Neon (Ne)	0.7
Hydrogen (H)	0.1	Helium (He)	0.138
Hydrogen cyanide (HCN)	0.9	Carbon monoxide (CO)	0.97
Acetylene (C_2H_2)	0.91	Ethylene (C_2H_4)	0.98
		Nitrogen (N)	0.967

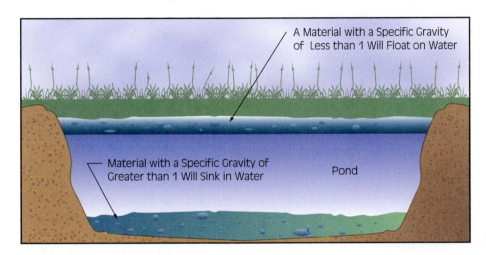

A Material with a Specific Gravity of Less than 1 Will Float on Water

Material with a Specific Gravity of Greater than 1 Will Sink in Water

Pond

FIGURE 2-89   Specific gravity.

materials floating on top of the water as opposed to those underwater. Carbon disulfide and 1,1,1-trichloroethane are two common materials that sink in water. Any material that sinks in water is especially troublesome if it has the potential to reach any groundwater, because remediation (cleanup) efforts are difficult and expensive.

Also of concern are materials that are water soluble, that is, have the ability to mix with water, not sink or float. Corrosives and many poisons are water soluble and are difficult to remove from a water source. Examples are provided in the following Firefighter Fact. Alcohol is also water soluble. Materials that are water soluble are difficult to extinguish.

FIREFIGHTER FACT

When dealing with water-soluble materials it is important to protect bodies of water, because the cleanup can be difficult or impossible if the material enters the water. In one incident more than 10,000 gallons (37,854 liters) of sodium hydroxide, a very corrosive material, entered a small stream. The creek had to be dammed, and a neutralizing agent was put into the stream. Any life existing in this mile of stream was killed by the sodium hydroxide, but the environmental damage could have spread for many more miles if it had not been neutralized. Once neutral, the water was released, which eventually led to a larger body of water, where no damage occurred. This took considerable resources and time to accomplish, and luckily the necessary neutralizing agent was located at the site of the spill.

In another incident a very small amount of a pesticide was sprayed near a pond. After a rainfall the pesticide ran into the pond, killing all of the fish. The only method of removing this pesticide is the addition of other chemicals that are also hazardous. It will be years before the pond is able to sustain life.

In another incident in Baltimore City a railcar was involved in an accident and released 18,000 gallons (68,137 liters) of hydrochloric acid, which eventually ran into a local stream. The addition of the acid actually raised the pH level of the stream to near acceptable levels, because this stream was already heavily contaminated from a number of other sources.

Note that nothing should be flushed into any body of water without the express permission of the local or state environmental agencies.

TABLE 2-12 pH of Common Materials

Material	pH	Material	pH
Water*	7	Sulfuric acid	1
Stomach acid	2	Gasoline	7
Orange juice†	2	Hydrochloric acid	0
Drain cleaner**	14	Pepsi‡	2
Potassium hydroxide	14	Household ammonia***	14

*Tap water is usually a 7, while rainwater in the Northeast can be 3–6 (acid rain).

†Citric acid is the main ingredient.

**The main ingredient is sodium hydroxide (lye).

‡Phosphoric acid is the main ingredient.

***This is a 5 percent solution of ammonium hydroxide and water.

Corrosivity

This book has already discussed tanks and containers that hold corrosives, as well as the concerns of dealing with corrosives. Corrosive is a term that is applied to both acids and bases, and is used to describe a material that has the potential to corrode or eat away skin or metal. Some examples are shown in **Table 2-12.** People deal with corrosives every day in that the human body is naturally acidic and they use many corrosive materials in their everyday lives. Acids are sometimes referred to as corrosives, whereas bases are also known as alkalis or caustics. The accurate way to describe a corrosive is to identify the material's pH, which provides some measure of corrosiveness. The designation pH is an abbreviation for a number of terms, including potential of hydrogen, power of hydrogen, and percentage of hydrogen ions. It is used to designate the corrosive nature of a material. Acids have hydronium ions, and bases have hydroxide ions. It is the percent (ratio) of these ions that makes up the pH number. Materials having a pH of 0 to 6.9 are considered acids, and materials with a pH of 7.1 to 14 are considered bases. A material having a pH of 7 is considered neutral. The pH scale is a logarithmic scale, meaning the movement from 0 to 1 is an increase of 10. The movement from 0 to 2 is an increase of 100.

When dealing with a corrosive response, one of the common methods to mitigate the release is to neutralize the corrosive. One thought is that water can be used to dilute and thereby neutralize the spill. This presents two major issues: one, the mixing of water (chemical reaction), and two, the runoff from the reaction. Corrosives and water can be a dangerous combination. One should never add a corrosive to water, as it presents great risk. There may be some spattering and heat generation.

If 1 gallon (3.8 liters) of an acid had a pH of 0, it would take 10 gallons (38 liters) of water to move the material to a pH of 1. To move it to a pH of 2 would take 100 gallons (380 liters) of water, and to change the pH to 6 would require 1 million gallons (3,785,412 liters) of water—all for a 1-gallon (3.8 liters) spill. Chemically neutralizing a corrosive spill is the better choice, but even that can present some issues. When neutralizing a strong acid, a weak base should be used to perform the neutralization. A street method of calculation for neutralization is that it takes more than 8,800 pounds of potash to bring 1,000 gallons (3,785 liters) of 50 percent sulfuric acid to neutral, which is more realistic than controlling 1 million gallons (3,785,412 liters) of runoff. These examples are provided for information only; the neutralization of corrosives is a Technician-level skill and can be very dangerous.

If the skin or eyes are exposed to a corrosive material, they should be immediately flushed with large quantities of water. This flushing should continue for at least twenty minutes uninterrupted. Some corrosive materials will cause immediate blindness and skin burns, and water should be used to prevent further injury.

Chemical Reactivity

Chemicals when they mix will have one of three types of reactions: exothermic, endothermic, or no reaction. The most common is the **exothermic reaction,** meaning the release of heat.

Material	Flash Point	Material	Flash Point (°F)
Gasoline	−45°F (−43°C)	Diesel fuel	> 100°F (38°C)
Isopropyl alcohol	53°F (11.6°C)	Motor oil	300–450°F (149–232°C)
Acetone	−4°F (−20°C)	Xylene	90°F (32°C)

TABLE 2-13 Flash Points of Some Common Materials

Fire is a rapid oxidation reaction (exothermic) accompanied by heat and light. When most chemicals mix and provide an exothermic reaction there usually is not a lot of light but there can be substantial heat. By mixing one tablespoon each of vinegar (acetic acid) and ammonia (base) at the same temperature, an immediate rise in temperature of 10°F (12°C) will occur. When handling oleum (concentrated sulfuric acid) and applying water to a spill, the resulting mixture will bubble, fume, boil, and heat to over 300°F (149°C) the instant the water hits the acid. An **endothermic reaction** is one in which the energy created by the reaction is absorbed and cooling occurs.

Flash Point

A flash point is described as the lowest temperature at which a fuel off-gases an ignitable mixture and that when introduced to a spark or flame will briefly ignite but not sustain burning, **Table 2-13.** This resulting flash fire will ignite just the vapors and self-extinguish once those vapors are burned up. The liquid itself does not burn; it is the mixture of vapors and air that ignites. Following closely behind the flash point is the fire point of a liquid. The fire point is the lowest temperature at which a fuel off-gases an ignitable mixture and that when introduced to a spark or flame will ignite and sustain burning. In the laboratory, scientists can replicate flash points and fire points; on the street, however, they are usually one and the same.

Autoignition Temperature

The autoignition temperature, sometimes referred to as ignition point, is the lowest temperature at which a fuel will off-gas an ignitable mixture and at which the fuel will self-ignite and continue to burn. Ignition points are much higher than flash points and represent a potential hazard level depending on the temperature. Depending on the context, the term **self-accelerating decomposition temperature (SADT)** may be used, which is essentially the same thing as

the autoignition temperature. Regardless of what it is called, it is a level to avoid.

Flammable Range

The two main areas within a flammable range are the lower explosive limit and upper explosive limit. The flammable range is the difference between the two extremes. A fire or explosion needs an ignition source, a fuel (vapors), and air. The proper vapor-to-air ratio is also required or ignition cannot occur. The **lower explosive** (flammable) **limit (LEL)** is the lowest amount of vapor mixed with air that can provide the proper mixture for a fire or explosion. The air monitor that is used to detect combustible gases is designed to read this level. The **upper explosive** (flammable) **limit (UEL)** is the highest amount of vapor mixed with air that will sustain a fire or explosion.

Each year many natural gas explosions occur throughout the United States. The LEL of natural gas is 5 percent, so if it is mixed with 95 percent air with an ignition source present an explosion or fire is possible, as shown in **Figure 2-90.** The UEL for natural gas is 15 percent, so 85 percent air is required to result in a fire or explosion. The flammable range for natural gas is 5 to 15 percent, and a fire or explosion can occur at any point in that range, as long as there is enough air to complete the mixture. With 4 percent methane and 96 percent air, there would not be any fire or explosion nor would there be at 16 percent natural gas and 84 percent air. Acetylene, which is a common gas, has a LEL of 2.5 percent and UEL of 100 percent, which means that when 2.5 percent LEL is reached, the potential for a fire exists. Having less than the LEL is referred to as too lean, and amounts in excess of the UEL are called too rich. Levels less than the LEL are much safer than levels above the UEL.

SAFETY

When confronted with a situation that has a level higher than the UEL, the situation is very dangerous.

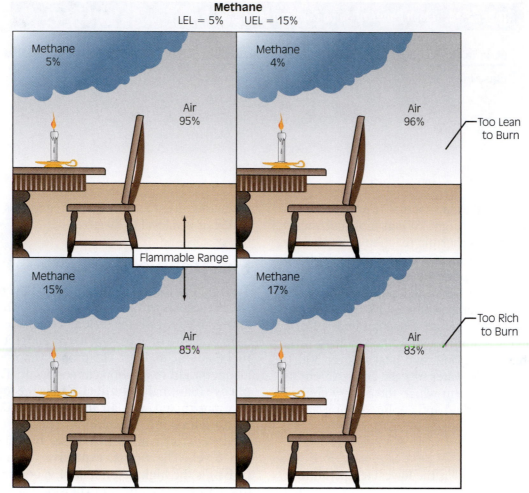

Methane
LEL = 5%   UEL = 15%

Methane 5% / Air 95%

Methane 4% / Air 96% — Too Lean to Burn

Flammable Range

Methane 15% / Air 85%

Methane 17% / Air 83% — Too Rich to Burn

To have a fire or explosion the lower explosive limit must be reached. Each gas has a flammable range in which there can be a fire or explosion. Below the LEL or above the UEL means there cannot be a fire.

FIGURE 2-90   Flammable range.

Ventilation should be carried out using non-spark-inducing devices, and great care should be taken to minimize any potential electrical arcs such as not using light switches or doorbells.

CONTAINERS AND PROPERTIES

The interrelationship with chemical and physical properties and the containers that hold hazardous materials is important. When there are chemical releases or incidents involving containers, knowing how the materials may react is important for responder safety. The lower the boiling point that a chemical has means that the container is going to be under more pressure and, if involved in a fire situation, the heat will cause an increase in pressure within the container. Materials that react or are corrosive can also create pressure within a container or, if the corrosive is placed in the wrong container, can cause the container to fail. When containers are under pressure, there is a good chance that the venting or rupture of the container will be violent. Understanding flash point and flammable range are important as this response to potentially flammable materials is common. The lower the flash point, the greater the fire risk, and the more likely the container will be under some pressure, since the vapor pressure is generally more forceful as the flash point decreases.

Radiation

We are subjected to radiation exposure in various forms every day. Our bodies have radioactive substances in our makeup. We eat various forms of radiation sources, and we breathe in radiation without any harm every day. Our exposure to these everyday

TABLE 2-14   Common Radioactive Sources

Name	Energy Emitted	Half-Life	Use	Decays to Form
Polonium-214	α	164 microseconds	Dust-removing brushes	Lead-210
Lead-214	β and γ	27 min	Industrial	Bismuth-214
Cobalt-60	β and γ	5.3 years	X-ray machines, industrial applications	Nickel-60
Cesium-137	β	30 years	Medical field, industrial imaging, check source	Barium-137
Radon-222	α	3.8 days	Naturally occurring gas	Polonium-218
Uranium-238	α	4.5 billion years	Ore	Thorium-234
Plutonium-239	α	24,100 years	Nuclear weapons	Uranium-235
Americium-241	α	433 years	Smoke detectors	Neptunium-237

radiation sources far exceeds those that would be found if we worked at a nuclear power plant. Through television, medical tests, and elevation we are subjected to levels of radiation that under normal circumstances cause us no harm. Some common radiation sources are listed in **Table 2-14.** To understand how radiation can hurt us, we must have an understanding of radioactivity. The basis of radioactivity is the makeup of the basic atom, specifically the nucleus. Having knowledge of radiation is important to understand not only how radiation affects the human body, but also how radiation monitors work.

An atom is comprised of electrons, neutrons, and protons. Protons and neutrons reside in the nucleus in the center of the atom, while electrons orbit the nucleus. Protons have a positive charge and determine the element or type of atom. Neutrons are the same size as protons but are neutral. Each element, with a given number of protons, can have several forms, termed **isotopes,** which are determined by the number of neutrons in the nucleus. The chemical properties of each isotope of an element are the same—you cannot chemically or physically identify one from the other. If there are too few or too many neutrons, the nucleus becomes unstable. **Radioisotopes,** isotopes whose nuclei are unstable, are radioactive, and emit radiation to become more stable. This emission of radiation is known as **radioactive decay** and usually is in the form of gamma radiation but may also be alpha, beta, or neutron radiation, as shown in Table 2-14.

An unstable material may become stable after one or two decays, but others may take many decay cycles. If a radioisotope decays by emission of an alpha or beta particle, the number of protons in the nucleus is changed, and the radioisotope becomes a different element. Uranium is the base for the development of **radon,** a common radioactive gas found in homes. Eventually, the radon will decay into lead, as shown in **Figure 2-91.** Each radioisotope has a constant **half-life.** Half-life is defined as the amount of time for half of a given radioactive source to decay. The **activity** of a source of radioactive material is a measure of the number of decays per second that occur within the source. Source activity is measured in **becquerel (Bq)** or **curies (Ci).** One becquerel is equal to one radioactive decay per second. One curie is equal to 37 gigabecquerels, or 37 billion disintegrations per second, in the same way that one kilometer is equal to 1,000 meters. The physical size of a radioactive source is not an indicator of radioactive strength or activity. One curie is also the activity of exactly one gram of radium-222.

Types of Radiation

Radiation is comprised of two basic categories: ionizing radiation and nonionizing radiation. Some examples of nonionizing radiation include radio waves, microwaves, infrared, visible light, and ultraviolet light. Alpha, beta, gamma, and X-rays are forms of ionizing radiation, and some characteristics of each are provided in **Figure 2-92.** There are two subcategories of ionizing radiation, one with energy and weight, and the other comprising just energy. Alpha and beta are known as particulates and have weight and energy. Gamma radiation has just energy and no weight.

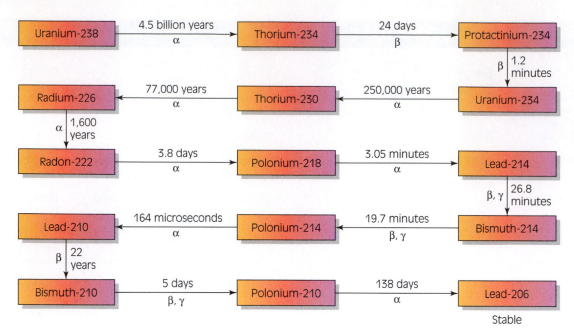

FIGURE 2-91   Half-life chart for uranium 238.

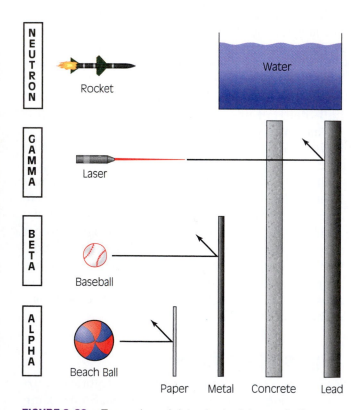

FIGURE 2-92   Examples of risks for ionizing radiation.

Alpha (α)

Alpha radiation consists of two neutrons and two protons, which carry a 2+ charge, and is identical to a helium nucleus. Alpha is a particulate and only has the ability to move a few feet in air. Its primary haz-

ard is through inhalation or ingestion. Street clothing or other protective clothing provides ample protection against alpha radiation.

Beta (β)

Beta particles are electrons (−) or positrons (+) and weigh 1/1,836 the weight of a proton. Beta radiation is one of two forms, low or high energy, but both are particulates. The low energy is comparable to alpha but has the ability to move a little farther in air and causes more damage. High-energy beta moves even farther and can cause greater harm. Beta can move several feet and higher levels of protective clothing are required, but structural firefighters' clothing offers some protection.

Gamma (γ)

Gamma rays come from the energy changes in the nucleus of an atom. Gamma is not a particulate but is airborne energy described as wavelike radiation that comes from within the source. These waves are sometimes called electromagnetic waves or electromagnetic radiation. Gamma can move a considerable distance. It is high energy and can cause internal damage to the body without causing external damage. Lead barriers offer some protection.

Neutron (n)

Neutron radiation is not common, but it is the basis for nuclear power plants and nuclear weapons. Neutrons are ejected from the nuclei of atoms during fission, when an atom splits into two smaller atoms. When fission occurs there is a release of energy in the form of heat, gamma radiation, and several neutrons. Fission can be spontaneous, in the case of californium-252, or induced, in the case of uranium-235 in nuclear power plants. Isotopes that can be induced to fission are called **fissile**.

Neutrons can travel great distances but are stopped by several feet of water, concrete, or hydrogen-rich materials. Neutron radiation is dangerous because it readily transfers its energy to water, and on average the human body is about 68 to 75 percent water. One unusual aspect to neutron radiation is its ability to activate nonradioactive isotopes, in other words, to make them radioactive. The nucleus of the targeted atom can capture a neutron, which makes it a different isotope. If the nucleus of the new isotope is unstable, the isotope will be radioactive. For example, when natural gold, gold-197, is bombarded with neutrons, it can be activated and become gold-198, which emits beta and gamma radiation. Activation is the reason that the materials used in a nuclear reactor become radioactive and must be disposed of as nuclear waste. Activation is accomplished only by neutron radiation.

X-rays

X-rays are comparable to gamma radiation waves, as they are wavelike, but X-rays are only dangerous when the X-ray machine in question is energized.

Toxic Products of Combustion

This is the one area in which firefighters suffer considerable chemical exposures. Any time a person is in smoke or breathes smoke, the body is being bombarded with toxic chemicals. Many toxic chemicals, such as carbon monoxide, carbon dioxide, hydrogen cyanide, hydrochloric acid, and phosgene, are produced in a fire. The worst type of chemical accident a firefighter can respond to is a house, car, or dumpster fire. Even brush fires are not exempt from the toxic products of combustion, because the field may have been sprayed with pesticides, herbicides, or other chemicals. Even burning wool or hay produces extremely toxic gases. Due to this constant exposure to these and other materials, it is important for firefighters to wear all of their protective clothing, especially the SCBA.

LESSONS LEARNED

The ability to recognize and identify the potential for hazardous materials to be present at an incident is important for the first responder. The numbers of tank trucks, tank cars, and containers can easily overwhelm the beginning student. At any incident, there is always a factor that relates to the recognition and identification process, whether it is the location, placards, container type, or physical senses. It could even be that sixth sense that alerts people to a potential problem. When that occurs, it is important to proceed with caution.

One of the important lessons for responders to remember is that they do not have to commit everything to memory, but they should know where to access hazardous materials information. It is not expected that each community have a railcar expert in their fire department. What is expected is that each fire department have a contact person available around the clock

to obtain that resource. It is possible that every material on this earth has the ability to cause harm in some fashion, but the chemical and physical properties play a factor in the type of harm that can be caused.

Materials that have low vapor pressures present little risk to the responders unless touched or eaten. Materials that have high vapor pressures do present a great risk to responders and to the community and should be treated with caution. Vapor pressure is only one of the terms with which responders should become familiar. Many responders have a fear of radiation, but this fear comes from a lack of education and experience. Understanding the harms from radiation is an important safety consideration. Local hazardous materials responders are a good resource and should be contacted early in an incident and whenever assistance dealing with hazardous materials is needed.

KEY TERMS

Aboveground Storage Tank (AST)   Tank that is stored above the ground in a horizontal or vertical position. Smaller quantities of fuels are often stored in this fashion.

Activity   A measure of the number of decays per second that occur with the source.

Becquerel (Bq)   A measure of radioactivity; the metric version of curies.

Boiling Point   The temperature to which a liquid must be heated in order to turn into a gas.

Building Officials Conference Association (BOCA)   A group that establishes minimum building and fire safety standards.

Bulk Tank   A large transportable tank, comparable to a tote, but considered to be the larger of the two.

Chemtrec   The Chemical Transportation Emergency Center, which provides technical assistance and guidance in the event of a chemical emergency; a network of chemical manufacturers that provide emergency information and response teams if necessary.

Cryogenic Gas   Any gas that exists as a liquid at a very cold temperature, always below $-150°F$ $(-101°C)$.

Curies (Ci)   The measure of activity level for radiation sources.

Deflagrates   Rapid burning, which in reality with regard to explosions can be considered a slow explosion, but is traveling at a lesser speed than a detonation.

Emergency Response Guidebook (ERG)   Book provided by the DOT that assists the first responder in making decisions primarily at transportation-related hazardous materials incidents.

Endothermic Reaction   A chemical reaction in which heat is absorbed, and the resulting mixture is cold.

Excepted Packaging   Type of packaging used to transport low-risk radioactive materials; the packaging is exempted from DOT regulations.

Exothermic Reaction   A chemical reaction that releases heat, such as when two chemicals are mixed and the resulting mixture is hot.

External Floating Roof Tank   Tank with the roof that covers the liquid within the tank exposed on the outside. The roof floats on the top of the liquid, which does not allow for vapors to build up.

First Responders   A group designated by the community as those who may be the first to arrive at a chemical incident. This group is usually composed of police officers, EMS providers, and firefighters.

Fissile   Isotopes that can be induced into fission, which creates a release of energy.

Frangible Disk   A type of pressure-relieving device that actually ruptures in order to vent the excess pressure. Once opened, the disk remains open; it does not close after the pressure is released.

Freezing Point   The temperature at which liquids become solids.

Gas   A state of matter in which the material moves freely about and is difficult to control. Steam is an example.

Half-Life   The amount of time for a given radiation source to decay by one half, which means it is emitting radioactivity.

Intermodal Containers   These are constructed in a fashion so that they can be transported by highway, rail, or ship. Intermodal containers exist for solids, liquids, and gases.

Internal Floating Roof Tank   Tank with a roof that floats on the surface of the stored liquid, but also has a cover on top of the tank to protect the top of the floating roof.

Isotope   A material that has a different form due to the number of neutrons that are in the nucleus.

Leaking Underground Storage Tank (LUST)   Describes a leaking tank that is underground.

Liquid   A state of matter that implies fluidity, which means a material has the ability to move as water would. There are varying states of being a liquid from moving very quickly to very slowly. Water is an example.

Lower Explosive Limit (LEL)   The lower part of the flammable range, and is the minimum required to have a fire or explosion.

Low Specific Activity (LSA)   Designation that indicates that a material is emitting low levels of radiation.

Material Safety Data Sheet (MSDS)   An information sheet that provides chemical-specific safety information.

Melting Point   The temperature at which solids become liquids.

Nuclear Regulatory Commission (NRC)   The government regulatory agency responsible for oversight of nuclear materials.

Ordinary Tank   A horizontal or vertical tank that usually contains combustible or other less hazardous chemicals. Flammable materials and other hazardous chemicals may be stored in smaller quantities in these types of tanks.

Oxidizer   Materials that readily release oxygen; by yielding oxygen, an oxidizer can easily cause or enhance the combustion of other materials. Oxidizers can dramatically increase the rate of burning when combustible material is ignited.

Persistence   An indication of the time that a material will remain as a liquid, and is related to vapor pressure.

Radioactive Decay   Process whereby as a material emits radiation, it decays, changing its form; for example, uranium eventually becomes lead after it decays.

Radioisotope   A radioactive form of an element unstable due to the number of neutrons.

Radon   A radioactive gas that is emitted from the earth, sometimes collecting in basement areas.

Relief Valve   A device designed to vent pressure in a tank, so that the tank itself does not rupture due to an increase in pressure. In most cases these devices are spring loaded so that when the pressure decreases the valve shuts, keeping the chemical inside the tank.

Reportable Quantity (RQ)   Both the EPA and DOT use the term. It is a quantity of chemicals that may require some type of action, such as reporting an inventory or reporting an accident involving a certain amount of the chemical.

Sea Containers   Shipping boxes that were designed to be stacked on a ship, then placed onto a truck or railcar.

Self-Accelerating Decomposition Temperature (SADT)   Temperature at which a material will ignite itself without an ignition source present. Can be compared to ignition temperature.

Solid   A state of matter that describes materials that may exist in chunks, blocks, chips, crystals, powders, dusts, and other types. Ice is an example.

Specification (Spec) Plates   All trucks and tanks have a specification plate that outlines the type of tank, capacity, construction, and testing information.

Spent Nuclear Fuel   Radioactive fuel that was used in a nuclear reactor.

States of Matter   Describe in what form matter exists, such as solids, liquids, or gases.

Sublimation   The ability of a solid to go to the gas phase without being liquid.

Surface Contaminated Object (SCO)   Materials that may be contaminated with radioactive waste and are usually low hazard.

Tote   A large tank usually 250 to 500 gallons (946–1893 liters), constructed to be transported to a facility and dropped for use.

Type A Container   A container that is designed to hold low-risk radioactive material and that offers moderate protection against accidents.

Type B Container   A container that is designed to hold high-risk radioactive materials and is a hardened transport case capable of withstanding severe accidents.

Underground Storage Tank (UST)   Tank that is buried under the ground. The most common are gasoline and other fuel tanks.

Upper Explosive Limit (UEL)   The upper part of the flammable range. Above the UEL, fire or an explosion cannot occur because there is too much fuel and not enough oxygen.

Vapor Pressure   The amount of force that is pushing vapors from a liquid. The higher the force the more vapors (gas) being put into the air.

REVIEW QUESTIONS

1. What are the nine hazard classes as defined by DOT?
2. An explosive placard is what color?
3. What three things on a placard indicate the potential hazards?
4. A DOT-406/MC-306 tank truck commonly carries what product?
5. A DOT-406 tank truck has a characteristic shape from the rear. What is it?
6. An MC-331 carries what type of gases?
7. What does the blue section of the NFPA 704 system refer to?
8. What are the locations of emergency shutoffs on an MC-331?
9. What are the four basic clues to recognition and identification for hazardous materials?
10. The placards shown in Table 2-1 are required for which quantities?
11. A tractor trailer carrying Division 1.1 materials is well involved in fire. What should be the firefighter's initial tactics?

12. When propane tanks are involved in a fire, what is a potential consequence?

13. What types of radiation are particles?

14. Which one type of radiation has the ability to activate other materials and make them radioactive?

15. What impact does half-life have with regard to a radiation source?

FOR FURTHER REVIEW

For additional review of the content covered in this chapter, including activities, games, and study materials to prepare for the certification exam, please refer to the following resources:

Firefighter's Handbook Online Companion
Click on our Web site at **http://www.delmarfire.cengage.com** for FREE access to games, quizzes, tips for studying for the certification exam, safety information, links to additional resources and more!

Firefighter's Handbook Study Guide
Order#: 978-1-4180-7322-0
An essential tool for review and exam preparation, this Study Guide combines various types of questions and exercises to evaluate your knowledge of the important concepts presented in each chapter of *Firefighter's Handbook*.

ENDNOTES

1. A tank truck placarded for gasoline could also haul diesel fuel or fuel oil without having to change placards.

2. The requirements for the subdivisions are presented in general. Other requirements may need to be met for a material to be assigned to a subdivision. Refer to 49 CFR 170–180 for more information.

ADDITIONAL RESOURCES

Bevelacqua, Armando, *Hazardous Materials Chemistry,* 2nd ed. Thomson Delmar Learning. Clifton Park, NY, 2006.

Bevelacqua, Armando and Richard Stilp, *Hazardous Materials Field Guide,* 2nd ed. Thomson Delmar Learning. Clifton Park, NY, 2006.

Bevelacqua, Armando and Richard H. Stilp, *Terrorism Handbook for Operational Responders,* 3rd ed. Delmar, Cengage Learning, Clifton Park, NY.

Hawley, Chris, *Hazardous Materials Air Monitoring and Detection Devices,* 2nd ed. Thomson Delmar Learning Clifton Park, NY, 2006.

Hawley, Chris, *Hazardous Materials Incidents,* 3rd ed. Thomson Delmar Learning. Clifton Park, NY, 2007.

Henry, Timothy V. *Decontamination for Hazardous Materials Emergencies.* Delmar Publishers. Albany, NY, 1998.

Hildebrand, Michal S. and Gregory Noll, *Composite Propane Cylinders* White Paper, Propane Education & Research Council, Washington, D.C., 2007.

Lesak, David, *Hazardous Materials Strategies and Tactics.* Prentice Hall, Englewood Cliffs, NJ, 1998.

Noll, Gregory, Michael Hildebrand, and James Yvorra, *Hazardous Materials: Managing the Incident,* 3rd ed. Red Hat Publishing, Chester, MD, 2005.

Schnepp, Rob and Paul Gantt, *Hazardous Materials: Regulations, Response and Site Operations.* Delmar Publishers, Albany, NY, 1998.

Stilp, Richard and Armando Bevelacqua, *Emergency Medical Response to Hazardous Materials Incidents.* Delmar Publishers, Albany, NY, 1996.

3

Hazardous Materials: Information Resources

I had been on the job for several years, and there was nothing unusual about the call when it first came in—nothing remarkable that would lead to any heightened sense of awareness. "Engine 29, check out the smell of smoke in the building at 211 Mason Street." It would be a simple food on the stove call or something to that effect.

When we arrived on the scene the local police department had already made initial entry to the facility (which turned out to be a restaurant) and reported a slight haze with a strange pungent odor but no visible fire. We made entry into the vacant facility and immediately noticed the haze and pungent odor, but because it was a restaurant we thought it was simply something to do with the food preparation process. As we made our way through the facility, we began to notice a strange taste in our mouths and several of those present began to experience irritation to the eyes and throat, yet still nothing clicked. It was, after all, simply a restaurant, perhaps some burnt wiring or bad cooking, no big deal. As we continued our investigation, several of those present began to complain of tightness in the chest, headaches, and nausea. I too was feeling strange but did not want to throw in the towel so another responder and I continued on.

Then suddenly as the haze began to clear via the doors and windows that had been opened for ventilation, we began to notice an abundance of dead insects and rodents throughout the facility. And now on the radio we heard the call for an ambulance as one of the responders who had been on scene for some time began to experience more than simply mild irritations. We began to realize that the haze was in fact some type of aerosolized insecticide/pesticide and that we had been exposed and contaminated. Now and only now did we begin to comprehend the error of our ways, so to speak, and implement the proper response and protocols.

As the weeks followed, I replayed the incident over and over and came to realize all of the mistakes that we had made, and how imperative it was to seek the proper training to ensure that I never put myself or those with me in this position again. Two of the responders present were unable to return to work for a period of time (in one case several months). They had to be treated for the effects of exposure to organophosphates. Both are fine today, but much has been written about the possible long-term effects of exposure to organophosphates, and one can only wonder what course these individuals' health could have taken had they been further exposed. And that I do, all too often.

Speaking only to my errors, there was no attempt to obtain the proper information from the appropriate resources or the owners. (After we realized we had no fire, we should have consulted the owners before further entry, because there was no life hazard.) Nor was subsequent contact made with the other tenants of this row structure who may have been able to provide information, or to even take basic actions to ensure the health and safety of those involved. The failure to properly research and gather information on this call could have resulted in much more severe health problems for all involved had the concentrations been higher or ventilation unsuccessful. But the fact that two responders were unnecessarily exposed still remains with me today.

—Street Story by Tom Creamer, Special Operations Coordinator (ret.),
City of Worcester Fire Department, Worcester, Massachusetts

LEARNING OBJECTIVES

After completing this chapter, the reader should be able to:

3-1. Explain the terms used on Material Safety Data Sheets.

3-2. Tell where MSDS are located.

3-3. Identify the standard information available on an MSDS.

3-4. List the type of information available on shipping papers.

3-5. Use the Emergency Response Guidebook (ERG).

3-6. List the types of assistance that can be provided by Chemtrec.

3-7. Explain the methods used to contact a shipper or emergency contact.

3-8. Describe other resources that may be available in the community.

INTRODUCTION

Chemical information is available through a variety of sources, including those carried on emergency apparatus and information sources that can be reached via telephone, fax, or computer. The shipper and the facility are required to maintain certain documents that will assist first responders in determining the chemical hazards they may face. Knowing what information is available and how to interpret this information is a valu-able tool. This chapter discusses the most common sources of information.

CAUTION

Specific and current chemical information is essential if the first responder is going to protect lives and property.

EMERGENCY RESPONSE GUIDEBOOK

The DOT's ERG is a well-known book for emergency responders, **Figure 3-1.** The DOT makes one copy available for every emergency response apparatus in the country.

The ERG is commonly referred to as the DOT book or the orange book. This book, which is published about every three years, contains information regarding the most commonly transported chemicals as regulated by the DOT. It is intended as a guidebook for first responders during the initial phase of a hazardous materials incident. It provides information regarding the potential hazards of responding to these materials and is one of the only books that provides specific evacuation recommendations. Although an excellent book for first responders, it does have limitations for the more advanced responder, who requires more specific chemical information.

The book consists of these major sections:

■ Placard information

■ **ADR/RID** marking system information

■ Listing by DOT identification number

■ Alphabetical listing by shipping name

■ Response guides

FIGURE 3-1 The DOT *Emergency Response Guidebook* should be found in every emergency vehicle in the United States. It provides chemical emergency response information that is valuable to the first responder.

RESIST RUSHING IN !
APPROACH INCIDENT FROM UPWIND
STAY CLEAR OF ALL SPILLS, VAPORS, FUMES AND SMOKE

HOW TO USE THIS GUIDEBOOK DURING AN INCIDENT INVOLVING DANGEROUS GOODS

ONE **IDENTIFY THE MATERIAL** BY FINDING ANY **ONE** OF THE FOLLOWING:

THE 4-DIGIT ID NUMBER ON A PLACARD OR ORANGE PANEL

THE 4-DIGIT ID NUMBER (after UN/NA) ON A SHIPPING DOCUMENT OR PACKAGE

THE NAME OF THE MATERIAL ON A SHIPPING DOCUMENT, PLACARD OR PACKAGE

IF AN **ID NUMBER** OR THE **NAME OF THE MATERIAL** CANNOT BE FOUND, SKIP TO THE NOTES BELOW.

TWO **LOOK UP THE MATERIAL'S 3-DIGIT GUIDE NUMBER** IN EITHER:

THE ID NUMBER INDEX..(the yellow-bordered pages of the guidebook)

THE NAME OF MATERIAL INDEX..(the blue-bordered pages of the guidebook)

If the guide number is supplemented with the letter "**P**", it indicates that the material may undergo violent polymerization if subjected to heat or contamination.

If the index entry is highlighted (in either yellow or blue), it is a TIH (Toxic Inhalation Hazard) material, a chemical warfare agent or a Dangerous Water Reactive Material (produces toxic gas upon contact with water). **LOOK FOR THE ID NUMBER AND NAME OF THE MATERIAL** IN THE TABLE OF INITIAL ISOLATION AND PROTECTIVE ACTION DISTANCES (the green-bordered pages). Then, if necessary, **BEGIN PROTECTIVE ACTIONS IMMEDIATELY** (see Protective Actions on page 298). If protective action is not required, use the information jointly with the 3-digit guide.

USE GUIDE 112 FOR ALL EXPLOSIVES EXCEPT FOR EXPLOSIVES 1.4 (EXPLOSIVES C) WHERE GUIDE 114 IS TO BE CONSULTED.

THREE **TURN TO THE NUMBERED GUIDE** (the orange-bordered pages) **AND READ CAREFULLY.**

NOTES **IF A NUMBERED GUIDE CANNOT BE OBTAINED BY FOLLOWING THE ABOVE STEPS,** AND A PLACARD CAN BE SEEN, LOCATE THE PLACARD IN THE TABLE OF PLACARDS (pages 16-17), THEN GO TO THE 3-DIGIT GUIDE SHOWN NEXT TO THE SAMPLE PLACARD.

IF A REFERENCE TO A GUIDE CANNOT BE FOUND AND THIS INCIDENT IS BELIEVED TO INVOLVE DANGEROUS GOODS, TURN TO **GUIDE 111** NOW, AND USE IT UNTIL ADDITIONAL INFORMATION BECOMES AVAILABLE. If the shipping document lists an emergency response telephone number, call that number. If the shipping document is not available, or no emergency response telephone number is listed, IMMEDIATELY CALL the appropriate **emergency response agency listed on the inside back cover of this guidebook.** Provide as much information as possible, such as the name of the carrier (trucking company or railroad) and vehicle number. AS A LAST RESORT, CONSULT THE TABLE OF RAIL CAR AND ROAD TRAILER IDENTIFICATION CHART (pages 18-19). IF THE CONTAINER CAN BE IDENTIFIED, REMEMBER THAT THE INFORMATION ASSOCIATED WITH THESE CONTAINERS IS FOR THE WORST CASE POSSIBLE.

Page 1

FIGURE 3-2 The first page of the ERG is a step-by-step listing of how to use the book. When not sure how to proceed, the responder should turn to this page.

■ Table of initial isolation and protective action distances
■ List of dangerous water-reactive materials

On the inside cover, the book provides an example of a shipping document used in truck transportation and provides information on how the shipping document should be written. It also provides an example of a placard with an identification number panel. The first page, which is shown in **Figure 3-2,** is an essen-tial page for responder safety because it outlines all of the actions the first responder should take. It provides a decision tree process by taking the reader through the incident step by step. It also has an important list-ing of the guides for explosives.

The next few pages of the ERG provide informa-tion on safety precautions and whom to call for assis-tance. The first responder should already know where the closest local assistance will be coming from and where to contact state assistance if required. The

HAZARD CLASSIFICATION SYSTEM

The hazard class of dangerous goods is indicated either by its class (or division) number or name. For a placard corresponding to the primary hazard class of a material, the hazard class or division number must be displayed in the lower corner of the placard. However, no hazard class or division number may be displayed on a placard representing the subsidiary hazard of a material. For other than Class 7 or the OXYGEN placard, text indicating a hazard (for example, "CORROSIVE") is not required. Text is shown only in the U.S. The hazard class or division number must appear on the shipping document after each shipping name.

Class 1 - Explosives

Division 1.1	Explosives with a mass explosion hazard
Division 1.2	Explosives with a projection hazard
Division 1.3	Explosives with predominantly a fire hazard
Division 1.4	Explosives with no significant blast hazard
Division 1.5	Very insensitive explosives with a mass explosion hazard
Division 1.6	Extremely insensitive articles

Class 2 - Gases

Division 2.1	Flammable gases
Division 2.2	Non-flammable, non-toxic* gases
Division 2.3	Toxic* gases

Class 3 - Flammable liquids (and Combustible liquids [U.S.])

Class 4 - Flammable solids; Spontaneously combustible materials; and Dangerous when wet materials/Water-reactive substances

Division 4.1	Flammable solids
Division 4.2	Spontaneously combustible materials
Division 4.3	Water-reactive substances/Dangerous when wet materials

Class 5 - Oxidizing substances and Organic peroxides

Division 5.1	Oxidizing substances
Division 5.2	Organic peroxides

Class 6 - Toxic* substances and Infectious substances

Division 6.1	Toxic*substances
Division 6.2	Infectious substances

Class 7 - Radioactive materials

Class 8 - Corrosive substances

Class 9 - Miscellaneous hazardous materials/Products, Substances or Organisms

* The words "poison" or "poisonous" are synonymous with the word "toxic".

Page 14

FIGURE 3-3 The DOT uses nine classes of hazardous materials, as described in Chapter 2, that are listed in the ERG. Occasionally the only information that is available is the hazard class.

DOT book provides a contact number for federal assistance, although responders should proceed by requesting local, state, and then federal assistance. The contact numbers and agencies are divided among Canada, the United States, and Mexico. It is important for the first responder to read the DOT book prior to an incident because it provides a large amount of background material that could not be effectively read during an emergency.

The book also provides a listing of the hazard class system, **Figure 3-3,** and provides a reference point for the hazard classes that may be listed on placards or shipping papers. This listing precedes the placard section, which is valuable if the only information available is the placard. The placard section, **Figure 3-4,** provides information about how to proceed at an incident where the only information is a placard. All of the possible current plac-

TABLE OF PLACARDS AND INITIAL
USE THIS TABLE ONLY IF MATERIALS CANNOT BE SPECIFICALLY IDENTIFIED BY

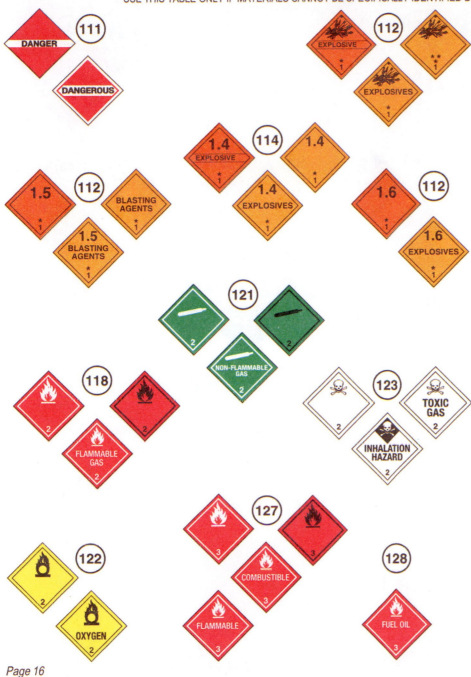

Page 16

FIGURE 3-4 If the only information that is available is a placard, the ERG can provide assistance. Listed below each placard is a guide number that provides response information based on the placard information.

ards are listed with their pictures, and the reader is referred to the accompanying guide page. Much like the placard section, included in the ERG are silhouettes of tank trucks, shown in **Figure 3-5,** and of rail cars, shown in **Figure 3-6.** Some international shipments use additional hazard markings as shown in **Figure 3-7.**

The yellow section, **Figure 3-8,** is a numerical listing by the identification, or ID, number. This is a number assigned by the DOT that identifies a material being shipped. This number takes two forms,

a North America (NA) number or a United Nations (UN) number, but it is always a four-digit number. Although one would think that this number would specifically identify a material, in some cases it may not. Depending on the material it may be lumped into a general category such as Hazardous Waste, Liquid, or NOS (Not Otherwise Specified), but in most cases it will identify a specific substance. The yellow section lists the ID number, shipping name, and guide reference number. If the material is highlighted as shown in **Figure 3-9,** then the reader must refer to the guide

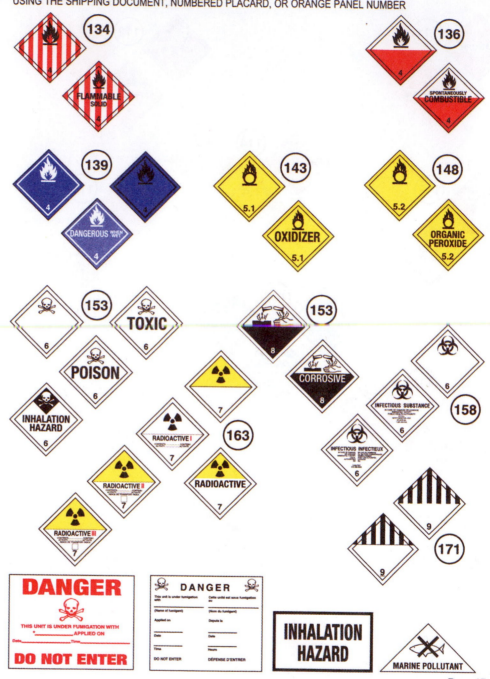

FIGURE 3-4   (Continued)

Page 17

page listed as well as the Table of Initial Isolation and Protective Action Distances, located at the back of the book in the green section.

The following abbreviations are used in the DOT ERG:

- NOS—not otherwise specified
- PIH—poison by inhalation
- P—polymerization hazard

- SCA—surface contaminated articles (radiation)
- LSA—low specific activity (radiation)
- PG I, II, or III—Packing Group I, II, or III
- ORM—other regulated material (listed as ORM A-E)
- LC_{50}—lethal concentration to 50 percent of the population
- LD_{50}—lethal dose to 50 percent of the population

ROAD TRAILER IDENTIFICATION CHART*

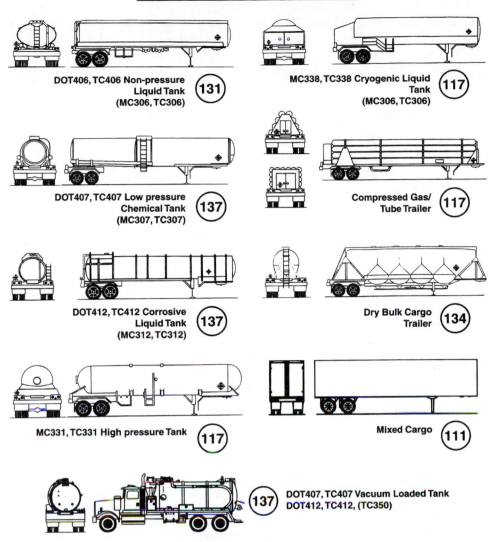

DOT406, TC406 Non-pressure Liquid Tank (MC306, TC306) (131)

MC338, TC338 Cryogenic Liquid Tank (MC306, TC306) (117)

DOT407, TC407 Low pressure Chemical Tank (MC307, TC307) (137)

Compressed Gas/ Tube Trailer (117)

DOT412, TC412 Corrosive Liquid Tank (MC312, TC312) (137)

Dry Bulk Cargo Trailer (134)

MC331, TC331 High pressure Tank (117)

Mixed Cargo (111)

DOT407, TC407 Vacuum Loaded Tank DOT412, TC412, (TC350) (137)

CAUTION: This chart depicts only the most general shapes of road trailers. Emergency response personnel must be aware that there are many variations of road trailers, not illustrated above, that are used for shipping chemical products. The suggested guides are for the most hazardous products that may be transported in these trailer types.

*   **The recommended guides should be considered as last resort if product cannot be identified by any other means.**

Page 19

FIGURE 3-5 Truck silhouettes are provided with the appropriate guide page for each of the truck styles.

The blue section of the book mirrors the yellow section except that it is alphabetical by shipping name, **Figure 3-10.** These shipping names are assigned by the DOT and for the most part are identical to the chemical names, but in some cases they can vary. A lot of names are assigned to chemicals, including the actual chemical name, synonyms, trade names, and shipping names.

SAFETY

Some chemicals can have more than sixty synonyms, which can create a confusing situation. Table 3-1 provides one example.

The chemical known by the DOT as chlordane is also called by the names given in **Table 3-1.** This

RAIL CAR IDENTIFICATION CHART*

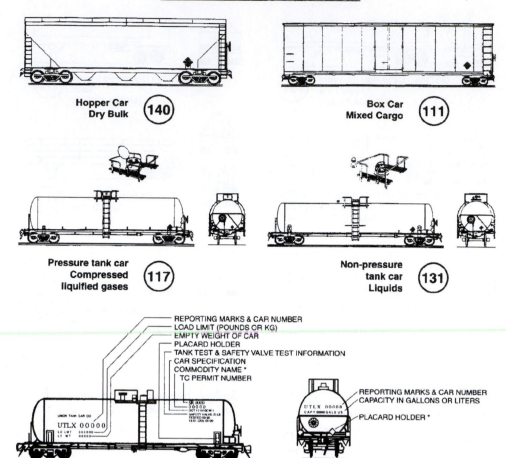

Hopper Car
Dry Bulk (140)

Box Car
Mixed Cargo (111)

Pressure tank car
Compressed
liquified gases (117)

Non-pressure
tank car
Liquids (131)

REPORTING MARKS & CAR NUMBER
LOAD LIMIT (POUNDS OR KG)
EMPTY WEIGHT OF CAR
PLACARD HOLDER
TANK TEST & SAFETY VALVE TEST INFORMATION
CAR SPECIFICATION
COMMODITY NAME *
TC PERMIT NUMBER

REPORTING MARKS & CAR NUMBER
CAPACITY IN GALLONS OR LITERS
PLACARD HOLDER *

CAUTION: Emergency response personnel must be aware that rail tank cars vary widely in construction, fittings and purpose. Tank cars could transport products that may be solids, liquids or gases. The products may be under pressure. It is essential that products be identified by consulting shipping documents or train consist or contacting dispatch centers before emergency response is initiated.

The information stenciled on the sides or ends of tank cars, as illustrated above, may be used to identify the product utilizing:

a.   the commodity name shown; or

b.   the other information shown, especially reporting marks and car number which, when supplied to a dispatch center, will facilitate the identification of the product.

*   **The recommended guides should be considered as last resort if product cannot be identified by any other means.**

Page 18

FIGURE 3-6 Rail car silhouettes are provided with the appropriate guide page for each of the rail car styles.

section lists the ID number, shipping name, and guide reference number. If the material is highlighted in blue, the reader should refer to the guide page listed as well as the Table of Initial Isolation and Protective Action Distances, located at the back of the book in the green section.

The middle of the book, the orange section, makes up the actual guide pages, **Figure 3-11.** A total of sixty-one guides is given for the more than 4,000 chemi-

cals listed by the DOT. It is this generalization that makes this book valuable to the initial responder but less so to an advanced responder, such as a hazardous materials technician. A hazardous materials technician needs more specific information, which the ERG does not provide.

Each guide takes up two pages and is divided into three major sections: potential hazards, public safety, and emergency response. The guide may also provide

HAZARD IDENTIFICATION CODES
DISPLAYED ON SOME INTERMODAL CONTAINERS

Hazard identification codes, referred to as "hazard identification numbers" under European and some South American regulations, may be found in the top half of an orange panel on some intermodal bulk containers. The 4-digit identification number is in the bottom half of the orange panel.

The hazard identification code in the top half of the orange panel consists of two or three digits. In general, the digits indicate the following hazards:

2 - EMISSION OF GAS DUE TO PRESSURE OR CHEMICAL REACTION

3 - FLAMMABILITY OF LIQUIDS (VAPORS) AND GASES OR SELF-HEATING LIQUID

4 - FLAMMABILITY OF SOLIDS OR SELF-HEATING SOLID

5 - OXIDIZING (FIRE-INTENSIFYING) EFFECT

6 - TOXICITY OR RISK OF INFECTION

7 - RADIOACTIVITY

8 - CORROSIVITY

9 - MISCELLANEOUS DANGEROUS SUBSTANCE

- Doubling of a digit indicates an intensification of that particular hazard (i.e. 33, 66, 88).
- Where the hazard associated with a material can be adequately indicated by a single digit, the digit is followed by a zero (i.e. 30, 40, 50).
- A hazard identification code prefixed by the letter "X" indicates that the material will react dangerously with water (i.e. X88).
- When 9 appears as a 2nd or 3rd digit, this may present a risk of spontaneous violent reaction.

FIGURE 3-7 Intermodal containers will have an orange panel near the placard like that shown in the figure. The four-digit number is the DOT ID number, and the one- or two-digit number above that is a classification code, providing information on the basic hazards.

Page 20

some additional information about a specific chemical. It is important for the responder to read the entire guide, and, if referred to the Table of Initial Isolation and Protective Action Distances, to read that section also.

The potential hazards section lists the predominant hazard on the top line. If fire is the major concern, then it will be listed on top; if health is the major concern, then it will be listed on top. The informa-

tion regarding the health effects provides the route of exposure and any major symptoms of exposure. It also details other potential health and environmental concerns with fire products and runoff. The fire section is not as detailed as the health section but does provide some assistance in identifying potential hazards. It provides information related to the flammability of the product and will state whether the vapors will stay low to the ground or will rise in air.

ID No.	Guide No.	Name of Material
1027	115	Cyclopropane
1027	115	Cyclopropane, liquefied
1028	126	Dichlorodifluoromethane
1028	126	Refrigerant gas R-12
1029	126	Dichlorofluoromethane
1029	126	Refrigerant gas R-21
1030	115	1,1-Difluoroethane
1030	115	Difluoroethane
1030	115	Refrigerant gas R-152a
1032	118	Dimethylamine, anhydrous
1033	115	Dimethyl ether
1035	115	Ethane
1035	115	Ethane, compressed
1036	118	Ethylamine
1037	115	Ethyl chloride
1038	115	Ethylene, refrigerated liquid (cryogenic liquid)
1039	115	Ethyl methyl ether
1039	115	Methyl ethyl ether
1040	119P	Ethylene oxide
1040	119P	Ethylene oxide with Nitrogen
1041	115	Carbon dioxide and Ethylene oxide mixture, with more than 9% but not more than 87% Ethylene oxide
1041	115	Carbon dioxide and Ethylene oxide mixtures, with more than 6% Ethylene oxide
1041	115	Ethylene oxide and Carbon dioxide mixture, with more than 9% but not more than 87% Ethylene oxide
1041	115	Ethylene oxide and Carbon dioxide mixtures, with more than 6 % Ethylene oxide

ID No.	Guide No.	Name of Material
1043	125	Fertilizer, ammoniating solution, with free Ammonia
1044	126	Fire extinguishers with compressed gas
1044	126	Fire extinguishers with liquefied gas
1045	124	Fluorine
1045	124	Fluorine, compressed
1046	121	Helium
1046	121	Helium, compressed
1048	125	Hydrogen bromide, anhydrous
1049	115	Hydrogen
1049	115	Hydrogen, compressed
1050	125	Hydrogen chloride, anhydrous
1051	117	AC
1051	117	Hydrocyanic acid, aqueous solutions, with more than 20% Hydrogen cyanide
1051	117	Hydrocyanic acid, liquefied
1051	117	Hydrogen cyanide, anhydrous, stabilized
1051	117	Hydrogen cyanide, stabilized
1052	125	Hydrogen fluoride, anhydrous
1053	117	Hydrogen sulfide
1053	117	Hydrogen sulfide, liquefied
1053	117	Hydrogen sulphide
1053	117	Hydrogen sulphide, liquefied
1055	115	Isobutylene
1056	121	Krypton
1056	121	Krypton, compressed
1057	115	Lighter refills (cigarettes) (flammable gas)
1057	115	Lighters (cigarettes) (flammable gas)
1058	120	Liquefied gas (nonflammable)

Page 26

FIGURE 3-8 The yellow pages are a numeric listing of the chemicals regulated by the DOT. This listing is by four-digit United Nations/North America identification number and also provides the shipping name and the emergency response guide page information.

Depending on the product, it may also describe the material's ability to float on water. Products shipped at an elevated temperature as well as those materials that have the ability to **polymerize** will have a notation.

One of the problems with the ERG is the fact that gasoline, which is highly flammable, is in the same guide that is provided for diesel fuel, which to the fire service is combustible. Although the DOT uses 141°F (60.5°C) as the difference between flammable and combustible, the fire service still uses 100°F (38°C) as the difference. A major difference is seen when handling a gasoline spill as opposed to a diesel fuel spill under normal conditions. By only using the guide as intended, the reader has no way of differentiating between the two products. Although

ID No.	Guide No.	Name of Material
2474	157	Thiophosgene
2475	157	Vanadium trichloride
2477	131	Methyl isothiocyanate
2478	155	Isocyanate solution, flammable, poisonous, n.o.s.
2478	155	Isocyanate solution, flammable, toxic, n.o.s.
2478	155	Isocyanate solutions, n.o.s.
2478	155	Isocyanates, flammable, poisonous, n.o.s.
2478	155	Isocyanates, flammable, toxic, n.o.s.
2478	155	Isocyanates, n.o.s.
2480	155	Methyl isocyanate
2481	155	Ethyl isocyanate
2482	155	n-Propyl isocyanate
2483	155	Isopropyl isocyanate
2484	155	tert-Butyl isocyanate
2485	155	n-Butyl isocyanate
2486	155	Isobutyl isocyanate
2487	155	Phenyl isocyanate
2488	155	Cyclohexyl isocyanate
2490	153	Dichloroisopropyl ether
2491	153	Ethanolamine
2491	153	Ethanolamine, solution
2491	153	Monoethanolamine
2493	132	Hexamethyleneimine
2495	144	Iodine pentafluoride
2496	156	Propionic anhydride
2498	129	1,2,3,6-Tetrahydrobenzaldehyde
2501	152	1-Aziridinyl phosphine oxide (Tris)
2501	152	Tri-(1-aziridinyl)phosphine oxide, solution

FIGURE 3-9 If the listing in the yellow pages or the blue pages is highlighted, as is the listing of ID number 2481, Ethyl isocyanate, then the reader must turn to the orange guide and to the green section. The green section is the Table of Initial Isolation and Protective Action Distances.

taking this route might be safer, because it assumes a worst-case scenario, there are times when this action is inappropriate and can lead to other problems, such as an unnecessary evacuation for a diesel fuel spill.

NOTE

Only the public safety section (orange pages) of the ERG provides recommendations about how far to isolate the scene and provides some initial strategic goals for the first arriving company. If the chemical is highlighted, then responders should use both the orange response guides and the information provided in the green section.

The public safety section provides information for the initial public protection options and key issues for the safety of the responders. It is this section as well as the Table of Initial Isolation and Protective Action Distances (green section), **Figure 3-12,** that make this book one of the necessary tools in the mitigation of a chemical release. The green section provides detailed isolation and evacuation information. This information relates to extremely hazardous substances, ones that can present a significant risk to the community. The orange section lists isolation distances for the emergency responders, and the green section is for the community. It also recommends that responders contact the emergency contact provided on the shipping papers or, if these are not available, then to use one of the contacts listed on the inside back cover. It provides some tactical objectives for handling radioactive substances and some additional considerations regarding these types of incidents.

The public safety section also lists personal protective equipment (PPE) recommendations and provides four basic PPE scenarios. It always recommends the use of positive-pressure SCBA, which is always a good recommendation when responding to chemical releases. The next level in severity is one in which chemical protective equipment is required because turnout gear will provide limited protection. In some cases it will state that turnout gear will provide limited protection, without any recommendation for other PPE. PPE is discussed in Chapter 4 of this text, and many of these same issues will be covered in that chapter. The last recommendation is the suggestion that in some cases chemical protective equipment is a good idea, but that turnout gear is acceptable for fire situations.

This section also provides information regarding initial evacuation distances for large spills and for fires. If the material is listed in the Table of Initial Isolation and Protective Action Distances, the reader will be referred to this section, located at the back of the book in the green section.

Name of Material	Guide No.	ID No.	Name of Material	Guide No.	ID No.
Allyl alcohol	131	1098	Aluminum phosphide pesticide	157	3048
Allylamine	131	2334	Aluminum powder, coated	170	1309
Allyl bromide	131	1099	Aluminum powder, pyrophoric	135	1383
Allyl chloride	131	1100	Aluminum powder, uncoated	138	1396
Allyl chlorocarbonate	155	1722	Aluminum processing by-products	138	3170
Allyl chloroformate	155	1722			
Allyl ethyl ether	131	2335	Aluminum remelting by-products	138	3170
Allyl formate	131	2336	Aluminum resinate	133	2715
Allyl glycidyl ether	129	2219	Aluminum silicon powder, uncoated	138	1398
Allyl iodide	132	1723			
Allyl isothiocyanate, inhibited	155	1545	Aluminum smelting by-products	138	3170
Allyl isothiocyanate, stabilized	155	1545	Amines, flammable, corrosive, n.o.s.	132	2733
Allyltrichlorosilane, stabilized	155	1724			
Aluminum, molten	169	9260	Amines, liquid, corrosive, flammable, n.o.s.	132	2734
Aluminum alkyl halides	135	3052			
Aluminum alkyl halides, liquid	135	3052	Amines, liquid, corrosive, n.o.s.	153	2735
Aluminum alkyl halides, solid	135	3052	Amines, solid, corrosive, n.o.s.	154	3259
Aluminum alkyl halides, solid	135	3461	2-Amino-4-chlorophenol	151	2673
Aluminum alkyl hydrides	138	3076	2-Amino-5-diethylaminopentane	153	2946
Aluminum alkyls	135	3051	2-Amino-4,6-dinitrophenol, wetted with not less than 20% water	113	3317
Aluminum borohydride	135	2870			
Aluminum borohydride in devices	135	2870			
			2-(2-Aminoethoxy)ethanol	154	3055
			N-Aminoethylpiperazine	153	2815
Aluminum bromide, anhydrous	137	1725	Aminophenols	152	2512
Aluminum bromide, solution	154	2580	Aminopyridines	153	2671
Aluminum carbide	138	1394	Ammonia, anhydrous	125	1005
Aluminum chloride, anhydrous	137	1726	Ammonia, anhydrous, liquefied	125	1005
Aluminum chloride, solution	154	2581	Ammonia, solution, with more than 10% but not more than 35% Ammonia	154	2672
Aluminum dross	138	3170			
Aluminum ferrosilicon powder	139	1395	Ammonia, solution, with more than 35% but not more than 50% Ammonia	125	2073
Aluminum hydride	138	2463			
Aluminum nitrate	140	1438			
Aluminum phosphide	139	1397	Ammonia solution, with more than 50% Ammonia	125	1005

Page 99

FIGURE 3-10 The blue pages are alphabetical by shipping name. These listings also provide the UN/NA number and the emergency response guide page.

SAFETY

The initial isolation distances listed in the orange section of the DOT book vary from a minimum of 30 feet (9 meters) to a maximum of 800 feet (243 meters). The minimums are derived for materials that are solid or present little risk to the responder. The higher the toxicity or the risk, the greater the isolation distance. The average distance is 330 feet (101 meters), which is what should be used for the establishment of an isolation zone for an unknown material, a distance that in an urban setting can present some significant problems. As more information becomes available, this isolation distance can increase or decrease as needed. When dealing with explosives the minimum distance should be 800 feet (243 meters) for the initial isolation distance. The isolation distances for persons downwind are expanded from the initial isolation distances in the orange section—from 0.1 to 7 miles (0.16 to 11 km)—in the green section.

TABLE 3-1   Chemical Names

Aspon-Chlordane	Octachlorodihydrodicyclopentadiene
Belt	1,2,4,5,6,7,8,8-Octachloro-2,3,3a,4,7,7a-hexahydro-4,7-methanoindene
CD 68	1,2,4,5,6,7,8,8-Octachloro-2,3,3a,4,7,7a-hexahydro-4,7-ethano-1H-indene
Chloordaan	1,2,4,5,6,7,8,8-Octachloro-3a,4,7,7a-hexahydro-4,7-methylene indane
Chlordan	Octachloro-4,7-methanohydroindane
g-Chlordan	Octachloro-4,7-methanotetrahydroindane
Chlorindan	1,2,4,5,6,7,8,8-Octachloro-4,7-methano-3a,4,7,7a-tetrahydroindane
Chlor kil	1,2,4,5,6,7,8,8-Octachloro-3a,4,7,7a-tetrahydro-4,7-methanoindane
Chlorodane	1,2,4,5,6,7,8,8-Octachloro-4,7,7a-Tetrahydro-4,7-methanoindane
Chlortox	1,2,4,5,6,7,8,8-Octachlor-3a,4,7,7a-Tetrahydro-4,7-endo-methano-indan (German)
Clordan	Octa-Klor
Clorodane	Oktaterr
Cortilan-Neu	Ortho-Klor
Dichlorochlordene	1,2,4,5,6,7,8,8-Ottochloro-3A,4,7,7A-Tetraidro-4,7-endo-methano-indano (Italian)
Dowchlor	RCRA Waste Number U036
ENT 9,932	SD 5532
ENT 25,552-X	Shell SD-5532
HCS 3260	Synklor
Kypchlor	Tat Chlor 4
M 140	Topichlor 20
NCI-C00099	Topiclor
Niran	Topiclor 20
1,2,4,5,6,7,8,8-Octachloor-3a,4,7,7a-tetrahydro-4,7-endo-methano-indaan (Dutch)	Toxichlor
1,2,4,5,6,7,10,10-Octachloro-4,7,8,9-tetrahydro-4,7-methyleneindane	Velsicol 1068

Excerpted from Richard J. Lewis, Sr., *Dangerous Properties of Industrial Materials,* 8th ed. Van Nostrand Reinhold Co., New York, 1994.

The emergency response section provides information regarding fires, spills, and first aid. The fire section is divided between small fires and large fires, and potential tactics that can be used for both. For both types of fires, there is a listing of what type of extinguishing agent may be needed such as water, foam, dry chemical, halon, or CO_2. Specific materials may also list special agents that may be needed. Listed under the large-fire section are some tactical considerations as well as some recommendations for specific substances. If the material is carried by tank truck or railcar, the book provides some

GUIDE 111 MIXED LOAD/UNIDENTIFIED CARGO ERG2004

POTENTIAL HAZARDS

FIRE OR EXPLOSION
- May explode from heat, shock, friction or contamination.
- May react violently or explosively on contact with air, water or foam.
- May be ignited by heat, sparks or flames.
- Vapors may travel to source of ignition and flash back.
- Containers may explode when heated.
- Ruptured cylinders may rocket.

HEALTH
- Inhalation, ingestion or contact with substance may cause severe injury, infection, disease or death.
- High concentration of gas may cause asphyxiation without warning.
- Contact may cause burns to skin and eyes.
- Fire or contact with water may produce irritating, toxic and/or corrosive gases.
- Runoff from fire control may cause pollution.

PUBLIC SAFETY

- CALL Emergency Response Telephone Number on Shipping Paper first. If Shipping Paper not available or no answer, refer to appropriate telephone number listed on the inside back cover.
- As an immediate precautionary measure, isolate spill or leak area for at least 100 meters (330 feet) in all directions.
- Keep unauthorized personnel away.
- Stay upwind.
- Keep out of low areas.

PROTECTIVE CLOTHING
- Wear positive pressure self-contained breathing apparatus (SCBA).
- Structural firefighters' protective clothing provides limited protection in fire situations ONLY; it may not be effective in spill situations.

EVACUATION
Fire
- If tank, rail car or tank truck is involved in a fire, ISOLATE for 800 meters (1/2 mile) in all directions; also, consider initial evacuation for 800 meters (1/2 mile) in all directions.

Page 170

FIGURE 3-11 The response guide in the orange section provides the responder with basic information regarding the health, fire, and public safety issues. Each part needs to be read completely before taking any action.

additional information on fighting those types of fires. In this section some helpful hints are given that apply in many cases involving fires and these types of containers. The hints include the following:

- Fight fire from a distance using unstaffed monitors.

- Withdraw immediately if the sound level from the venting safety device rises or the tank begins to discolor.
- Cool containers with flooding quantities of water until well after the fire is out.

The spill or leak section lists some general tactical objectives and provides some specific information on

ERG2004   MIXED LOAD/UNIDENTIFIED CARGO   **GUIDE 111**

EMERGENCY RESPONSE

FIRE
CAUTION: Material may react with extinguishing agent.
Small Fires
- Dry chemical, CO_2, water spray or regular foam.

Large Fires
- Water spray, fog or regular foam.
- Move containers from fire area if you can do it without risk.

Fire involving Tanks
- Cool containers with flooding quantities of water until well after fire is out.
- Do not get water inside containers.
- Withdraw immediately in case of rising sound from venting safety devices or discoloration of tank.
- ALWAYS stay away from tanks engulfed in fire.

SPILL OR LEAK
- Do not touch or walk through spilled material.
- ELIMINATE all ignition sources (no smoking, flares, sparks or flames in immediate area).
- All equipment used when handling the product must be grounded.
- Keep combustibles (wood, paper, oil, etc.) away from spilled material.
- Use water spray to reduce vapors or divert vapor cloud drift. Avoid allowing water runoff to contact spilled material.
- Prevent entry into waterways, sewers, basements or confined areas.

Small Spills • Take up with sand or other non-combustible absorbent material and place into containers for later disposal.

Large Spills • Dike far ahead of liquid spill for later disposal.

FIRST AID
- Move victim to fresh air. • Call 911 or emergency medical service.
- Give artificial respiration if victim is not breathing.
- **Do not use mouth-to-mouth method if victim ingested or inhaled the substance; give artificial respiration with the aid of a pocket mask equipped with a one-way valve or other proper respiratory medical device.**
- Administer oxygen if breathing is difficult.
- Remove and isolate contaminated clothing and shoes.
- In case of contact with substance, immediately flush skin or eyes with running water for at least 20 minutes.
- Shower and wash with soap and water.
- Keep victim warm and quiet.
- Effects of exposure (inhalation, ingestion or skin contact) to substance may be delayed.
- Ensure that medical personnel are aware of the material(s) involved and take precautions to protect themselves.

Page 171

FIGURE 3-11 (Continued)

some substances. It is important to be aware that this section could lead responders into an action that may be beyond their level of training if not used correctly. As an example, most of the guides recommend that the reader can stop a leak if it can be done without risk. The intention is for a responder trained to the operations level to be able to close a remote shutoff valve if available. Most readers may incorrectly follow the recommendation and try to stop the leak regardless of the method, a tactic that requires a hazardous materials technician to accomplish safely. Another area of concern is the suggestion that a water spray be used to knock down vapors. If necessary to save lives, using a water spray is an acceptable tactic.

Page 320

TABLE OF INITIAL ISOLATION AND PROTECTIVE ACTION DISTANCES

ID No.	NAME OF MATERIAL	SMALL SPILLS (From a small package or small leak from a large package)				LARGE SPILLS (From a large package or from many small packages)			
		First ISOLATE in all Directions		Then PROTECT persons Downwind during-		First ISOLATE in all Directions		Then PROTECT persons Downwind during-	
		Meters (Feet)		DAY Kilometers (Miles)	NIGHT Kilometers (Miles)	Meters (Feet)		DAY Kilometers (Miles)	NIGHT Kilometers (Miles)
2204 2204	Carbonyl sulfide Carbonyl sulphide	30 m	(100 ft)	0.1 km (0.1 mi)	0.6 km (0.4 mi)	300 m	(1000 ft)	3.0 km (1.9 mi)	8.1 km (5.0 mi)
2232 2232	Chloroacetaldehyde 2-Chloroethanal	30 m	(100 ft)	0.2 km (0.1 mi)	0.3 km (0.2 mi)	90 m	(300 ft)	0.8 km (0.5 mi)	1.6 km (1.0 mi)
2334	Allylamine	30 m	(100 ft)	0.1 km (0.1 mi)	0.5 km (0.3 mi)	120 m	(400 ft)	1.1 km (0.7 mi)	2.5 km (1.5 mi)
2337	Phenyl mercaptan	30 m	(100 ft)	0.1 km (0.1 mi)	0.1 km (0.1 mi)	60 m	(200 ft)	0.4 km (0.2 mi)	0.6 km (0.4 mi)
2382 2382	1,2-Dimethylhydrazine Dimethylhydrazine, symmetrical	30 m	(100 ft)	0.1 km (0.1 mi)	0.2 km (0.1 mi)	60 m	(200 ft)	0.6 km (0.4 mi)	1.2 km (0.8 mi)
2407	Isopropyl chloroformate	30 m	(100 ft)	0.1 km (0.1 mi)	0.3 km (0.2 mi)	90 m	(300 ft)	0.7 km (0.5 mi)	1.5 km (0.9 mi)
2417 2417	Carbonyl fluoride Carbonyl fluoride, compressed	30 m	(100 ft)	0.2 km (0.1 mi)	1.1 km (0.7 mi)	90 m	(300 ft)	1.0 km (0.6 mi)	3.6 km (2.3 mi)
2418 2418	Sulfur tetrafluoride Sulphur tetrafluoride	60 m	(200 ft)	0.7 km (0.4 mi)	3.2 km (2.0 mi)	500 m	(1600 ft)	4.7 km (2.9 mi)	10.6 km (6.6 mi)
2420	Hexafluoroacetone	30 m	(100 ft)	0.3 km (0.2 mi)	1.3 km (0.8 mi)	800 m	(2500 ft)	7.2 km (4.5 mi)	11.0+ km (7.0+ mi)
2421	Nitrogen trioxide	30 m	(100 ft)	0.1 km (0.1 mi)	0.5 km (0.3 mi)	60 m	(200 ft)	0.4 km (0.3 mi)	1.9 km (1.2 mi)
2437	Methylphenyldichlorosilane (when spilled in water)	30 m	(100 ft)	0.1 km (0.1 mi)	0.1 km (0.1 mi)	30 m	(100 ft)	0.3 km (0.2 mi)	1.1 km (0.7 mi)
2438	Trimethylacetyl chloride	30 m	(100 ft)	0.1 km (0.1 mi)	0.2 km (0.1 mi)	60 m	(200 ft)	0.5 km (0.3 mi)	0.8 km (0.5 mi)
2442	Trichloroacetyl chloride	30 m	(100 ft)	0.2 km (0.2 mi)	0.8 km (0.5 mi)	120 m	(400 ft)	1.2 km (0.8 mi)	2.2 km (1.4 mi)
2474	Thiophosgene	90 m	(300 ft)	0.8 km (0.5 mi)	2.4 km (1.5 mi)	360 m	(1200 ft)	3.6 km (2.3 mi)	6.8 km (4.2 mi)
2477	Methyl isothiocyanate	30 m	(100 ft)	0.1 km (0.1 mi)	0.2 km (0.1 mi)	60 m	(200 ft)	0.5 km (0.3 mi)	1.0 km (0.7 mi)
2480	Methyl isocyanate	60 m	(200 ft)	0.5 km (0.3 mi)	1.9 km (1.2 mi)	600 m	(1800 ft)	5.4 km (3.3 mi)	11.0+ km (7.0+ mi)

FIGURE 3-12   The Initial Isolation and Protective Action Distances table provides recommended isolation distances for the materials that were highlighted in the yellow or blue sections. The table is divided into small spills and large spills.

SAFETY

The use of water indiscriminately to knock down "vapors" can create many other problems. These problems include increasing the spill size, creating runoff problems, possible chemical reactions, and adding an additional hazard to other responders who will try to mitigate the leak. If necessary, as in the case of knocking down anhydrous ammonia vapors that may affect a neighborhood, it is an acceptable tactic. To flow water into the back of a trailer that has a leaking drum, in which the contents have not been identified, is inappropriate and could cause further problems.

The first-aid section provides basic medical treatment and some basic decontamination recommendations for chemical burns. The information is basic and if confronted with a patient, the responder should contact the local hospital or Chemtrec for further assistance.

The Table of Initial Isolation and Protective Action Distances is the green section in the back of the book, **Figure 3-12.** It provides specific isolation and evacuation distances for the materials that were highlighted in the yellow or blue section. An example of how to use these isolation distances is shown in **Figure 3-13** using ethyl isocyanate (ID 2481). This section is further subdivided between small spills and large spills, and both are divided between day and night distances. The criteria used to establish these distances used the following information: the DOT incident database (HMRIS), typical package size, typical flow rate from a ruptured package, and the release rate of vapors from a spill. The DOT chose the average day as being warm and sunny with a temperature of 95°F (35°C). They chose sixty-one cities and did a five-year study of the weather to establish a pattern. It was during this study that it was determined that materials will travel farther at night than during the day. They also chose to use an evacuation distance such

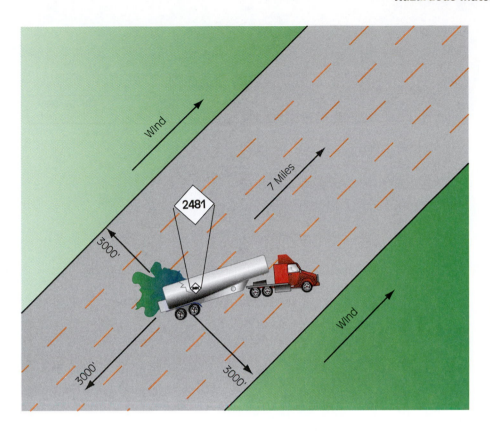

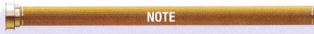

FIGURE 3-13 A large spill of ethyl isocyanate (ID 2481) is listed as having an isolation distance of 1,700 feet in all directions. In such a case, responders need to protect those persons downwind for 7+ miles.

that in 90 percent of the incidents, the distance used would be too large. But using this scenario, in 10 percent of the incidents the distance would be too small.

NOTE

The distances are a guide for the first thirty minutes of an incident and use a typical day as defined by the DOT. The ERG is a guide to get an evacuation started, and as soon as possible air monitoring needs to be established to definitively define the evacuation distances.

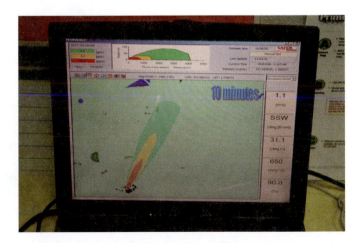

FIGURE 3-14 Software is available that can plot toxic gas cloud plumes, which can help determine isolation areas and guide evacuation decisions.

In addition, an emergency response software package known as CAMEO uses a vapor-cloud modeling program known as ALOHA. This modeling program is generally referred to as a plume dispersion model, an example of which is shown in **Figure 3-14,** and can assist with evaluating potential downwind evacuations, using real-time weather data.

The DOT defines a small spill as a leaking container smaller than a 55-gallon (208 liters) drum or a leak from a small cylinder, whereas a large spill is defined as coming from a container larger than 55 gallons (208 liters) or a large leak from a cylinder. These are further explained in that a leak from several small

containers may be considered a large leak. A small leak from a large container may also be considered a small leak.

The last section included in the green section is the List of Dangerous Water-Reactive Materials, **Figure 3-15.** This section provides the evacuation distances for these materials if they contact water. The distances vary from 0.3 (0.5 km) to 6 miles (9.7 km). Another helpful item that this section

TABLE OF WATER-REACTIVE MATERIALS WHICH PRODUCE TOXIC GASES

Materials Which Produce Large Amounts of Toxic-by-Inhalation (TIH) Gas(es) When Spilled in Water

ID No.	Guide No.	Name of Material	TIH Gas(es) Produced
1716	156	Acetyl bromide	HBr
1717	155	Acetyl chloride	HCl
1724	155	Allyltrichlorosilane, stabilized	HCl
1725	137	Aluminum bromide, anhydrous	HBr
1726	137	Aluminum chloride, anhydrous	HCl
1728	155	Amyltrichlorosilane	HCl
1732	157	Antimony pentafluoride	HF
1745	144	Bromine pentafluoride	HF Br₂
1746	144	Bromine trifluoride	HF Br₂
1747	155	Butyltrichlorosilane	HCl
1752	156	Chloroacetyl chloride	HCl
1754	137	Chlorosulfonic acid	HCl
1754	137	Chlorosulfonic acid and Sulfur trioxide mixture	HCl
1754	137	Chlorosulphonic acid	HCl
1754	137	Chlorosulphonic acid and Sulphur trioxide mixture	HCl
1754	137	Sulfur trioxide and Chlorosulfonic acid	HCl
1754	137	Sulphur trioxide and Chlorosulphonic acid	HCl
1758	137	Chromium oxychloride	HCl
1763	156	Cyclohexyltrichlorosilane	HCl
1766	156	Dichlorophenyltrichlorosilane	HCl
1767	155	Diethyldichlorosilane	HCl
1769	156	Diphenyldichlorosilane	HCl
1771	156	Dodecyltrichlorosilane	HCl
1777	137	Fluorosulfonic acid	HF

Chemical Symbols for TIH Gases:

Br₂ Bromine · Cl₂ Chlorine · HBr Hydrogen bromide · HCl Hydrogen chloride · HCN Hydrogen cyanide · HF Hydrogen fluoride · HI Hydrogen iodide · H₂S Hydrogen sulfide · H₂S Hydrogen sulphide · NH₃ Ammonia · PH₃ Phosphine · SO₂ Sulfur dioxide · SO₂ Sulphur dioxide · SO₃ Sulfur trioxide · SO₃ Sulphur trioxide

Use this list only when material is spilled in water. Page 345

FIGURE 3-15 List of Dangerous Water-Reactive Materials from the green section of the ERG (1 of 4 pages).

provides is the chemical that this material makes when it contacts water, an unusual piece of information for this type of text.

The last pages are filled with definitions, glossary, and explanations. The inside back cover provides a listing of additional emergency contacts for the United States, Canada, and Mexico. Further discussion regarding these agencies is found later in this chapter.

The firefighter should consider these general reminders when using the DOT ERG:

- It is an excellent source for evacuation and isolation distances.
- It is an excellent source for water-reactive hazards.
- It is an excellent first responder document, one that will keep the responders safe.

- Is a great starting point for the first thirty minutes of the incident.
- The specific chemical information is limited, and in some cases too general for a hazardous materials team, although it is well suited for the first responder level.
- Specific identification of the product is essential to responder and citizen safety.

Using the DOT Emergency Response Guide (ERG)

JPR 3-1 outlines specific steps in using the ERG when responding to an incident involving dangerous goods. Please note that this JPR sequence is a basic guideline, and users should reference the first page of the ERG (as shown in **Figure 3-2**) for further information. Users should also practice using the ERG prior to an incident so that they are familiar with its contents and fully prepared to effectively mitigate the threat of hazardous materials.

1. Resist rushing in! Approach incidents from an upwind direction and stay clear of all spills, vapors, fumes, and smoke.
2. Identify the material by finding any one of the following:
   - 4-digit ID number on placard/ID panel, **JPR 3-1A.**
   - 4-digit ID number on shipping document or package, **JPR 3-1B.**
   - Name of material on shipping document, placard, or package, **JPR 3-1C.**
3. Look up the material's 3-digit guide number in either:
   - ID number index
   - Name of material index
4. Turn to the numbered guide, **JPR 3-1D.**

MATERIAL SAFETY DATA SHEETS

Material Safety Data Sheets, or MSDS, as they are commonly called, are a result of the hazard communication standard, which is OSHA regulation 29 CFR 1910.1200, which became effective in 1994. As stated in Chapter 1, this regulation requires employers who use chemicals above the household quantity to create MSDS. They are also required to develop a hazard communication plan, label all chemical containers, and provide training to employees on an annual basis. This training is based on the chemical hazards the employees will face in the workplace. An MSDS is required to have a variety of information. The amount, quality, and order of information is determined by the manufacturer. The original intent of the MSDS was to protect employees working at the facility, not emergency responders. Although many of the sheets do have applicability and serve a useful purpose, in some cases they may not have the necessary emergency response information. The information that may be found on an MSDS is listed in **Table 3-2.**

The quality of information varies from MSDS to MSDS and from manufacturer to manufacturer. The early MSDS had a lot of good information; however, as litigation has increased, the typical MSDS provides a worst-case scenario, which results in an ex-

tremely conservative approach to the handling of the chemical.

Given the choice of using the DOT ERG or an MSDS, the firefighter should rely more on the technical information on the MSDS. The DOT ERG places many chemicals into a few general categories. Although it may be conservative, the MSDS is specific to the chemical. It is important through the preplanning process that responders learn about the chemicals from a facility representative. Responders should work through the MSDS and the ERG and determine prior to the incident which information is most beneficial.

Although many attempts have been made to modify the format of the MSDS and to improve the quality of information, the MSDS has remained the same since inception. There are a couple of recommended formats, and a new format exists for the European chemical industry. It can be anticipated that the United States will have to conform to this format. The information provided in **Table 3-2** is from this format. One of the improvements with this new MSDS format is a section specifically on emergency response procedures and considerations for firefighters. A sample MSDS is provided in **Figure 3-16.**

Using the MSDS Wisely

It is always recommended that responders use more than one source of information. The EPA issued a safety alert in June 1999 on the use of the MSDS, and the majority of that bulletin is reproduced here.

> **CAUTION**
>
> "A critical consideration when choosing a response strategy is the safety of emergency responders. Adequate information about on-site chemicals can make a big difference when choosing a safe response strategy. This information must include name, toxicity, physical and chemical characteristics, fire and reactivity hazards, emergency response procedures, spill control, and protective equipment. Generally, responders rely primarily on Material Safety Data Sheets (MSDS) maintained at the facility. However, MSDS may not provide sufficient information to effectively and safely respond to accidental releases. This Alert is designed to increase awareness of MSDS limitations, so that first responders can take proper precautions, and identify additional sources of chemical information, which could help prevent death or injury."

Determining an Action Plan Using the MSDS

A Material Safety Data Sheet can be a very effective tool when responding to an incident at a facility that houses chemicals. It can cut down on the

JOB PERFORMANCE REQUIREMENT   3-1
Using the DOT Emergency Response Guide (ERG)

A   What clues are available, and what portions of the DOT ERG could be used? *(Courtesy of Maryland Department of the Environment Emergency Response Division)*

B   What clues are available, and what portions of the DOT ERG could be used? *(Courtesy of Maryland Department of the Environment Emergency Response Division)*

response time and avoid injury or even death of those involved in the incident. When responding to such a facility, you would be well advised to follow these procedures:

1. Review the MSDS to determine the chemical threat and whether there are other products in the chemical that are of concern.

2. Don the appropriate level PPE for the chemical threat, particularly in cases where the material has escaped its container, **JPR 3-2A.**

3. If the material has released into the building, follow procedures for evacuating and securing the building.

4. Determine which extinguishing agents are required for the incident and identify the container in which the material will be found, **JPR 3-2B.**

Accidents and How the MSDS Relates

In May 1997 a massive explosion and fire occurred at an agricultural chemical packaging facility in eastern Arkansas. Prior to the explosion, employees observed smoke in a back warehouse and evacuated. The facility called local responders and asked for help to control smoldering inside a pesticide container. The local fire department rapidly responded and reviewed the smoldering product's MSDS. The MSDS lacked information on decomposition temperatures or explosion hazards. The firefighters decided to investigate the building. While they were approaching, a violent explosion occurred.

Fragments from a collapsing cinder block wall killed three firefighters and seriously injured a fourth.

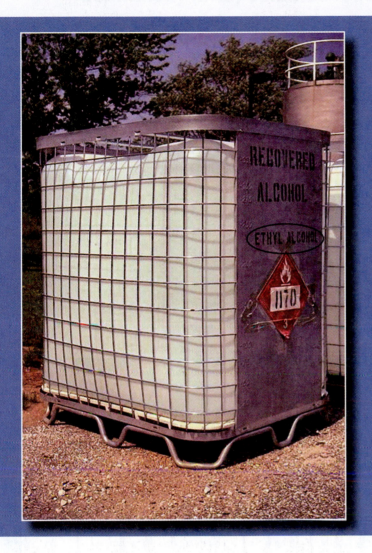

C   What clues are available, and what portions of the DOT ERG could be used?

In April 1995, an explosion and fire at a manufacturing facility in Lodi, New Jersey, caused the death of five responders. The explosion occurred while the company was blending aluminum powder, sodium hydrosulfite, and other ingredients. Even though the material was water reactive, the MSDS for the product advised the use of a "water spray . . . to extinguish fire." The recommendation in the MSDS for "small fires" was to flood with water; however, "small fire" was not defined, the amount of water necessary was not specified, and no information dealt with how to respond to large fires (which can occur during blending processes).

The MSDS only described the hazards associated with the product. In this case, responders needed information on the hazards associated with the reactivity during the blending process (which was significantly different from the product).

Emergency responders should note that the chemical information provided on an MSDS usually presents the hazards associated with that particular product. Once the product is placed in a process some factors may change, resulting in the increase, decrease, or elimination of hazards. These factors may include reactions with other chemicals and changes in temperature, pressure, and physical/chemical characteristics.

MSDS in the Workplace

In 1988 the Occupational Safety and Health Administration (OSHA) required facilities storing or using hazardous chemicals to comply with the Hazard Communication Standard. This standard requires employers to provide employees with an MSDS for every hazardous chemical present on site and to train those

JOB PERFORMANCE REQUIREMENT 3-1

Using the DOT Emergency Response Guide (ERG) (Continued)

D An example DOT ERG listing, which consists of two pages.

GUIDE 127   FLAMMABLE LIQUIDS (POLAR/WATER-MISCIBLE)   ERG2004

POTENTIAL HAZARDS

FIRE OR EXPLOSION
- **HIGHLY FLAMMABLE: Will be easily ignited by heat, sparks or flames.**
- Vapors may form explosive mixtures with air.
- Vapors may travel to source of ignition and flash back.
- Most vapors are heavier than air. They will spread along ground and collect in low or confined areas (sewers, basements, tanks).
- Vapor explosion hazard indoors, outdoors or in sewers.
- Those substances designated with a "P" may polymerize explosively when heated or involved in a fire.
- Runoff to sewer may create fire or explosion hazard.
- Containers may explode when heated.
- Many liquids are lighter than water.

HEALTH
- Inhalation or contact with material may irritate or burn skin and eyes.
- Fire may produce irritating, corrosive and/or toxic gases.
- Vapors may cause dizziness or suffocation.
- Runoff from fire control may cause pollution.

PUBLIC SAFETY

- **CALL Emergency Response Telephone Number on Shipping Paper first. If Shipping Paper not available or no answer, refer to appropriate telephone number listed on the inside back cover.**
- As an immediate precautionary measure, isolate spill or leak area for at least 50 meters (150 feet) in all directions.
- Keep unauthorized personnel away.
- Stay upwind.
- Keep out of low areas.
- Ventilate closed spaces before entering.

PROTECTIVE CLOTHING
- Wear positive pressure self-contained breathing apparatus (SCBA).
- Structural firefighters' protective clothing will only provide limited protection.

EVACUATION
Large Spill
- Consider initial downwind evacuation for at least 300 meters (1000 feet).
Fire
- If tank, rail car or tank truck is involved in a fire, ISOLATE for 800 meters (1/2 mile) in all directions; also, consider initial evacuation for 800 meters (1/2 mile) in all directions.

Page 202

employees to properly recognize the hazards of the chemicals and to handle them safely. An MSDS normally provides information on the physical/chemical characteristics and first-aid procedures. This information is valuable for employees to safely work with the chemical. However, the content for the MSDS on emergency response procedures, fire, and reactive hazards may be insufficient for local responder use in an emergency situation. Vagueness, technical jargon, understandability, product versus process

ERG2004   FLAMMABLE LIQUIDS   **GUIDE**
(POLAR/WATER-MISCIBLE)   **127**

D   (Continued)

EMERGENCY RESPONSE

FIRE
CAUTION: All these products have a very low flash point: Use of water spray when fighting fire may be inefficient.

Small Fires
- Dry chemical, CO_2, water spray or alcohol-resistant foam.

Large Fires
- Water spray, fog or alcohol-resistant foam.
- Use water spray or fog; do not use straight streams.
- Move containers from fire area if you can do it without risk.

Fire involving Tanks or Car/Trailer Loads
- Fight fire from maximum distance or use unmanned hose holders or monitor nozzles.
- Cool containers with flooding quantities of water until well after fire is out.
- Withdraw immediately in case of rising sound from venting safety devices or discoloration of tank.
- ALWAYS stay away from tanks engulfed in fire.
- For massive fire, use unmanned hose holders or monitor nozzles; if this is impossible, withdraw from area and let fire burn.

SPILL OR LEAK
- ELIMINATE all ignition sources (no smoking, flares, sparks or flames in immediate area).
- All equipment used when handling the product must be grounded.
- Do not touch or walk through spilled material.
- Stop leak if you can do it without risk.
- Prevent entry into waterways, sewers, basements or confined areas.
- A vapor suppressing foam may be used to reduce vapors.
- Absorb or cover with dry earth, sand or other non-combustible material and transfer to containers.
- Use clean non-sparking tools to collect absorbed material.

Large Spills
- Dike far ahead of liquid spill for later disposal.
- Water spray may reduce vapor; but may not prevent ignition in closed spaces.

FIRST AID
- Move victim to fresh air.   • Call 911 or emergency medical service.
- Give artificial respiration if victim is not breathing.
- Administer oxygen if breathing is difficult.
- Remove and isolate contaminated clothing and shoes.
- In case of contact with substance, immediately flush skin or eyes with running water for at least 20 minutes.   • Wash skin with soap and water.
- In case of burns, immediately cool affected skin for as long as possible with cold water. Do not remove clothing if adhering to skin.
- Keep victim warm and quiet.
- Ensure that medical personnel are aware of the material(s) involved and take precautions to protect themselves.

Page 203

concerns, and missing information on an MSDS may increase the risk to emergency responders. MSDS are provided by manufacturers, importers, and/or distributors. MSDS chemical hazard information can vary substantially depending on the provider. Sometimes this discrepancy is due to different testing procedures. However, whoever prepared the MSDS is responsible for ensuring the accuracy of the hazard information. **Table 3-3** summarizes information from various MSDS for the chemical azinphos methyl and

TABLE 3-2   Information Included on an MSDS

Information	Note
Chemical product and company identification	The name the chemical company uses to identify the product is used near the top of the MSDS. Information related to the manufacturer, such as the address and other contact information, is provided near the top of the first sheet. The MSDS is required to have a twenty-four-hour emergency contact number, although in most cases this number is Chemtrec's.
Chemical composition	The ingredients of the chemical are listed here. If the mixture is a trade secret, there is a provision to exclude this information. If this information is needed for medical reasons, the manufacturer is to provide this information to the physician. In some cases the true identity of the material may never be known, and only the hazards may be provided.
Hazards identification	In this section, an emergency overview is provided for both an employer and emergency responders. This section is lengthy and includes potential health effects, first-aid, and firefighting measures. The exposure levels are typically included with the health effects but may be listed separately. New to MSDS will be a section on accidental releases. This section will be beneficial to emergency responders. Handling and storage considerations, as well as engineering controls including proper PPE, are also provided in this section.
Physical and chemical properties	Specific chemical information such as boiling points, vapor pressures, flash points, and many other specific items are included in this section.
Stability and reactivity	Some chemicals become unstable after a period of time or as a result of poor storage conditions, or they may react with other chemicals. Any information regarding this type of detail is provided here.
Toxicology information	The long-term health effects and other concerns regarding acute and chronic exposures are provided in this section.
Ecological information	Information regarding any potential environmental effects is listed here.
Disposal considerations	Although in many cases this section states "follow local regulations," this is the section that outlines any regulatory requirements for disposal.
Transport information	Information regarding the DOT regulations is listed here—typically the hazard class and UN identification number.
Regulatory information	Any other regulations that apply to the use, storage, or disposal of the chemical are listed here. If the chemical is covered by any of the other regulations, this information is listed in this section.

illustrates how different sources can provide varied and conflicting information.

SHIPPING PAPERS

Besides the use of placards when chemicals are transported the carrier is required to provide shipping papers for the cargo. The shipping papers generally provide the following information:

- Shipping company
- Destination of the packages
- Emergency contact information
- The number and weight of the packages
- The proper shipping name of the materials and the hazard class of the materials
- Special notations for hazardous materials

If the vehicle is not carrying hazardous materials, there is no requirement to have any specific information, which at times can be confusing for first responders. When carrying hazardous materials the shipping papers may include a packing group (PG) number listed as a I, II, or III. The lower the number, the worse the chemical is, and it may have special shipping requirements.

Other information may include a reportable quantity (RQ) for a hazardous material. Some chemicals have a threshold of reporting when spilled much like

AIR PRODUCTS

Material Safety Data Sheet
Version 1.2
Revision Date 08/12/2003

MSDS Number 300000000099
Print Date 10/01/2003

1. PRODUCT AND COMPANY IDENTIFICATION

Product name	: Nitrogen
Chemical formula	: N2
Synonyms	: Nitrogen, Nitrogen gas, Gaseous Nitrogen, GAN
Product Use Description	: General Industrial
Company	: Air Products and Chemicals,Inc 7201 Hamilton Blvd. Allentown, PA 18195-1501
Telephone	: 800-345-3148
Emergency telephone number	: 800-523-9374 USA 01-610-481-7711 International

2. COMPOSITION/INFORMATION ON INGREDIENTS

Components	CAS Number	Concentration (Volume)
Nitrogen	7727-37-9	100 %

Concentration is nominal. For the exact product composition, please refer to Air Products technical specifications.

3. HAZARDS IDENTIFICATION

Emergency Overview

High pressure gas.
Can cause rapid suffocation.
Self contained breathing apparatus (SCBA) may be required.

Potential Health Effects

Inhalation	: In high concentrations may cause asphyxiation. Asphyxiation may bring about unconsciousness without warning and so rapidly that victim may be unable to protect themselves.
Eye contact	: No adverse effect.
Skin contact	: No adverse effect.
Ingestion	: Ingestion is not considered a potential route of exposure.
Chronic Health Hazard	: Not applicable.

Exposure Guidelines

Air Products and Chemicals,Inc — 1/7 — Nitrogen

Material Safety Data Sheet
Version 1.2
Revision Date 08/12/2003

MSDS Number 300000000099
Print Date 10/01/2003

Primary Routes of Entry	: Inhalation
Target Organs	: None known.
Symptoms	: Exposure to oxygen deficient atmosphere may cause the following symptoms: Dizziness. Salivation. Nausea. Vomiting. Loss of mobility/consciousness.

Aggravated Medical Condition

None.

Environmental Effects

Not harmful.

4. FIRST AID MEASURES

General advice	: Remove victim to uncontaminated area wearing self contained breathing apparatus. Keep victim warm and rested. Call a doctor. Apply artificial respiration if breathing stopped.
Eye contact	: Not applicable.
Skin contact	: Not applicable.
Ingestion	: Ingestion is not considered a potential route of exposure.
Inhalation	: Remove to fresh air. If breathing is irregular or stopped, administer artificial respiration. In case of shortness of breath, give oxygen.

5. FIRE-FIGHTING MEASURES

Suitable extinguishing media	: All known extinguishing media can be used.
Specific hazards	: Upon exposure to intense heat or flame, cylinder will vent rapidly and or rupture violently. Product is nonflammable and does not support combustion. Move away from container and cool with water from a protected position. Keep containers and surroundings cool with water spray.
Special protective equipment for fire-fighters	: Wear self contained breathing apparatus for fire fighting if necessary.

6. ACCIDENTAL RELEASE MEASURES

Personal precautions	: Evacuate personnel to safe areas. Wear self-contained breathing apparatus when entering area unless atmosphere is proved to be safe. Monitor oxygen level. Ventilate the area.
Environmental precautions	: Do not discharge into any place where its accumulation could be dangerous. Prevent further leakage or spillage if safe to do so.

Air Products and Chemicals,Inc — 2/7 — Nitrogen

Material Safety Data Sheet
Version 1.2
Revision Date 08/12/2003

MSDS Number 300000000099
Print Date 10/01/2003

Methods for cleaning up	: Ventilate the area.
Additional advice	: If possible, stop flow of product. Increase ventilation to the release area and monitor oxygen level. If leak is from cylinder or cylinder valve, call the Air Products emergency telephone number. If the leak is in the user's system, close the cylinder valve, safely vent the pressure, and purge with an inert gas before attempting repairs.

7. HANDLING AND STORAGE

Handling

Protect cylinders from physical damage; do not drag, roll, slide or drop. Do not allow storage area temperature to exceed 50°C (122°F). Only experienced and properly instructed persons should handle compressed gases. Before using the product, determine its identity by reading the label. Know and understand the properties and hazards of the product before use. When doubt exists as to the correct handling procedure for a particular gas, contact the supplier. Do not remove or deface labels provided by the supplier for the identification of the cylinder contents. When moving cylinders, even for short distances, use a cart (trolley, hand truck, etc.) designed to transport cylinders. Leave valve protection caps in place until the container has been secured against either a wall or bench or placed in a container stand and is ready for use. Use an adjustable strap wrench to remove over-tight or rusted caps. Before connecting the container, check the complete gas system for suitability, particularly for pressure rating and materials. Before connecting the container for use, ensure that back feed from the system into the container is prevented. Ensure the complete gas system is compatible for pressure rating and materials of construction. Ensure the complete gas system has been checked for leaks before use. Employ suitable pressure regulating devices on all containers when the gas is being emitted to systems with lower pressure rating than that of the container. Never insert an object (e.g. wrench, screwdriver, pry bar, etc.) into valve cap openings. Doing so may damage valve, causing a leak to occur. Open valve slowly. If user experiences any difficulty operating cylinder valve discontinue use and contact supplier. Close container valve after each use and when empty, even if still connected to equipment. Never attempt to repair or modify container valves or safety relief devices. Damaged valves should be reported immediately to the supplier. Close valve after each use and when empty. Replace outlet caps or plugs and container caps as soon as container is disconnected from equipment. Do not subject containers to abnormal mechanical shocks which may cause damage to their valve or safety devices. Never attempt to lift a cylinder by its valve protection cap or guard. Do not use containers as rollers or supports or for any other purpose than to contain the gas as supplied. Never strike an arc on a compressed gas cylinder or make a cylinder a part of an electrical circuit. Do not smoke while handling product or cylinders. Never re-compress a gas or a gas mixture without first consulting the supplier. Never attempt to transfer gases from one cylinder/container to another. Always use backflow protective device in piping. When returning cylinder install valve outlet cap or plug leak tight. Never use direct flame or electrical heating devices to raise the pressure of a container. Containers should not be subjected to temperatures above 50°C (122°F). Prolonged periods of cold temperature below -30°C (-20°F) should be avoided.

Storage

Full containers should be stored so that oldest stock is used first. Containers should be stored in a purpose build compound which should be well ventilated, preferably in the open air. Stored containers should be periodically checked for general condition and leakage. Observe all regulations and local requirements regarding storage of containers. Protect containers stored in the open against rusting and extremes of weather. Containers should not be stored in conditions likely to encourage corrosion. Containers should be stored in the vertical position and properly secured to prevent toppling. The container valves should be tightly closed and where appropriate valve outlets should be capped or plugged. Container valve guards or caps should be in place. Keep containers tightly closed in a cool, well-ventilated place. Store containers in location free from fire risk and away from sources of heat and ignition. Full and empty cylinders should be segregated. Do not allow storage temperature to exceed 50°C (122°F). Return empty containers in a timely manner.

Air Products and Chemicals,Inc — 3/7 — Nitrogen

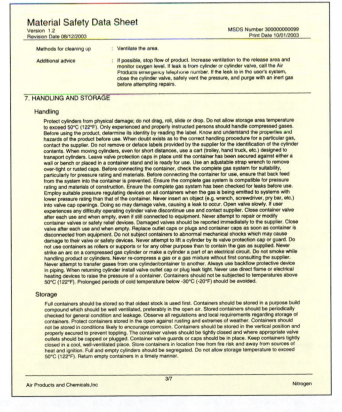

FIGURE 3-16 An example of an MSDS. (*Courtesy of Air Products and Chemicals, Inc. Presented for illustrative purposes only*)

Material Safety Data Sheet

Version 1.2
Revision Date 08/12/2003

MSDS Number 300000000099
Print Date 10/01/2003

Technical measures/Precautions

Containers should be segregated in the storage area according to the various categories (e.g. flammable, toxic, etc.) and in accordance with local regulations. Keep away from combustible material.

8. EXPOSURE CONTROLS / PERSONAL PROTECTION

Personal protective equipment

Respiratory protection	: Self contained breathing apparatus (SCBA) or positive pressure airline with mask are to be used in oxygen-deficient atmosphere. Air purifying respirators will not provide protection. Users of breathing apparatus must be trained.
Hand protection	: Sturdy work gloves are recommended for handling cylinders. The breakthrough time of the selected glove(s) must be greater than the intended use period.
Eye protection	: Safety glasses recommended when handling cylinders.
Skin and body protection	: Safety shoes are recommended when handling cylinders.
Special instructions for protection and hygiene	: Ensure adequate ventilation, especially in confined areas.
Remarks	: Simple asphyxiant.

9. PHYSICAL AND CHEMICAL PROPERTIES

Form	: Compressed gas.
Color	: Colorless gas
Odor	: No odor warning properties.
Molecular Weight	: 28 g/mol
Relative vapor density	: 0.97 (air = 1)
Density at 70 °F (21 °C)	: 0.075 lb/ft3 (0.0012 g/cm3) Note: (as vapor)
Specific Volume at 70 °F (21 °C)	: 13.80 ft3/lb (0.8615 m3/kg)
Boiling point/range	: -320.8 °F (-196 °C)
Critical temperature	: -232.6 °F (-147 °C)
Melting point/range	: -346.0 °F (-210 °C)
Water solubility	: 0.02 g/l

4/7
Air Products and Chemicals,Inc   Nitrogen

Material Safety Data Sheet

Version 1.2
Revision Date 08/12/2003

MSDS Number 300000000099
Print Date 10/01/2003

10. STABILITY AND REACTIVITY

Stability	: Stable under normal conditions.
Hazardous decomposition products	: None.

11. TOXICOLOGICAL INFORMATION

No known toxicological effects from this product.

12. ECOLOGICAL INFORMATION

Ecotoxicity effects

Aquatic toxicity	: No data available.
Toxicity to other organisms	: No data available.
Mobility	: No data available.
Bioaccumulation	: No data available.

Further information

No ecological damage caused by this product.

13. DISPOSAL CONSIDERATIONS

Waste from residues / unused products	: Contact supplier if guidance is required. Return unused product in orginal cylinder to supplier.
Contaminated packaging	: Return cylinder to supplier.

14. TRANSPORT INFORMATION

CFR

Proper shipping name	: Nitrogen, compressed
Class	: 2.2
UN/ID No.	: UN1066

IATA

Proper shipping name	: Nitrogen, compressed
Class	: 2.2
UN/ID No.	: UN1066

IMDG

Proper shipping name	: NITROGEN, COMPRESSED
Class	: 2.2
UN/ID No.	: UN1066

5/7
Air Products and Chemicals,Inc   Nitrogen

Material Safety Data Sheet

Version 1.2
Revision Date 08/12/2003

MSDS Number 300000000099
Print Date 10/01/2003

CTC

Proper shipping name	: NITROGEN, COMPRESSED
Class	: 2.2
UN/ID No.	: UN1066

Further Information

Avoid transport on vehicles where the load space is not separated from the driver's compartment. Ensure vehicle driver is aware of the potential hazards of the load and knows what to do in the event of an accident or an emergency.

15. REGULATORY INFORMATION

OSHA Hazard Communication Standard (29 CFR 1910.1200) Hazard Class(es)
Compressed Gas

Country	Regulatory list	Notification
USA	TSCA	Included on Inventory.
EU	EINECS	Included on Inventory.
Canada	DSL	Included on Inventory.
Australia	AICS	Included on Inventory.
South Korea	ECL	Included on Inventory.
China	SEPA	Included on Inventory.
Philippines	PICCS	Included on Inventory.
Japan	ENCS	Included on Inventory.

EPA SARA Title III Section 312 (40 CFR 370) Hazard Classification:
Sudden Release of Pressure Hazard

US. California Safe Drinking Water & Toxic Enforcement Act (Proposition 65)
This product does not contain any chemicals known to State of California to cause cancer, birth defects or any other harm.

16. OTHER INFORMATION

NFPA Rating

Health	: 0
Fire	: 0
Instability	: 0
Special	: SA

HMIS Rating

Health	: 0
Flammability	: 0
Physical hazard	: 3

Prepared by   : Air Products and Chemicals, Inc. Global EH&S Product Safety Department

For additional information, please visit our Product Stewardship web site at

http://www.airproducts.com/productstewardship/

6/7
Air Products and Chemicals,Inc   Nitrogen

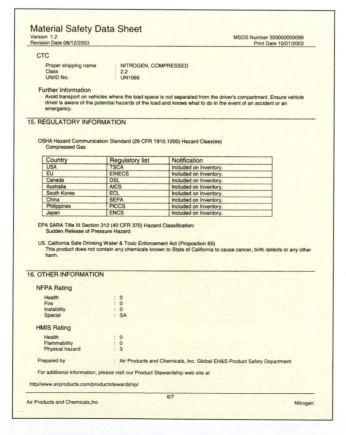

FIGURE 3-16   (Continued) *(Courtesy of Air Products and Chemicals, Inc. Presented for illustrative purposes only)*

JOB PERFORMANCE REQUIREMENT 3-2
Determining an Action Plan Using the MSDS

A A number of clues are present to assist with the identification of this hazardous material. *(Courtesy of Maryland Department of the Environment Emergency Response Division)*

B This incident provides an opportunity to visualize a reactive materials release. *(Courtesy of Cambria County, Pennsylvania, Emergency Services)*

in storage situations and SARA section 304. The RQ will be listed on the shipping papers. If the material is spilled and exceeds the quantity listed, the driver must report the spill to the **National Response Center (NRC).**

The driver/operator of the vehicle is supposed to keep the shipping papers with the vehicle at all times and should be able to provide them upon request.

> **CAUTION**
>
> Problems occur in accidents where there are multiple shipping papers from a variety of pickups. If the driver has kept all of the previous shipping papers from other days, it can be confusing sorting out what the actual cargo is.

If the driver provides information regarding the cargo, and the shipping papers agree with that information, the results should be confirmed with the shipping company. If all three are in agreement as to what is being shipped, it is probably accurate, but this is not guaranteed. A driver who picked up some illegal waste is not going to put it on a shipping paper, nor placard the vehicle, and probably will not admit to having it. It is obvious that the shipping company will not know about it, so until visually confirmed

with appropriate levels of PPE, one must never assume what a cargo consists of.

Mode of Transportation

In highway transportation the shipping papers are called a bill of lading, or most commonly just shipping papers. The papers are supposed to be within arm's reach of the driver, most commonly in a pouch on the driver's door. On tank trucks a duplicate set of papers may be located in a tube attached near the landing gear on the driver's side. In most cases the driver will leave the papers in the truck, and they usually need to be retrieved. If the driver had many pickup stops there will be multiple papers, and some companies put each individual item on a ticket. The hazardous materials are sometimes color coded, as in the case of United Parcel Service (UPS), which uses red tabs to identify hazardous materials packages.

For rail, the shipping papers are called the **consist** or **waybill** and are in the control of the engineer, or the conductor if the train has one. There may be two sets of papers, one listing the contents of each car and one that is by car number, starting with the first car past the engine being number 1 (in some cases car number 1 is the last car). In a derailment this numbering

TABLE 3-3   Comparison of MSDS Data for Azinphos Methyl—AZM

MSDS Note	MSDS–A	MSDS–B	MSDS–C	MSDS–D	CAMEO
Hazard Rating	Health—2 Flammable—0 Reactivity—0	None listed	Health—3 Flammable—2 Reactivity—2	Health—4 Flammable—0 Reactivity—0	Health—3 (extremely hazardous) Flammable—2 (ignites when moderately heated) Reactivity—2 (violent chemical change possible)
Reactivity Hazards	Stable under normal conditions. Hazardous polymerization will not occur.	Depends on the characteristics of dust; decomposes under influence of acids and bases.	Stable material. Unstable above 100°F sustained temperature. Hazardous polymerization will not occur.	Releases toxic, corrosive, flammable, or explosive gases. Polymerization will not occur.	Will decompose.
Incompatibility	High Temperatures, oxidizers, alkaline substances	Acids and bases	Heat, moisture	Heat, flames, sparks, and other ignition sources	Heat, UV light
Fire Hazards	Vapors from fire are hazardous.	Combustible, gives off irritating or toxic fumes (or gases) in a fire.	Decomposes above 130°F with gas evolution and dense smoke. Dust explosion hazard for large dust cloud.	Containers may rupture or explode if exposed to heat.	Decomposes giving off ammonia, hydrogen, and CO.

system usually goes out the window but in some cases may be useful. Most railcars are identified well with car numbers, and as long as they are upright can usually be identified.

Rail uses a number known as a **Standard Transportation Commodity Code (STCC),** usually referred to as a "stick" number. It is a seven-digit number, and if it starts with a "49," the material is a hazardous material. Out of all of the modes of transportation, in rail incidents the engineer is not likely to surrender the papers and will want to accompany the papers anywhere they go. Rail rules require that the engineer be in possession of the papers at all times. If they are lost, engineers face serious personal penalties. There is a computer system in place called Operation Respond that tracks rail shipments. If the car number is known, the computer can identify what the cargo is in a particular car. The system is being slowly implemented in several cities and only involves a handful of carriers at this time.

On a ship the papers are called the **dangerous cargo manifest,** or **DCM** for short. These papers are in the control of the ship's captain or the master who is second in command of the ship. Regardless of the location of the ship and the crew complement, there is always someone in charge of the ship. There may be a considerable number of crew aboard a ship, but there may be only one person who speaks English. The most difficult time to obtain information regarding cargo is when the ship is loading and unloading, especially for container ships. Although the papers may be on board for all of the containers, it can be some time before the crew actually knows where each container is and what the actual cargo is. One of the major problems is determining what is in each individual container, and this information may be limited or not available at all. Luckily each container is well marked with identification information, and after some time the source can be traced to get information.

It is interesting to note that one of the biggest hazards from ships comes from the fuel on board. Even a cruise ship can have up to a million gallons of fuel on board. When dealing with ships, special training is required, because they are extremely hazardous to operate on, and responders can get easily lost and/or fall into deep holds in the ship.

In air, the shipping papers are called **Air Bills** and are in the control of the pilot and usually stored in the cockpit. The type of aircraft (passenger or cargo) will determine the type and amount of materials that can be sent via air. The most common aircraft involved with chemical spills are cargo aircraft, and in many cases exact identification is very time consuming and difficult.

FACILITY DOCUMENTS

Each facility that has chemicals above consumer quantities is supposed to have MSDS as described earlier. In SARA Title III, facilities that are covered by these requirements are also required to submit a Tier 2 form, a listing of all chemicals on site that have an MSDS, and a site plan. If they are using extremely hazardous substances (EHS) they are also required to have an emergency plan. Facilities may also be covered by other regulations and may be required to have other plans in place. Each facility should be able to provide upon request, relatively quickly and without any hassles, an MSDS for a given material. Many facilities will leave a binder of MSDS at the gate or with the security guard. Twenty-four-hour staff should have full access to information as well as emergency contact numbers. The size of the facility will determine the amount of staff available to track this process. The SARA reports and information are updated annually and must be submitted by March 1 of each year. The MSDS may not be revised annually but the facility must ensure their accuracy. Responders should meet with the facility to review these documents and to view the facility.

COMPUTER RESOURCES

Many of the chemical information texts are also available electronically typically as a CD-ROM disk or are available online. Many first responders use the **Computer-Aided Management for Emergency Operations (CAMEO) program** for chemical information. CAMEO is a software package that combines chemical response information with emergency planning capability. The information within CAMEO is easily accessed and can be used by first responders. CAMEO also has the ability to provide a vapor cloud model, known as plume dispersion, as shown in **Figure 3-17.** When the local data and leak information are input, a program known as Aerial Location of Hazardous Atmospheres (ALOHA) can determine the worst-case scenario for the vapor cloud travel.

There are also CDs that have MSDS on them, and many companies, especially chemical distributors, and universities have their MSDS on the Internet. A simple Internet search of "Material Safety Data Sheets" provides several thousand results of possible MSDS locations. Other programs provide chemical information, but in most cases they are specifically designed for a hazardous materials team and may be above the first responder level. One of the greatest advantages of computer software is the ability to search for a chemical by its name and synonyms quickly.

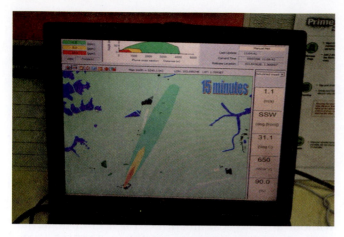

FIGURE 3-17 A toxic gas cloud projection model known as Aloha is part of the hazardous materials software program CAMEO.

CHEMTREC

The Chemical Transportation Emergency Center, or **Chemtrec,** as it is called, is an information service provided by the American Chemistry Council (ACC). A group of chemical manufacturers established the association for several purposes, but one of the outgrowths is the Chemtrec service, which is a free service to emergency responders, paid for by the chemical manufacturers through an annual fee. Although other services also provide chemical information, Chemtrec is the largest and has been providing this service longer than any other company. When a company joins the Chemtrec system, it provides MSDS to Chemtrec as well as emergency contact information. Chemtrec can be considered a large phone book, so if a responder has a company name or chemical, Chemtrec can provide a contact name and number. Many shippers use Chemtrec as their emergency contact point.

If Chemtrec does not have an MSDS on file, they do have other chemical information databases. One of the big advantages is their ability to contact the manufacturer directly, and they will conference call with a responder so that accurate information can be obtained right from the product specialist. If Chemtrec does not have a specific manufacturer's name, they can connect the responder with other specialists who may be able to provide technical assistance. Chemtrec is well connected when dealing with chemical injuries and exposures and can provide medical information as well as provide a contact for further information. One thing Chemtrec does not do is make any regulatory notifications—that is the responsibility of the shipper. When calling Chemtrec, the responder should have the following information available:

- *Caller's name and phone number.* Most responders provide the dispatch phone number, because the on-scene cell phone may be tied up.
- *Name of the shipper or manufacturer.* If this information is not known, Chemtrec will contact a manufacturer for assistance, until the manufacturer is identified.
- *Shipping paper information.* This includes truck or railcar number, the carrier name, the consignee (receiver) name, type of incident, and local conditions.

The Canadian equivalent is called CANUTEC and stands for Canadian Transportation Emergency Center and provides the same services as Chemtrec. In Mexico the Emergency Transportation System for the Chemical Industry is known as SETIQ and provides the same service as Chemtrec and CANUTEC. All three services' phone numbers and other emergency contact numbers are provided in the DOT ERG.

REFERENCE AND INFORMATION TEXTS

Many texts are available from a variety of sources that provide chemical information. In the Additional Resources section at the end of the chapter there is a listing of common texts used by hazardous materials teams, like those shown in **Figure 3-18.** One thing to consider is that every piece of apparatus should carry several reference sources. If a hazardous materials team is available in a community, responders probably do not have to have any additional sources of information, because contacting the hazardous materials team by radio or phone is quick and easy. If responders are in an area where the hazardous materials team is not immediately available or it faces considerable travel time, then additional sources of information would be recommended.

The DOT ERG is fairly easy to use, but many of the other reference sources can be complicated and may require some knowledge of chemistry terms, many of which may be above the first-responder level. With some study of the preliminary information contained in each text, most responders can readily access the information they need. The other difference is that these texts are chemical specific and the information is for that material only. The reference texts are also slanted toward the group that develops the text. These books are generally used by hazardous materials technicians but would be recommended for first responders as well. Each book has a different perspective, and there is no one book that is the "only" book

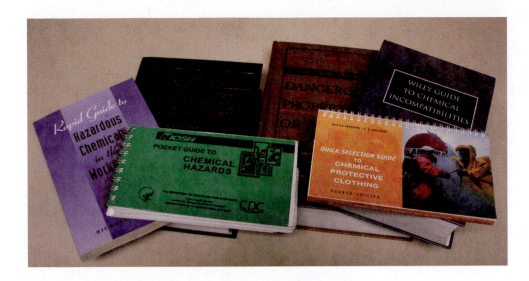

FIGURE 3-18 Several common hazardous materials reference sources, including the NIOSH pocket guide and the Sax guide to Dangerous Properties to Industrial Materials.

to use. There is a book, the *NIOSH Pocket Guide to Chemical Hazards* shown in **Figure 3-18,** which comes closest to being the "one" book used by hazardous materials technicians. When operating at a hazardous materials release, it is best to check more than one source of information and always confirm the information with other sources.

The National Institutes of Occupational Safety and Health (NIOSH) is the research arm to OSHA. They do not issue regulations, but recommend regulatory issues to OSHA. They conduct studies to improve worker safety and when the research demonstrates a need for new regulations or revisions to existing regulations, they provide this information to OSHA. When there are injuries or fatalities in the workplace, they conduct investigations and, again, if there is a need for new or revised regulations in this category, they also provide this information to OSHA.

They do provide a considerable amount of research resources to the fire service and provide a number of sources of information. The *NIOSH Pocket Guide* provides information on the commonly used industrial chemicals. The *Pocket Guide* provides information on exposure levels, chemical and physical properties, and protective equipment recommendations. It is one of the few publications that provides **ionization potentials (IP),** which are used to determine whether a **photoionization detector (PID)** will determine if a chemical is present. A PID uses an ultraviolet lamp as the mechanism to ionize gas samples. In order to be detected by a PID, the gas must have an ionization potential of less than the lamp strength in the PID. Most PIDs have a lamp strength of 10.6 electron volts (eV). When using the *NIOSH Pocket Guide,* you can determine which gases have IPs below 10.6 eV and therefore can be read by the PID.

Another commonly used book is the *Chemical Hazards Risk Information System (CHRIS),* which is published by the U.S. Coast Guard. It provides information on chemical hazards, chemical and physical properties, and water spills. The *CHRIS* book is actually a three-volume set, and it has a great synonym index. Between the *NIOSH Pocket Guide* and the *CHRIS* book, most of the common chemicals are covered.

One book that provides the most detail, and covers more than 23,000 chemicals, is *Sax's Dangerous Properties of Industrial Materials,* as shown in **Figure 3-18.** This book is invaluable to hazardous materials teams and has considerable toxicological information, as well as chemical and physical properties. It covers trade names, synonyms, CAS numbers, and DOT numbers in the index, which is its own volume. There are a number of reference texts that are available to first responders and hazardous materials teams, many of which are listed in the suggested readings section at the end of this chapter. The books covered here are the basic texts and only represent a small number of texts that should be used to determine chemical response information.

A variety of texts is recommended because responders cannot always anticipate the type of material that they may run across. If a responder is in an area where the hazardous materials team has an extended response time, it is recommended that several persons in that department be trained to the technician level, so as to be able to provide some technical assistance to the incident commander. To function as a hazardous materials team, several persons must be trained, but one trained person could provide chemical-specific information and act as a liaison after the arrival of the hazardous materials team.

INDUSTRIAL TECHNICAL ASSISTANCE

Each community usually has a technical specialist in a given field. As an example, in a facility that ships and receives railcars, there is usually someone within that facility who is knowledgeable regarding railcars. In the event of an emergency in that community, it may be wise to use the resources of that person. Almost every town that has some type of industrial facility has a technical specialist within their community.

Many areas of the country, including Baltimore, Houston, and Baton Rouge, have industrial mutual aid groups that are designed to assist each other and the community in the event of a chemical release. In Baltimore this group is called the South Baltimore Industrial Mutual Aid Plan (SBIMAP) and provides technical specialists who assist on the scene and also provides specialized equipment throughout a three-state area. It has been in existence for many years,

and one hazardous materials team uses the services of a chemist each time it responds.

Each industrial facility usually has a person responsible for safety and health, and that person is usually a good technical resource. Within the facility there may be chemists or chemical engineers who can provide chemical information. When dealing with chemical exposures and toxicology, many facilities have industrial hygienists who work with those issues daily. These people should be contacted prior to a large incident, to find out where they live and what their availability is. When approached, many of these professionals are more than willing to assist their community.

NOTE

Accurate and quick information is essential to every hazardous materials incident. The more information that can be obtained, the easier the incident is to resolve.

LESSONS LEARNED

First responders should be starting the information process by trying to obtain as much information as possible, so they can make safe and accurate tactical decisions. Using the recognition and identification skills, a considerable amount of information can be obtained. This information, combined with the use of reference sources, can help provide a significant amount of useful data. If waiting for a hazardous materials team, the more information that can be obtained prior to the arrival of the team will make the response task easier and less stressful. The more information that can be obtained through

the use of the DOT ERG and, more important, the *NIOSH Pocket Guide,* the safer the response. Following the recommendations in the DOT ERG, the initial responders would be taking a conservative approach, which is appropriate. As more information is available, other texts, computer programs, and technical assistance can provide more specific information that can aid in a more incident-specific response. As essential as this information is, first responders should not take any risks attempting to get this information, such as entering a hazard area to get shipping papers.

KEY TERMS

ADR/RID   Abbreviations used by the international shipping community and established by a European Agreement by the United Nations Economic and Social Council Committee of Experts on Transporting Dangerous Goods. The U.S. DOT is a voting member of this shipping body, which governs Regulations Concerning the International Transport of Dangerous Goods by Rail (RID) and Road (ADR).

Air Bill   The term used to describe the shipping papers used in air transportation.

Chemtrec   The Chemical Transportation Emergency Center, which provides technical assistance and guidance in the event of a chemical emergency. This network of chemical manufacturers provides emergency information and response teams if necessary.

Computer-Aided Management for Emergency Operations (CAMEO) Program   A computer program that combines a chemical information database with emergency planning software. It is commonly used by hazardous materials teams to determine chemical information.

Consist   The shipping papers that list the cargo of a train. The listing is by railcar, and the consist lists all of the cars.

Dangerous Cargo Manifest (DCM)   The shipping papers for a ship, which list the hazardous materials on board.

Ionization Potential (IP)   The ability of a gas or vapor to be ionized. It is most commonly used to determine whether a photoionization detector can detect a gas or vapor.

National Response Center (NRC)   The location that must be called to report a spill if it is in excess of the reportable quantity.

Photoionization Detector (PID)   A detection device that can detect toxic gases or vapors and is commonly used by hazardous materials response teams.

Polymerize   A chain reaction in which the material quickly duplicates itself and, if contained, can be very explosive.

Standard Transportation Commodity Code (STCC)   A number assigned to chemicals that travel by rail.

Waybill   A term that may be used in conjunction with consist, but is a description of what is on a specific railcar.

REVIEW QUESTIONS

1. What are three common methods of obtaining chemical information?
2. The DOT ERG is intended to be useful for how long at an incident?
3. If given a four-digit UN identification number, which resource provides quick initial information?
4. What mode of transportation uses STCC numbers?
5. Are MSDS required in a home for household quantities of hazardous materials?
6. Can Chemtrec be used for medical information?
7. Which section of the DOT book is used if you have a shipping name?
8. Which reference book provides information related to suggested isolation distances?
9. What advantage does an industrial contact provide at a chemical spill?
10. What cargo probably will not be listed on the shipping papers?

FOR FURTHER REVIEW

For additional review of the content covered in this chapter, including activities, games, and study materials to prepare for the certification exam, please refer to the following resources:

Firefighter's Handbook Online Companion
Click on our Web site at **http://www.delmarfire.cengage.com** for FREE access to games, quizzes, tips for studying for the certification exam, safety information, links to additional resources and more!

Firefighter's Handbook Study Guide
Order#: 978-1-4180-7322-0
An essential tool for review and exam preparation, this Study Guide combines various types of questions and exercises to evaluate your knowledge of the important concepts presented in each chapter of *Firefighter's Handbook*.

RESPONSE REFERENCES

American Association of Railroads, *American Association of Railroads Hazardous Materials Action Guides.* Washington D.C., 2006. Provides basic to advanced response information; one of the few guides that actually provides cleanup and mitigation strategies; lists common materials transported by rail.

American Association of Railroads, *Emergency Handling of Hazardous Materials in Surface Transportation.* Washington, D.C., 2005. Covers materials commonly transported on rail, which applies to highway incidents as well.

American Conference of Governmental Industrial Hygienists, *ACGIH Threshold Limit Values,* 2007 ed., Cincinnati, OH. Provides the threshold limit values (TLV) for all of the chemicals that have been studied by the ACGIH; targets common industrial chemicals.

Ash, Michael and Irene Ash (eds.), *Gardener's Chemical Synonym and Trade Names,* 11th ed., Milne, Hoboken, NJ, 1999. Listing of common synonyms for industrial materials and household products; since some chemicals have more than 60 different synonyms, this text can be invaluable; no response information.

Brethrick, L. *Brethricks Handbook of Reactive Substances,* 6th ed., Butterworths, Boston, MA, 2000. One of the few sources of information on what happens when chemicals combine; not slanted toward any grouping of chemicals.

Chemical Hazard Risk Information System (CHRIS), U.S. DOT/U.S. Coast Guard, Washington D.C., 2001, www.chrismanual .com. Provides detailed chemical information, is one of the best reference texts, and is free.

Compressed Gas Association, *Handbook of Compressed Gases,* 4th ed., Van Nostrand Reinhold, New York, 1999. Provides extensive detail on common compressed gases; includes detailed information on each type of cylinder.

Crop Protection Handbook, Meister Publishing, Willoughby, OH, 2003. Formerly *Farm Chemicals Handbook;* definitive source for pesticide, herbicide, and insecticide information; no other text is as up-to-date or in-depth.

Fosberg, K. and S. Z. Mansdorf, *Quick Selection Guide to Chemical Protective Clothing,* 4th ed., Van Nostrand Reinhold, New York, 2003. Provides chemical compatibility information on several hundred chemicals; listings are for the most common types of protective clothing.

Lewis, Richard J., Sr., *Dangerous Properties of Industrial Materials,* 11th ed., 2005, John Wiley & Sons Publishing, Hoboken, NJ. One of the best and most detailed reference sources. Three-volume text with extensive cross-referencing and information on 26,000 chemicals.

NIOSH Pocket Guide to Chemical Hazards, National Institute of Occupational Safety and Health, Washington D.C., 2005, http://www.cdc.gov/niosh/npg. Covers most occupational chemicals and most chemicals that responders encounter. Is an excellent source of information and has some detailed chemical and physical properties. Both hard copy and electronic versions are available. The electronic version is free, and single copies of the text are also free.

ADDITIONAL RESOURCES

Bevelacqua, Armando and Richard Stilp, *Hazardous Materials Field Guide,* 2nd ed. Thomson Delmar Learning, Clifton Park, NY, 2006.

Bevelacqua, Armando and Richard Stilp, *Terrorism Handbook for Operational Responders,* 3rd ed. Delmar, Cengage Learning, Clifton Park, NY, 2003.

Callan, Michael, *Street Smart Haz Mat Response.* Red Hat Publishing, Chester, MD, 2001.

Hawley, Chris, *Hazardous Materials Air Monitoring and Detection Devices,* 2nd ed. Thomson Delmar Learning, Clifton Park, NY, 2006.

Hawley, Chris, *Hazardous Materials Incidents,* 2nd ed. Thomson Delmar Learning, Clifton Park, NY, 2004.

Hawley, Chris, Gregory G. Noll, and Michael S. Hildebrand, *Special Operations for Terrorism and HazMat Crimes.* Red Hat Publishing, Chester, MD, 2002.

Noll, Gregory G., Michael S. Hildebrand, and James Yvorra, *Hazardous Materials: Managing the Incident,* 3rd ed. Red Hat Publishing, Chester, MD, 2005.

Schnepp, Rob and Paul Gantt, *Hazardous Materials: Regulations, Response, and Site Operations.* Delmar Publishers, Albany, NY, 1998.

4

Hazardous Materials: Personal Protective Equipment

Chemical protective clothing has come a long way since I was introduced to chemical hazards one evening in 1974. I had responded to a dumpster fire early one morning. We arrived to see a thick, red-brown cloud issuing from the waste container. Obviously, but unfortunately not to us, we had a hazardous materials situation, not a fire. Undaunted, we went about our tasks, and, after half an hour, we eventually extinguished the fire. However, we all had burning eyes, noses, throats, and severe irritation to our wrists and necks. We called for the assistant chief on duty. He suggested our ailments were from the diesel fumes issuing from the fire truck. We thought otherwise but went back to the station, finished our tours, and went home. The next morning I was visited by a man from the chemical company who wanted to make sure we were all right. He explained to me that the dumpster had contained discarded nitric acid bottles, and that some had still been full. They had started the fire. He was more concerned about our exposure, however, since we had not worn chemical protective clothing or even SCBA. (Hey, the fire was outside! Remember those days?) He went on to explain that nitric acid had "a delayed reaction to any acute exposure." This, he told me, would have affected the lungs. They might have developed fluid or edema, creating chemical pneumonia. In addition, my throat might have swollen and made it difficult to breathe. I remember saying to him, "You're probably glad I answered the door!" He said that, actually, he was.

Over the years I have looked back on that incident, and I know it is what got me into hazardous materials. It was the driving force behind my signing up for one of the first National Fire Academy Chemistry of Hazardous Materials courses. I know now that chemical protective clothing, and at the very least SCBA, should always be worn in any smoke, vapor, or cloud situation, but hindsight is the best sight, I guess.

A final note for anyone who might be concerned about me. Do not worry; everything works out in the end. A few months later I went to an acetylene-tank fire. I was acting officer and had decided to extinguish the fire with a sodium-bicarbonate extinguisher. Upon closer examination I noticed the fire was out. I turned to my engine crew to tell them not to charge the extinguisher. Sadly, one overanxious fellow discharged the unit right into my face—and me, once again, without SCBA! With my mouth wide open, I got a lung full of baking soda. So we can all relax, because I am balanced—chemically neutral—again!

—*Street Story by Mike Callan, President, Callan and Company, Middlefield, Connecticut*

LEARNING OBJECTIVES

After completing this chapter, the reader should be able to:

4-1. Describe the causes of harm.

4-2. Explain the health hazards associated with chemical releases.

4-3. Discuss various chemical-related health terms.

4-4. Identify the various levels of PPE.

4-5. Demonstrate the use of SCBA and other respiratory protection at chemical releases.

4-6. Demonstrate the use of firefighting protective clothing at chemical releases.

4-7. Explain the signs and symptoms of heat stress.

INTRODUCTION

Personal protective equipment, or PPE as it is commonly called, takes on many different shapes and versions. One example is provided in **Figure 4-1.** Even standard firefighter protective clothing (FFPC) has many variations and ensembles. When dealing with hazardous materials these configurations are endless.

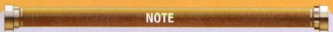

NOTE
The use of PPE is essential to the health and safety of the first responders.

The failure to use PPE or to use it properly may cause an injury or have fatal effects.

SAFETY
Many firefighter injuries can be prevented by the proper use of PPE, most of which only takes a few seconds to put in place.

The best respiratory protection for the first responder above all else is to use the self-contained breathing apparatus (SCBA). SCBA offers a substantial amount of chemical protection and should be considered the minimum protection against chemical spills. Firefighter turnout gear offers limited protection against some chemicals, but it is not intended to be used for chemical spill response. Firefighter turnout gear is not tested or approved for chemical spills, and, although it may offer some protection in some chemical environments, it should only be used for immediate life-threatening rescue situations where there will be limited time in the hazard area.

As the need to be in the hazard area increases, so does the need for the proper PPE. This chapter outlines specialized types of PPE, but specific hands-on training is required prior to the use of this PPE. In general, hazardous materials technicians and specialists wear chemical-protective clothing, but persons trained to the operations level may be required to don chemical-protective clothing to perform decontamination operations or other patient-related activities after specific PPE training is provided.

HEALTH HAZARDS

Health issues are a serious concern with hazardous materials emergencies. They can affect every responder and may even be carried home. They can affect the responder immediately, or it may be years before they take a toll. Protecting the body from hazardous materials is easily accomplished by wearing some form of protective clothing and having a basic understanding of toxicology and how chemicals cause harm.

Toxicology

Toxicology is the study of poisons and their effect on the body, and people who study the effect of poisons on the body are known as toxicologists. Although toxicologists are typically found in the medical community, most industrial facilities have industrial hygienists on staff whose responsibility is to protect the workers' health and safety. Their primary focus is on the chemical hazards that exist within the facility. These people have extensive training in toxicology and chemical exposures and are great resources to the emergency services. Because the world of toxic exposures can be complicated and in emergency situations information is needed quickly, a quick consultation with an industrial hygienist may make the incident easier to resolve.

Types of Exposures

Being exposed to a chemical may present a risk, the level of which is typically spoken of in terms of the chemical's potential *hazard*. The hazard that a

FIGURE 4-1 PPE is used to protect the wearer from a variety of hazards, but no one type of PPE protects the wearer from all types of hazards. This Level A fully encapsulated suit protects the wearer from chemical hazards but offers little protection from heat, flame, or mechanical hazards. All PPE adds heat stress, and this garment adds significant heat stress on the wearer. Decisions on which PPE to wear are difficult to make and require additional training.

chemical presents is an indicator as to the potential harm it can cause. There are two types of exposures, acute and chronic, both of which can have serious health effects.

NOTE

An acute exposure is a quick, one-time exposure to a chemical.

Typically, an **acute** exposure is one where the body is subjected to a large dose over a short period of time. By definition, this period of time can be from sec-

onds to 72 hours. It should be noted that after an acute exposure the body, depending on the product and its hazards, may cause injuries ranging from minimal to major and potentially leading to death. Overall, the human body does well with short-duration exposures and recovers from them, but all chemical exposures should be avoided. A simple example of an acute exposure is that of a nonsmoker who decides to smoke one cigarette. This one-time exposure for most people is not harmful, nor would it cause any long-term health effects.

An example of a chronic exposure, however, is that of the person who smokes three packs of cigarettes a day for twenty years. This person has received doses of cigarette smoke several times a day for a long period of time and is likely to have health problems associated with this chronic exposure. (Note that abnormalities do exist: The person who smoked one cigarette may develop lung problems from that one acute exposure, and the person who chain-smokes for twenty years may never develop health problems associated with the chronic exposure.)

SAFETY

Hazardous materials incidents have the ability to impact humans, the environment, and property. In some cases, such as with dioxin, the risk to humans, the environment, and property can last for many years. Through immediate effects or long-term or recurring (**chronic**), effects, hazardous materials can present a risk. Proper identification of potential hazards and the wearing of proper protective clothing are essential for responder safety.

Types of Hazards

In the realm of hazardous materials, *hazard* is defined as the category of risk that can be inflicted by exposure or contamination with a chemical. There are several methods used to identify possible hazards at a chemical release. The most common method in use today is known by the acronym **TRACEMP,** which stands for thermal, radiation, asphyxiation, chemical, **etiological,** mechanical hazards, and potential psychological harm. Each of the individual hazards has additional hazards that fit within that classification. Much like the **risk-based response** theory, the use of TRACEMP assigns a chemical to a risk category so that tactical decisions can be based on that classification. The subcategories within TRACEMP are given in **Table 4-1.**

TABLE 4-1	Subcategories Within TRACEMP
Thermal	Both heat and cold hazards fit into this category. If a flammable liquid ignites, it is classified as a thermal hazard. If liquefied gas contacts a person's skin, it could cause frostbite and a thermal (cold) burn. When liquefied gases are released into small spaces or rooms, the temperature can drop dramatically, presenting hypothermia and cold stress concerns.
Radiation	Types of radiation include non-ionizing and ionizing. Non-ionizing types include microwave and infrared. Ionizing types include alpha, beta, gamma, and neutron radiation and X rays. Alpha and beta are particles, while gamma, neutron, and X ray are forms of energy.
Asphyxiation	Both simple and chemical asphyxiants fit into this category.
Chemical	This category has a number of subgroups, including poisons and corrosives. The poisons may be referred to as toxics—and some chemicals are highly toxic. There are specific levels of exposure that determine into which category a chemical fits. Also within this category are convulsants, irritants, sensitizers, and allergens. The reaction to a particular chemical varies from person to person, much in the way some people are allergic to bee stings while others are not. Chemicals affect some people and not others.
Etiological	Bloodborne pathogens and biological materials exist within this category.
Mechanical	Although not chemical in nature, mechanical hazards exist within the hazard area of a chemical spill, including the standard slip, trip, and fall hazards that one should always be concerned about. Other examples of mechanical hazards would be getting hit from blast particles, such as from a bomb or BLEVE, or a drum falling on a responder.
Psychological	When exposed to traumatic or disaster situations, some responders may suffer immediate, short-term or long-term psychological stress. Post-Traumatic Stress Syndrome (PTSS) is one example.

SAFETY

In situations where liquefied gases are released, their low temperatures will drop the temperature of the room. In some cases, the drop can be dramatic. In one situation where anhydrous ammonia had been released and there was liquid ammonia on the floor, the hazardous materials team's boots would freeze to the floor. Their SCBA regulators were starting to freeze and several team members had frostbitten hands. In addition to these cold hazards, the vaporization of the ammonia resulted in a vapor cloud that obstructed visibility and the crews were not able to see their team members an arm's length away, not to mention the impact of the fogged up faceshields.

Categories of Health Hazards

Within the TRACEMP categories, there are terms that responders should understand, because MSDSs or industrial contacts may describe some chemicals as fitting into one or more of these categories. One of the most commonly used terms is **carcinogen**, which refers to a material with cancer-causing po-

tential. There are two classifications of carcinogens, known and suspected, with the majority being suspected carcinogens. According to the Chemical Abstracts Service (CAS), a division of the American Chemical Society (ACS), a group that tracks chemicals, there are 30 million chemicals in existence today. Each day the ACS adds 4,000 more chemicals to their listing. The National Toxicology Program under the Public Health Service of the U.S. Health and Human Services Administration issues an annual report on carcinogens. The eleventh annual report lists 54 chemicals that are known to cause cancer and 184 chemicals that are suspected of causing cancer. When dealing with chemical spill response, the risk of getting cancer always causes great fear. Firefighters are exposed to a large number of chemicals, many of them known cancer-causing agents, but if firefighters wear their SCBA these exposures are unlikely to cause problems. For older firefighters who worked in earlier years when SCBA was not used as extensively as it is today, cancer is still a leading cause of death, and many retired firefighters have not had the chance to enjoy retirement due to an early death.

SAFETY

Wearing SCBA and avoiding other off-duty activities that involve cancer-causing materials can prevent the development of cancer in most persons.

Another term that is commonly used is **irritant,** which is self-explanatory. An irritant is not corrosive but mimics the effect of a corrosive material in that it can cause irritation of the eyes and possibly the respiratory tract. One notable characteristic of an exposure to irritants is that the effects are easily reversed by exposure to fresh air. Mace and pepper spray are classified as irritants and may be called incapacitating agents by the military.

Sensitizer is a term used to describe a chemical that causes an effect that is in reality an allergic reaction, typically occurring through skin contact. In most cases an employee can work with a chemical for years and suffer no effects and then one day suffer a severe reaction to the material. Skin reddening, hives, itching, and difficulty breathing are possible symptoms when dealing with a sensitizing agent. Some persons, however, can become sensitive to a chemical after one exposure.

Allergens are chemicals that produce symptoms much like those of sensitizers but are the result of a reaction with an individual's immune system. In many cases, an individual may have been allergic to a substance but was unaware of it until the chemical was encountered.

A **convulsant** chemical is one that has the ability to cause convulsions upon exposure, typically by ingestion. There are a number of drugs that can cause a seizure-like response. Organophosphate pesticides make up one group of chemicals that can cause seizure-like activities. When someone is exposed to an organophosphate pesticide (OPP), the brain reacts to the chemicals and seizurelike activity results. The group of chemical warfare agents known as nerve agents is comparable to OPP compounds and causes the same seizure-like activity.

Some chemicals only affect one or more organs and are described as target organ hazards, or they may affect a body system, such as the central nervous system. The effects depend on the individual, dose, concentration, and length of exposure. Some of the target organ descriptions are provided in **Table 4-2.**

Radiation Hazards

Radiation has the ability to cause a number of health problems up to and including death, which can occur within weeks. Exposure to radiation can cause radiation sickness and other illnesses, some of which are shown in **Table 4-3.** Coming in direct contact with or being close to a radiation source can cause significant burns. The level of radiation and the dose determine

TABLE 4-2   Target Organs and Systems

Name	Target Organ or System
Central nervous system (CNS) chemicals	Affect the central nervous system and can cause short-term or long-term effects. Commonly short-term memory is lost after exposure to a CNS hazard material. Many of the hydrocarbons cause CNS effects, and people sometimes purposely expose themselves to a CNS agent to receive a "high" from the exposure. In the long term, the brain cells are damaged, never to recover. Neurotoxins essentially cause the same effects.
Peripheral nervous system (PNS) chemicals	Much like CNS chemicals, the PNS chemicals affect the body's ability to move in a coordinated fashion. Exposure to a PNS chemical causes a disruption of the brain's ability to move messages to the other body systems.
Hepatoxins	These types of materials affect the liver and if the exposure is high enough can cause severe damage to the liver.
Nephrotoxins	These materials adversely affect the kidneys.
Reproductive toxins	These toxins affect the ability to reproduce and can cause birth defects. These types of toxins can stay within the body, so they can have adverse effects on a pregnancy even if the exposure occurred a considerable time before the pregnancy.
Mutagens	An exposure to a mutagen may not cause any harm to the people who received the exposure, but the effect could be transmitted to their offspring. Mutagens cause damage to the genetic system and can cause mutations that can become hereditary.
Teratogens	These materials can affect an unborn child. The effects, however, do not happen at a cellular level and would not be passed along from generation to generation.

TABLE 4-3   Health Effects from Radiation

Exposure (REM)	Exposure (Sieverts)	Health Effect	Time to Onset
5–10	0.05–0.1	Changes in blood chemistry	
50	0.5	Nausea	Hours
50–100	0.5–1	Headache	
55	0.55	Fatigue	
70	0.7	Vomiting	
75	0.75	Hair loss	2–3 weeks
90	0.9	Diarrhea	
100	1	Hemorrhage	
100–200	1–2	Mild radiation poisoning	Within hours
200–300	2–3	Severe radiation poisoning	Within hours
400–600	4–6	Acute radiation poisoning, 60% fatal	Within an hour
600–1,000	6–10	Destruction of intestinal lining, internal bleeding, and death	15 minutes
1,000–5,000	10–50	Damage to central nervous system, loss of consciousness, and death	Less than 10 minutes
5,000–8,000	50–80	Immediate disorientation and coma, death	Under a minute
More than 8,000	More than 80	Immediate death	Immediate

the damage to the body. The primary issue with radiation is the fact that it can kill without any warning, since it is colorless and odorless and the danger can not be seen. Detection devices can easily detect radiation levels and warn responders when they are entering a potentially dangerous environment. Examples of radiation action levels are provided in **Table 4-4.**

Understanding the level of radiation danger can be confusing. Equally as confusing is the process to calculate a radiation dose. There are three measurements that can be used to describe a radiation dose: absorbed dose, equivalent dose, and effective dose. The **absorbed dose** is a measure of energy transferred to a material by radiation. The absorbed dose is measured in units called gray (Gy) or rad (radiation absorbed dose). One gray is equal to 1 joule of energy absorbed by 1 kilogram of material and is the SI (System International) unit for absorbed dose. One gray is

also equal to 100 rad, an older unit. To understand the impact on humans, we need to expand our thoughts on dose measurements and consider the effect of the absorbed energy on the body. The basic unit of equivalent dose in the United States is the roentgen, which is a value that is provided for the amount of ionization in air caused by X-rays or gamma radiation. Using the roentgen and applying the quality factor we can determine the dose and impact on humans, which is known as roentgen equivalent man or REM (also abbreviated as simply R). The value of 1 roentgen and of 1 REM are equal to each other and are used interchangeably. The rest of the world uses sieverts (Sv) to describe the equivalent dose to man. To further define how the dose impacts humans, another factor known as a weighting factor can be added. Target organs have a higher factor than target bone, for example. Radiation monitors measure in three scales:

TABLE 4-4   Radiation action levels and assorted doses

Dose	Cause or Type	Note
Average Exposures to Radiation		
0.01 R annually	Chest X-ray	Per-year exposure
0.2 R annually	Radon in the home	Per-year exposure
0.081 R annually	Living at high elevations (Denver)	Per-year exposure
1.4 R	Gastrointestinal series	
Action Levels		
1 mR/hr	Isolation zone (public protection level)	Recommended exposure limit for normal activities
5 R	Emergency response	For all activities
10 R	Emergency response	Protecting valuable property
25 R	Emergency response	Lifesaving or protection of large populations
>25 R	Emergency response	Lifesaving or protection of large populations. Only on a voluntary basis for persons who are aware of the risks involved.

(1) REM (R) or sieverts (S), (2) milliREM (mR) or millisieverts (mSv), and (3) microREM (μR) or microsieverts (μSv). The dose of radiation is expressed by a time factor, typically an hour. The conversions for these values are provided in **Table 4-5.**

Radiation Protection

Radiation doses should be kept "As Low As Reasonably Achievable." This is known in the nuclear industry as ALARA and is the cornerstone of radiation safety. The three factors that can influence radiation dose are time, distance, and shielding—concepts that should be applied to all types of chemical exposures as well. To minimize their dose, individuals should stay near the radiation source for as little time as possible, stand as far away from the source as possible, and place as much shielding between them and the source as possible. Naturally occurring background radiation occurs on the level of microREM (or microsieverts) per hour, such that when using a radiation detection device an accurate device will be reading this background radiation as well as any additional levels. When levels above the normal background are encountered, responders should use caution. Background radiation readings do vary depending on the location, elevation, terrain, building type, and a number of other factors. It is when the radiation dose changes scales, such as going from microREM to milliREM, that there is a

potential for health effects. Going from microREM to REM levels indicates severe danger. When encountering radiation levels that are significantly above background and have changed from microREM to milliREM levels or to REM levels, responders should move to an area of lower radiation dose.

Routes of Exposure

The four primary routes of exposure are respiratory, absorption, ingestion, and injection, **Figure 4-2.** The route that is the most commonly associated with causing health effects, both acute and chronic, is the respiratory system route, as shown in **Figure 4-3.**

> **SAFETY**
>
> In almost all cases the respiratory system requires some type of protection. For emergency services workers this is the easiest system to protect because they have easy access to SCBA, should be familiar with it, and are comfortable with its use.

The respiratory system can be affected by gases, vapors, and solid materials such as dust and other particles. In many cases the chemicals may not have any effect on the respiratory system itself, but may enter the body through the respiratory system and affect other organs or body systems. When dealing with

TABLE 4-5   Radiation Equivalents and Conversions

Term or Amount	Equal To
rad	gray (Gy) or absorbed dose (AD)
REM	sievert (Sv) or dose equivalent (DE)
1 µR	0.01 µSv
100 µR	1 µSv
1 mR	10 µSv
100 mR	1 mSv
1 REM	10 mSv
100 REM	1 Sv
curie (Ci)	becquerel (Bq)
27 picocuries (pCi)	1 disintegration/sec (d/sec) or 1 becquerel
1 pCi	37 mBq
1 µCi	37 kBq
27 µCi	1 MBq
1 mCi	37 MBq
27 mCi	1 GBq
1 Ci	37 GBq
27 Ci	1 TBq
1 pCi	2.22 dpm
1 Bq	60 dpm
1 dpm	0.45 pCi

respiratory hazards, there are two categories of asphyxiants, simple and chemical. Although the end result is usually the same, the manner in which a person is killed does differ significantly. Simple asphyxiants simply exclude the oxygen in the air and push it out of the area. It is not an adverse chemical reaction but simply a matter of something other than oxygen occupying the space in the body where the oxygen should be. Normal oxygen levels are 20.9 percent in air, and the body starts to develop difficulty breathing at less than 19 percent. A person will start to have serious problems at less than 16.5 percent oxygen. Gases such as

nitrogen, halon, and carbon dioxide (CO_2) will move oxygen out of the area and cause people in the area to have difficulty breathing. If the concentrations are high enough, death could result. Chemical asphyxiants work in a different manner. They cause a chemical reaction within the body and will not allow it to use the readily available oxygen. The most common chemical asphyxiant is carbon monoxide (CO). When CO is in the air in sufficient quantities, it enters the bloodstream through the lungs. It binds with the hemoglobin in the blood, forming carboxyhemoglobin. Because hemoglobin has an attraction for CO about 225 times that of oxygen, it will not allow the oxygen molecules to bind with the blood, which causes severe health problems and often death.

The other common route of entry is via skin absorption, because the skin is the body's largest organ. Although some chemicals can cause damage to the skin and may irritate it, this does not mean that it is toxic by skin absorption. The number of chemicals that are toxic by skin absorption is relatively low, but precautions should be taken to minimize contact, and, if at all possible, have no skin contact with chemicals. Skin contact with chemicals can cause burns, rashes, or drying of the skin. The only way to provide skin protection is to wear proper protective clothing that will not allow the chemicals to get onto the skin. Firefighter turnout gear will slow the process down but will not prevent the eventual migration of the chemical to the skin. Effective decontamination is required to ensure that all of the chemical is cleaned from the skin.

The other route of entry is ingestion, which is more common than one would think. The use of SCBA generally prevents this route of exposure, at least until the rehabilitation phase of the incident.

CAUTION

After fighting a fire, responders are typically covered in soot and other debris. Most firefighters do not decontaminate themselves prior to eating or drinking any refreshments that may be available to them at the scene. Without proper cleaning, they will typically ingest some of these products of combustion, none of which is healthy to eat.

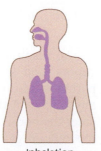  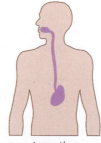 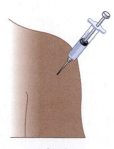

FIGURE 4-2   Routes of exposure.   Inhalation   Absorption   Ingestion   Injection

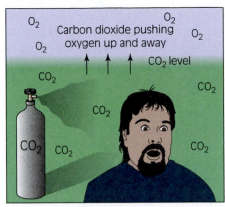

Simple Asphyxiant

Chemical Asphyxiant

FIGURE 4-3 Respiratory system route of exposure.

TABLE 4-6   Routes of Exposure/Harm by Hazard Class

Hazard Class	Primary Route of Entry	Possible Route of Entry	Harm
Explosives	Inhalation, ingestion, and absorption	Injection	Thermal, etiological, mechanical and psychological
Gases	Inhalation	Absorption	Thermal, asphyxiation, chemical, and psychological
Flammable Liquids	Inhalation, ingestion, and absorption	Injection	Thermal, asphyxiation, chemical, and psychological
Flammable Solids	Ingestion, and absorption	Inhalation and injection	Thermal, chemical, and psychological
Oxidizers and Organic Peroxides	Inhalation, ingestion, and absorption	Injection	Thermal, asphyxiation, chemical, mechanical, and psychological
Toxic or Infectious Materials	Inhalation, ingestion, and injection	Absorption	Asphyxiation, chemical, etiological, and psychological
Radioactive Materials	Inhalation, ingestion, and absorption	Injection	Thermal, radiation, and psychological
Corrosive Materials	Inhalation, ingestion, and absorption	Injection	Chemical and psychological
Miscellaneous Dangerous Goods	Inhalation, ingestion, and absorption		Thermal, radiation, asphyxiation, chemical, etiological, mechanical, and psychological

One other route of exposure is injection, although it is not considered to be one of the major routes. Many emergency services workers are exposed to hazardous materials via this route, and for this population it is probably one of the leading routes, after inhalation. The most common material that emergency services workers are exposed to is body fluids or what is referred to as bloodborne pathogens. Other methods of injection are through being near a high-pressure line when it breaks, such as a hydraulic rescue tool fluid line that would inject hydraulic fluid into a person's body. Other than standard infection control practices,

there is little protection for these types of exposures except to properly wear full PPE when working in and around situations where exposure to these fluids is possible. Some examples of the routes of exposure by hazard classes are provided in **Table 4-6.**

Factors that affect the rate of exposure, regardless of the route, are basic items such as temperature, pulse, and respiratory rate. The higher each of these items is in an individual, the more likely it is that the chemical will have some effect. The damage that chemicals have is based on the equation *Effect = Dose × Concentration × Time.* This equation relates to acute and chronic exposures. A small dose, that is, a small concentration, for a small period of time is not likely to have an adverse effect on a normal human. A large dose at a high concentration over a long period of time will in most cases have an effect. People exhibiting normal vital signs who are exposed to a chemical may not have any effects. But if they jog around the block prior to being exposed to a material, they may be affected.

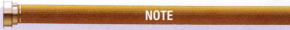

NOTE

The increased temperature of the body will allow for faster absorption into the body, and the accompanying increased respiratory rate will cause more chemicals to enter the respiratory system.

The increased pulse rate allows for the chemicals to be spread throughout the body faster. In some confined space incidents, where chemicals may have played a factor in injuries and deaths, increased vital signs play a role. In most cases the first victim may be unconscious and therefore the body's system has slowed down, and in some cases the victim went down due to a lack of oxygen. When a person recognizes that a coworker has gone down, the vital signs increase and when trying to perform a rescue the person may actually be exposed to a higher level of the chemical than the coworker he or she is trying to rescue. This is one of the reasons why in some cases the rescuer dies, and the original victim ends up surviving.

EXPOSURE LEVELS

In industry, monitoring for exposures is commonplace and is usually a preventive action, but in the emergency services it can be an afterthought, typically after an incident has occurred. Several different types of exposure values have been issued by a variety of agencies, some of which can be very confusing. Exposure values have been established for the commonly used chemicals and for a variety of situations. The key to preventing exposures is to monitor for hazardous materials and to wear appropriate PPE. The one key agency involved with exposure values is the Occupational Safety and Health Administration (OSHA). The exposure values that they set are the ones that must be followed by all industries, including the fire service because it too is considered an industry.[1] Another organization that issues exposure values is the American Conference of Governmental Industrial Hygienists (ACGIH), a group that advocates worker safety and conducts a lot of studies regarding chemical exposures. The National Institute of

Occupational Safety and Health (NIOSH), a research arm of OSHA, issues recommendations for exposure levels. OSHA is the only agency that provides legally binding exposure values; all of the others are recommendations. In some OSHA regulations the employer is required to use the lowest published exposure values, which in many cases are not OSHA's own values. When dealing with emergency situations, it is always recommended that responders follow the lowest published values.

The exposure values are based on an average male and are for an industrial application. The exposure values are typically based on an eight-hour day, forty-hour workweek with a sixteen-hour break between exposures. The values that are issued for the various substances are typically conservative; in a given population it is not atypical to find someone who is sensitive to a chemical at a lower value than the rest of the group. Each value has an extra margin for safety built in, ranging from 1 to 10,000 times the actual value. The exposure values are typically listed in parts per million (ppm) or as milligrams per meter cubed (mg/m^3). Explanations of these terms are provided in **Figure 4-4**.

The most common exposure values are expressed in ppm, but for some materials (typically solids) the values may be expressed as mg/m^3. The values are generally for a period of time, usually eight hours. OSHA refers to this eight-hour exposure value as the **permissible exposure limit (PEL)**. The ACGIH refers to this eight-hour exposure as the **threshold limit value (TLV)**. Both are average exposures over the

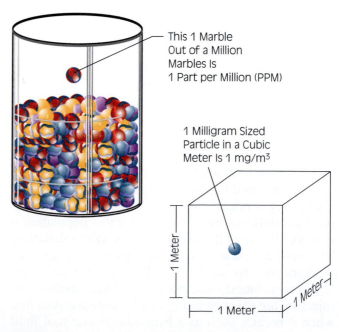

This 1 Marble Out of a Million Marbles Is 1 Part per Million (PPM)

1 Milligram Sized Particle in a Cubic Meter Is 1 mg/m^3

1 Meter

1 Meter

1 Meter

FIGURE 4-4   Explanation of units of measure.

eight-hour period. A worker can be exposed to more than the PEL or TLV as long as at the end of the day the exposure value is less than the PEL or TLV. In some cases these exposure values are called time weighted averages (TWAs), and they may be expressed as the OSHA-TWA or the ACGIH-TWA, which is an eight-hour daily average exposure. NIOSH issues **recommended exposure limits (RELs),** which are for a ten-hour day, forty-hour workweek. These exposure values are outlined in graphic form in **Figure 4-5.**

Other values that may be listed for chemicals are the **ceiling levels,** generally referred to as PEL-C or TLV-C. These provide an amount that is the highest level to which an employee can be exposed. When figuring an average, there are times when employees are going to be exposed to chemicals at a level higher than the PEL or TLV, but there are also times when they will be exposed to less than those levels. As long as their exposure average is less than the PEL or TLV, they are acceptable. Certain chemicals will cause effects at levels that may be obtained through worker exposure and would not be considered safe, but the overall exposure would fall below the PEL or TLV. To avoid unsafe levels, the safety organizations may attach a ceiling level to an exposure value, and that is the highest level that the employee can be exposed to, regardless of what the end average is.

Another exposure value is known as the **short-term exposure limit (STEL).** This value is assigned to a fifteen-minute exposure. An employee can be

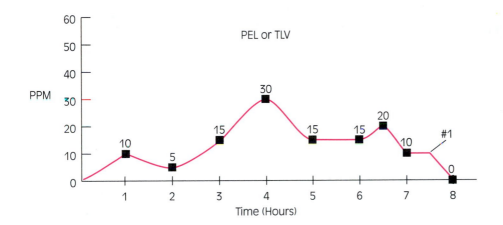

#1 PEL or TWA	
Hour	Exposure
1	10
2	5
3	15
4	30
5	15
6	15
7	20
8	10
Total 8	120 Avg Exposure 15 ppm

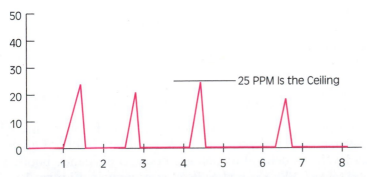

STEL: Allows four exposures of 15 minutes in length, with a minimum of an hour break in between. The ceiling is the highest exposure, no matter the time.

FIGURE 4-5 Exposure values.

TABLE 4-7 Exposure Values

Chemical	PEL (PPM)	IDLH (PPM)	LCt_{50} (ppm) (t = 3 mins.)	ICt_{50} (ppm) (t = 3 mins.)
Sarin (a chemical warfare agent)	0.000017	0.03	12	8
Mustard (a chemical warfare agent)	0.0005	0.0005	231	21.5
Acetone	1000	2500	N/A	N/A
Acrolein	0.1	2	N/A	N/A
Ammonia	50 (ST)	300	N/A	N/A
Chlorine	1	10	N/A	N/A
Ethion (a pesticide)	0.02	N/A	N/A	N/A
Hydrogen sulfide	20 (CEILING)	100	N/A	N/A
Carbon monoxide	50 (35 NIOSH)	200 (CEILING)	N/A	N/A

Note: N/A, not applicable.

exposed at this level for fifteen minutes and then is required to take an hour break from the exposure. The employee can do this four times a day without any adverse effects. NIOSH is also using an excursion value that is coupled with a time limit, five to thirty minutes typically. At this level, an employee can enter an environment one time and not suffer any effects.

The last value can be confusing because it is called the immediately dangerous to life or health (IDLH) value. One would think that being exposed to a chemical at the IDLH level would mean that death may be imminent. In reality this is a value that is the maximum airborne concentration that an individual could escape and not suffer any adverse effects. The actual definition does not match the legal definition. At the IDLH level, emergency responders need to be using SCBA as an absolute minimum, and if the chemical is toxic by skin absorption then a fully encapsulated gastight suit (Level A) must be used.

Other values that a responder may see are called **lethal doses (LD$_{50}$)** or **lethal concentrations (LC$_{50}$)**. The LD is for solids and liquids, and the LC is for gases. In most cases animal studies provide these values, but some are derived from human studies, suicides, and murders. The 50 attached to the LD or LC means 50 percent of the exposed population. In the studies a certain number of test subjects were exposed to a low level of chemicals. After a period of time, the test subjects were studied for any adverse effects. Another group was exposed to a higher level of the chemical. When the subjects were exposed to a certain level and 50 percent of the test subjects died, this established the LD$_{50}$ or LC$_{50}$. With terrorism, emergency responders are having to use some mili-

tary data, much of which is based on LD$_{50}$ or LC$_{50}$ type studies. The military values are generally expressed as **LCt$_{50}$** or lethal concentration to 50 percent of the population, with "t" representing time, usually expressed in minutes. The military also uses **ICt$_{50}$**, which is the incapacitating concentration to 50 percent of the population in a certain amount of time. **Table 4-7** lists exposure values for some chemicals.

Although the exposure values can be confusing it is important to know what each of the values means, and emergency responders should be aware of the exposures they receive during incidents. It is generally recommended for emergency responders to use the TLVs or PELs as the point at which SCBA should be utilized. Emergency responders are faced with a lot of chemical exposures, and at these levels their safety can be ensured if proper PPE is worn.

SAFETY

Air monitoring of these exposure levels is a good way to ensure responder safety.

TYPES OF PERSONAL PROTECTIVE EQUIPMENT

The most common type of PPE for the firefighter is firefighter protective clothing (FFPC). Full FFPC is defined as helmet, hood, coat, pants, boots, gloves, PASS, and SCBA, as shown in **Figure 4-6**. To be fully protected, all of this equipment must be in use. All of the snaps, zippers, and closures must be used to offer optimal protection. FFPC offers protection

10,000 times greater than a person who is not wearing SCBA. To offer this protection, the SCBA must be fitted properly to the person and must be a positive pressure device. If these two factors are not followed then the protection factor can be considerably less than the 10,000. A positive pressure SCBA has an airflow in the face piece all the time, and it maintains positive air pressure inside the mask to keep contaminants out. For firefighting or chemical spill response, it is imperative that the SCBA be a positive pressure unit and that it be activated automatically without any intervention on the part of the wearer.

SAFETY

A firefighter's chance for survival at a hazardous materials incident is dramatically improved when wearing SCBA, and it should be considered the minimum when dealing with chemical spills.

Although various types of SCBA equipment are available, the most common for chemical spill response is a sixty-minute type, which on an average allows twenty to thirty minutes of work time for a hazardous materials team member. When determining the use of SCBA, one has to consider the time it takes to enter the hazard area, working time, time to leave the area, and time to be decontaminated and undressed. In general, thirty- and forty-five-minute air supplies are inadequate for spill response.

Some teams and persons who deal with waste sites may use **supplied air respirators (SARs),** which have some advantages over SCBA. They do not have the weight of the SCBA, but do restrict movement somewhat due to the hoseline. Some logistical issues are associated with the use of SARs, but for long-term incidents they are of assistance.

Some response teams use **air-purifying respirators (APRs)** for minor spills, as shown in **Figure 4-7.** Although they offer some advantages and are common in industry they are not commonly used within the emergency services. With terrorism and bloodborne pathogen issues becoming more commonplace, however, the use of APRs will increase.

APRs also require **fit testing** just like the SCBA to ensure that the respirator fits and will offer the wearer protection. If a responder chooses to use an APR at an incident, a decision flowchart must be followed. In general, the responder must know the chemical, and it must have good warning properties, such as a characteristic odor, and the responder has to identify the amount in the air. The amount of oxygen in the area must be verified and good levels must be maintained. In addition, a variety of cartridges can be used, and when dealing with a variety of chemicals a large stock

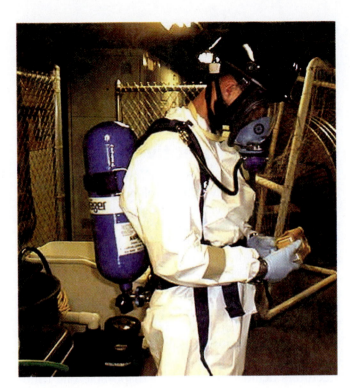

FIGURE 4-6 The use of SCBA offers tremendous protection against heat and chemical hazards. Responders who enter any environment that may have a chemical present should always use SCBA.

against heat and water. The exact amount of protection depends on the type of gear used. FFPC is not certified for chemical contact, nor should it be used for chemical protection. Some of the latest style gear is certified to protect against bloodborne pathogens, but to be sure, the firefighter should check for the NFPA certification.

SAFETY

Wearing protective clothing can present its own significant risk, and both OSHA and the NFPA require the use of the buddy system when entering potentially hazardous environments. When entering IDLH atmospheres, it is required that at least two firefighters enter together, and a backup team must be in place. When one partner has to leave the hazard area, the other partner should always leave at the same time. There should never be a situation where responders are in a hazard area alone.

Self-Contained Breathing Apparatus

With regard to chemical exposures, SCBA offers a protection factor of 10,000. What this means is that a person wearing SCBA has a survivability rating

FIGURE 4-7 This police officer is wearing an air purifying respirator (APR), which is filtering contaminants from the air.

of these cartridges must be maintained. With the exception of waste sites, minor releases, and possibly some decontamination work, SCBA is much easier to use.

Problems with SCBA include extra weight, fatigue, lack of full visibility, lack of mobility, contribution to heat stress, need to refill air supply, and other limiting factors. It is for this reason that many hazardous materials teams choose the other types of respiratory protection. The type of SCBA used will determine the amount of weight added to the responder, and in general the longer the work time the higher the weight of the apparatus. This added weight adds to the overall heat stress on the user. When using SCBA the wearer has limited vision because the face piece does not allow for a full spectrum of vision. When a chemical protective suit is added this limited field of vision gets even smaller. The use of SCBA in some applications such as confined spaces limits wearers as to how they can move about the confined space. For people who may be slightly claustrophobic with SCBA, there

is an additional layer of stress placed on them when donning chemical protective clothing. The users of SCBA and chemical protective clothing need to be medically cleared to function with this type of PPE and should receive periodical medical exams, according to the applicable regulations.

Chemical Protective Clothing

There are four basic levels of chemical protective clothing, but these levels are broken down into further components. The levels are assigned letters to signify their protection levels. Both OSHA and the EPA use Level A, B, C, and D, with Level A being the highest level of chemical protection.

Since the establishment of these levels, many changes have been made to PPE styles and types. When hazardous materials teams first started they had reusable protective clothing. It was used, cleaned, tested, and then reused. But the integrity of the suit became questionable after each use, so the suit manufacturers decided to switch to disposable one-time-use garments. Today, most teams use disposable suits, although some teams maintain at least one type of reusable garment, usually a Teflon fabric suit.

All chemical protective clothing must be checked prior to use for compatibility with the chemical that has been spilled. No matter the type of chemical protective clothing chosen, OSHA requires a PPE program that describes the use of protective clothing and the potential hazards employees may face. Part of the program consists of the maintenance, testing , inspection, and storage procedures. Many forms of protective clothing have specific requirements for storage and some types of clothing have finite shelf lives, some as short as five years. Level A suits require testing generally once a year or after a use. These suits are tested using a pressure test, which ensures their integrity.

CAUTION

No one suit fits all chemicals.

Some of the suits that are NFPA certified do come close, but there are a couple of chemicals that they are not to be used with. Chemical compatibility is based on the **permeation** of a chemical through the fabric of the suit. Permeation is the movement of a chemical through the fabric on the molecular level, **Figure 4-8.** In an acceptable suit, no damage to the fabric should result from a permeation test, nor should anything be seen visually. Imagine wrapping garlic in plastic wrap; initially there is no smell, but after a few hours the scent of garlic has permeated the refrigerator

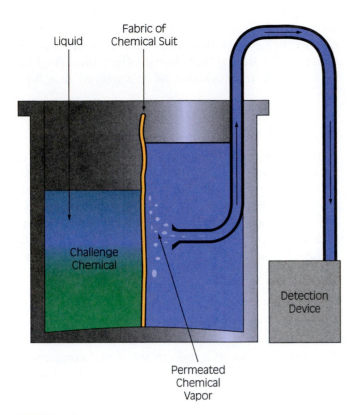

FIGURE 4-8 When conducting a permeation test, the fabric splits a test container, and a measurement device is used to see if the chemical goes through the fabric.

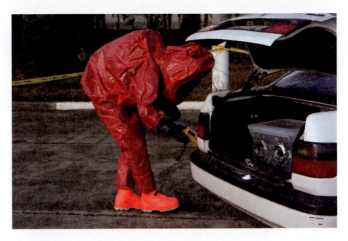

FIGURE 4-9 This Level A ensemble is a gas/vapor-tight garment that protects against most chemicals. Although the protection offered by this ensemble is very high, it is the most stressful to wear.

as if it had not been wrapped at all. Compatibility charts are provided by the manufacturer for the chemicals against which they have tested the fabric.

FIREFIGHTER FACT

Permeation is based on an 8-hour day or 480 minutes. When a department's inventory has more than one suit fabric, it is best to choose a fabric that is given a value of >8 hours, because that means that in tests the fabric showed no breakthrough in 8 hours. After the fabric reaches 8 hours, the test is usually halted, unless specific times are indicated by the manufacturer.

Most fabrics are tested against a battery of chemicals as specified by the American Society for Testing and Materials (ASTM) or the NFPA. This battery of chemicals represents the majority of the chemical families that responders are most likely to encounter.

Two other areas that are of concern when dealing with PPE are degradation and penetration. The degradation of a suit involves the physical destruction of the suit, leaving a hole or damage. Penetration is the movement of chemicals through natural openings such as zippers, glove/suit interface points, and other areas of the suit. It does not involve damage but is an area where chemicals may enter the suit.

Level A Ensemble Protective Clothing

The **Level A ensemble** as shown in **Figure 4-9,** is thought of as providing the highest level of protection against chemical exposure. It is a fully **encapsulated suit** and sometimes is called an encapsulated suit instead of a Level A suit.

NOTE

While it is true that a Level A ensemble offers the maximum level of chemical protection, it is also the leader as far as causing heat stress and physical and psychological stress on the responder wearing the garment.

To be considered a Level A ensemble the suit must have attached gloves and attached boots, and the zipper must be gastight. The suit is designed not to allow any gases to penetrate the garment, and by accomplishing this, it becomes liquid tight as well. Because materials cannot get inside the suit, including air, and gases cannot escape, the person wearing the suit needs to have an SCBA on the whole time. The suit does have relief valves that vent the exhaled air after a certain pressure buildup.

The requirement to use a Level A ensemble within the HAZWOPER regulation is when a firefighter is entering an atmosphere above the IDLH value and the chemical is toxic by skin absorption. There are several occasions, however, where the use of a Level A ensemble would be recommended for some chemicals that do not meet that definition. On occasions where a responder may be potentially covered with a toxic or corrosive material over the whole body, a Level A ensemble would be advisable.

STREETSMART TIP

When wearing a Level A ensemble prehydration is highly recommended, and dressing should take place in a cool quiet area, preferably in the shade.

During emergency operations all members must be monitored. Lack of full visibility and heat stress are major concerns when using a Level A ensemble. The Level A ensemble typically consists of the following components:

- Encapsulated suit with attached gastight gloves and boots
- Inner and outer gloves
- Hard hat
- Communication system
- Cooling system
- SCBA
- PBI/Nomex coveralls
- Overboots

To be an NFPA 1991 (Encapsulated Suit Specifications) certified Level A suit, the suit must have some flash fire resistance. To assist in compliance, newer suits use a blended fabric that offers chemical resistance as well as the flash resistance. Older style suits use a flash suit overgarment that is made of aluminized PBI/Kevlar fabric. This flash protection is not intended for firefighting, but it offers three to thirteen seconds of protection when involved in a flash fire.

Level B Ensembles

Within the **Level B ensemble** family, there is a lot of variety in suit types, as shown in **Figures 4-10A–C.** Although the EPA and OSHA acknowledge two basic types, a large number of styles are available. The two basic types are coverall style and encapsulated, but even these have subvarieties. The encapsulated style of Level B suits is similar to the Level A style, but does not have attached gloves, nor is it vapor-tight. It typically has attached booties, and the SCBA is worn on the inside. Some manufacturers provide glove ring assemblies that allow for the gloves to be preattached during storage. A variety of fabrics are available for the Level B ensemble style of suits, and compatibility is important. The Level B encapsulated suit is the workhorse of hazardous materials teams; it is the most common suit used.

The encapsulated Level B suit is sometimes referred to as a Bubble B or a B plus suit. These suits have some of the same heat stress issues associated with them as do the Level A ensembles, even though

(A)

(B)

(C)

FIGURE 4-10 The photos here all represent Level B suits: (A) a coverall style for law-enforcement officers; (B) an encapsulated style, which is not gastight; and (C) a military-designed two-piece garment worn by tactical officers. The respiratory protection is a rebreather style, which provides a four-hour air supply.

they are lighter. Other styles of Level B suits include a two-piece garment consisting of a jacket and pants, usually bib-overall style. The coverall style Level B suit may have attached booties or a hood, but there are a number of styles for a variety of uses. The one item that makes the Level B suit different from the lower levels of PPE is the use of SCBA. A Level A ensemble is a gastight suit, whereas the Level B ensemble is intended for splash protection and the SCBA offers the respiratory protection.

A Level B ensemble consists of these components:

- Level B suit
- Hard hat
- Inner and outer gloves
- SCBA
- Communication system
- Outer boots
- Nomex/PBI coveralls

Level C Ensemble

A Level C ensemble, as shown in **Figure 4-11,** incorporates the use of an air-purifying respirator within the ensemble. For obvious reasons an APR cannot be used within an encapsulated suit, but it can be used with the other styles. A Level C ensemble can be a coverall or a two-piece garment. A Level C ensemble is used where splashes may occur, but where respiratory hazards are minimal and are covered by the use of an APR.

When using an APR, an extensive listing of requirements must be met. A Level C ensemble consists of these components:

- Level C suit
- Air-purifying respirator
- Hard hat
- Inner and outer gloves
- Outer boots
- Nomex/PBI coveralls

Level D Ensemble

A Level D ensemble, shown in **Figure 4-12,** is actually regular work clothing. It is used when respiratory protection is not required and splashes are not a concern. The Level D ensemble provides no chemical protection, but does offer protection against other workplace hazards. A level D ensemble consists of these items:

- Work clothes
- Hard hat
- Chemical/work gloves

FIGURE 4-11 This Level C ensemble protects against splash hazards and low toxicity materials. It is the use of the air purifying respirator that makes this ensemble Level C.

- Safety glasses
- Safety shoes/boots

High-Temperature Clothing

The two basic types of high-temperature clothing are proximity and fire entry gear. A set of proximity gear is shown in **Figure 4-13.** The most common use for this type of gear is in airport firefighting and flammable liquids firefighting. The entry suits may also be used in high-temperature applications within industry, such as steel making.

High-temperature gear is usually identified by its characteristic aluminized outer shell. This shell is usually attached to a PBI/Kevlar fabric that offers a higher heat resistance than normal structural firefighting clothing. This proximity gear is named for its ability to allow the wearer to get close to the burning liquid. It offers protection for temperatures up to 300 to 400°F (149 to 204°C).

FIGURE 4-12 The Level D ensemble is normal work clothing, this example includes Nomex coveralls, chemical gloves, safety glasses, and steel-toe shoes.

Fire entry gear is designed to allow the wearer to enter a fully involved fire area for a period of thirty to sixty seconds over the life of the suit. The thought process was that it would allow for the rescue of trapped victims, which firefighters now know would not be alive if that type of garment is needed. There are certain applications for this gear involving industrial incidents where high temperatures would be found, but these are rare and most departments no longer carry this type of gear. Fire entry gear can be used in temperatures ranging up to 2,000°F (1,093°C).

Low-Temperature Clothing

When dealing with cryogenics the responder must wear protective clothing that protects the wearer against very cold temperatures. Propane, once it is released from a cylinder to the atmosphere, can reach temperatures well below 0°F (−17°C). Cryogenics by their definition are at least −150°F (−101°C) and are usually colder. Standard protective clothing is not effective against these types of materials. Responders should take precautions against cold stress, and prevent hypothermia. Although no full protective clothing ensemble is available for dealing with cold materials, various types of gloves, **Figure 4-14,** and gauntlets and aprons are available. Layering of clothing can offer some additional protection. Unfortunately in addition to the cold, cryogenics usually present other hazards such as fire, corrosion, toxicity, and asphyxiation.

Limitations of Personal Protective Equipment

There are four basic limitations to protective clothing, and they apply across the board from EMS infection control gear, to structural firefighter clothing, to the fully encapsulated Level A ensembles. Most of these issues are only thought of when dealing with chemical protective clothing, but more thought and emphasis should be placed on these issues when using any type of protective clothing. The four major issues are

FIGURE 4-13 These firefighters are wearing proximity firefighting suits that offer a high level of thermal resistance. They are typically used by air crash rescue crews and flammable liquid firefighters. *(Courtesy of Tech Sgt. Brian E. Christiansen, USAF)*

FIGURE 4-14 Cryogenic gloves are used when dealing with very cold refrigerated liquids. They offer protection from a range of −150° to −450°F.

heat stress, mobility, visibility, and communications problems.

STREETSMART TIP

Stress is a leading killer of firefighters, and heat stress plays a major factor in many of these deaths.

Firefighter clothing, although getting lighter, places tremendous stress on the wearer. The type of vapor barrier used in the FFPC will determine the amount of heat stress a person will encounter, and not all vapor barriers are created equal. W. L. Gore and Associates has developed a new vapor barrier that can be used in an emergency response environment. The clothing was designed for both firefighters and law enforcement. In addition to the firefighter's style, a law enforcement SWAT style coverall is also available. The clothing is known as GORE CHEMPAK Ultra Barrier Fabric. It is designed for a variety of urban search and rescue (USAR) and hazardous materials

FIGURE 4-15 This type of PPE is very useful for USAR situations, offers chemical protection, and is well-suited for WMD response. *(Courtesy of W. L. Gore and Associates)*

situations, as shown in **Figure 4-15.** Given the basic risk categories just discussed, this protective clothing is flame resistant and chemical resistant, which is a unique characteristic. It meets the NFPA 1994 Standard on Protective Clothing Ensemble for Chemical/Biological Terrorism Incidents, and the NFPA 1992 Standard on Liquid Splash Protective Ensembles and Clothing for Hazardous Materials Emergencies.

One of the best advantages for this clothing is its weight and mobility; it offers a high level of protection but in a more form-fitting profile. For both firefighters and law enforcement personnel, this presents an advantage over firefighter's protective clothing or Nomex SWAT clothes. Other than for active or structural firefighting, this protective clothing has benefits and offers protection against WMD (weapons of mass destruction) agents. The gloves made of GORE CHEMPAK Ultra Barrier Fabric are some of the most tactile gloves made for hazardous situations to a degree previously unheard of in chemical protective garments.

DuPont has also developed a new type of protective clothing that has some unique capabilities as well. The main features of the Tychem ThermoPro garment are its chemical and flash fire resistance. In other clothing for hazardous materials response, the combination of chemical and flash fire resistance was not an option; you had to choose one protective stance and take some risk with the other. Now you can have both features with the ThermoPro suit. It meets the NFPA 1992 standard and the NFPA 2112 Standard on Flame Resistant Garments for the Protection of Industrial Personnel Against Flash Fire. This garment is also designed for law enforcement applications and WMD response.

It is important for emergency responders to be able to recognize heat stress and its degrees of seriousness. Heat stress can lead to heat stroke, a condition that is almost always fatal. When wearing a Level A ensemble the temperature inside the suit can easily exceed 100°F (38°C), even on a cold day. When using PPE of any type, responders should take care to prehydrate and take their time when doffing their PPE—immediate removal of their PPE can shock the body, **Figure 4-16,** and cause serious health issues. This is important to remember when removing PPE from a hazardous materials responder, who is usually zipped up inside a very warm environment and cannot control the undressing progress.

FIGURE 4-16   Any type of PPE adds stress to the wearer, even in cold weather. Hydration is the key to survival.

The progression of heat stress is dependent on the amount of work being performed and the physical ability of the responders. To best combat any heat-related emergency, responders should hydrate and take frequent breaks, and until acclimated take extra precautions. When an incident occurs on a day in which the weather conditions would not be considered normal, personnel should pay extra attention to their activity levels. Early recognition is important for the health and safety of the responders. The levels and their warning signs are listed in **Table 4-8.**

> **CAUTION**
>
> Dehydration is a major factor when operating at a fire or a chemical spill. Frequent hydration and frequent rest breaks are important.

TABLE 4-8	Health Hazards and Common Warning Signs
Heat Exhaustion	This is the real first step toward dehydration, and it is fairly common. The root cause is excess sweating, resulting in a loss of body fluid. Symptoms include dizziness, headache, nausea, diarrhea, and vomiting. Heat exhaustion affects not only hazardous materials responders but firefighters as well. Responders should have had enough fluid that they feel the need to urinate. Heat exhaustion that is not treated can lead to heat stroke, which is a very serious and often fatal condition.
Heat Stroke	This is a serious emergency and can be fatal to upward of 80 percent of the patients. In simple terms, it means the body's ability to regulate its temperature has failed and is no longer functioning. The symptoms include unconsciousness, hot and dry skin, seizures, confusion, and disorientation. Being dehydrated when approaching heat stroke levels only complicates the process and actually speeds up the patient's deterioration. Heat stroke should be prevented because once the body shuts down, the end result is usually fatal.
Cold Stress	Situations with liquefied gases or cryogenic gases present situations where cold stress could be a factor. When skin comes in contact with or in proximity to cold vapors or liquids, it can be severely damaged. The symptoms of cold stress include discoloration of the skin, numbness, and pain. The more severe the damage, the darker the skin will become. Skin severely damaged will be black from the exposure.

LESSONS LEARNED

The routes of entry and type of harm that chemicals present vary and chemicals usually present multiple hazards. Having an understanding of these health effects and routes of exposure is important for responder safety. Most chemicals present chronic hazards, but acute hazards are also to be avoided. The use of protective clothing is important for the various hazards that responders may face. The SCBA offers a high level of protection and should be the absolute minimum for respiratory protection. If a responder cannot confirm the absence of hazardous materials through the use of air monitors, then a basic level of protective clothing should be used. The decision to use chemical protective clothing is not an easy one because many factors must be examined prior to its use. With the ensuing heat stress that accompanies the wearing of chemical protective clothing, cases can arise in which the protective clothing may be more dangerous than the chemical hazard. It is for this reason the responders must use effective risk assessment to determine the true hazard—and then dress for that hazard. Many responders have a fear of radiation, but radiation is no different than other hazardous materials; one just has to understand the hazards. With radiation and with chemicals, understanding dose is paramount to remaining safe. We are exposed to radiation every day, and just as with chemicals, the dose makes the poison.

KEY TERMS

Absorbed Dose   A measure of the amount of radiation transferred to a material.

Acute   A quick one-time exposure to a chemical.

Air-Purifying Respirator (APR)   Respiratory protection that filters contaminants out of the air, using filter cartridges. Requires the atmosphere to have sufficient oxygen, in addition to other regulatory requirements.

Allergen   A material that causes a reaction by the body's immune system.

Carcinogen   A material that is capable of causing cancer in humans.

Ceiling Level   The highest exposure a person can receive without suffering any ill effects. It is combined with the PEL, TLV, or REL as a maximum exposure.

Chronic   A continual or repeated exposure to a hazardous material.

Convulsant   A chemical that has the ability to cause seizure-like activity.

Encapsulated Suit   A chemical suit that covers the responder, including the breathing apparatus. Usually associated with Level A clothing, which is gas and liquid tight, but there are some Level B styles that are fully encapsulated but not gas or liquid tight.

Etiological   A form of a hazard that includes biological, viral, and other disease-causing materials.

Fit Testing   A test that ensures the respiratory protection fits the face and offers maximum protection.

ICt$_{50}$ The incapacitating level for time to 50 percent of the exposed group. It is a military term that is often used in conjunction with LCt$_{50}$.

Irritant   A material that is irritating to humans, but usually does not cause any long-term adverse health effects.

LCt$_{50}$ The lethal concentration for time to 50 percent of the group. Same as the LC$_{50}$, but adds the element of time. It is a military term.

Lethal Concentration (LC$_{50}$)   A value for gases that provides the amount of chemical that could kill 50 percent of the exposed group.

Lethal Dose (LD$_{50}$)   A value for solids and liquids that provides the amount of a chemical that could kill 50 percent of the exposed group.

Level A Ensemble (of Protective Clothing)   Fully encapsulated chemical protective clothing. It is gas and liquid tight and offers protection against chemical attack.

Level B Ensemble (of Protective Clothing)   A level of protective clothing that is usually associated with splash protection. Level B requires the use of SCBA. Various clothing styles are considered Level B.

Permeation   The movement of chemicals through chemical protective clothing on a molecular level; does not cause visual damage to the clothing.

Permissible Exposure Limit (PEL)   An OSHA value that regulates the amount of a chemical that a person can be exposed to during an eight-hour day.

Recommended Exposure Limit (REL)   An exposure value established by NIOSH for a ten-hour day, forty-hour workweek. Similar to the PEL and TLV.

Risk-Based Response   An approach to responding to a chemical incident by categorizing a chemical into a fire, corrosive, or toxic risk. Use of a risk-based approach can assist the responder in making tactical, evacuation, and PPE decisions.

Sensitizer   A chemical that after repeated exposures may cause an allergic-type effect on some people.

Short-Term Exposure Limit (STEL)   A fifteen-minute exposure to a chemical followed by a one-

hour break between exposures. Only allowed four times a day.

Supplied Air Respirator (SAR)   Respiratory protection that provides a face mask, air hose connected to a large air supply, and an escape bottle. Typically used for waste sites or confined spaces.

Threshold Limit Value (TLV)   An exposure value that is similar to the PEL, but is issued by the ACGIH. It is based on an eight-hour day.

TRACEMP   An acronym for the types of hazards that exist at a chemical incident: thermal, radiation, asphyxiation, chemical, etiological, mechanical, and potential psychological harm.

REVIEW QUESTIONS

1. What route of entry is the easiest to protect against?
2. What material are emergency responders commonly exposed to through injection?
3. After twenty years of smoking, what type of exposure has a person received?
4. What type of exposure is a one-time event?
5. What are the three common routes of exposure?
6. What six hazards potentially exist at a chemical release?
7. Which adverse medical effect mentioned in the chapter can become hereditary?
8. Nitrogen is what type of asphyxiant?
9. What exposure value uses a fifteen-minute time limit?
10. What is the term *ceiling* used for with regard to exposure values?
11. Which level of chemical protective clothing offers a high level of protection against toxic chemicals that are absorbed through the skin?
12. What is a major concern as the levels of protective clothing increase?
13. Which type of respiratory protection offers the highest level of protection?

FOR FURTHER REVIEW

For additional review of the content covered in this chapter, including activities, games, and study materials to prepare for the certification exam, please refer to the following resources:

Firefighter's Handbook Online Companion
Click on our Web site at **http://www.delmarfire.cengage.com** for FREE access to games, quizzes, tips for studying for the certification exam, safety information, links to additional resources and more!

Firefighter's Handbook Study Guide
Order#: 978-1-4180-7322-0
An essential tool for review and exam preparation, this Study Guide combines various types of questions and exercises to evaluate your knowledge of the important concepts presented in each chapter of *Firefighter's Handbook*.

ENDNOTE

1. Not all emergency services workers are covered by the Federal OSHA regulations. Some are covered by their state OSHA or are not covered at all by OSHA regulations. This varies from state to state.

ADDITIONAL RESOURCES

Basics of Nuclear Radiation, Technical Note TN-176, RAE Systems, San Jose, CA, 2005.

Bevelacqua, Armando, *Hazardous Materials Chemistry,* 2nd ed. Thomson Delmar Learning, Clifton Park, NY, 2006.

Bevelacqua, Armando and Richard Stilp, *Hazardous Materials Field Guide,* 2nd ed. Thomson Delmar Learning, Clifton Park, NY, 2006.

Bevelacqua, Armando and Richard Stilp, *Terrorism Handbook for Operational Responders,* 3rd ed. Delmar, Cengage Learning, Clifton Park, NY, 2003.

Hawley, Chris, *Hazardous Materials Air Monitoring and Detection Devices,* 2nd ed. Thomson Delmar Learning, Clifton Park, NY, 2006.

Hawley, Chris, *Hazardous Materials Incidents,* 3rd ed. Thomson Delmar Learning, Clifton Park, NY, 2007.

Henry, Timothy V., *Decontamination for Hazardous Materials Emergencies.* Delmar Publishers, Albany, NY, 1998.

Lesak, David, *Hazardous Materials Strategies and Tactics.* Prentice-Hall, Englewood Cliffs, NJ, 1998.

Noll, Gregory, Michael Hildebrand, and James Yvorra, *Hazardous Materials: Managing the Incident,* 3rd ed. Red Hat Publishing, Chester, MD, 2005.

Radiation Basics Technical Note, Health Physics Society, McLean, VA, 2005.

Schnepp, Rob and Paul Gantt, *Hazardous Materials: Regulations, Response and Site Operations.* Delmar Publishers, Albany, NY, 1998.

Stilp, Richard and Armando Bevelacqua, *Emergency Medical Response to Hazardous Materials Incidents.* Delmar Publishers, Albany, NY, 1996.

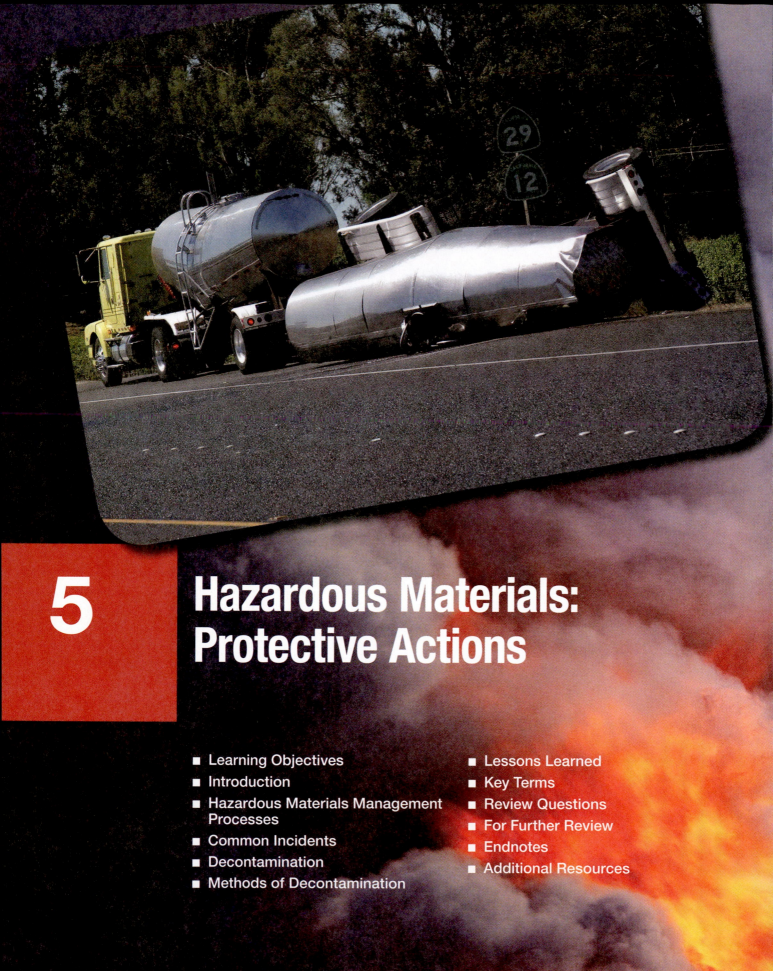

5

Hazardous Materials: Protective Actions

It was the start of what we call a normal day on the job, until later in the morning when an alarm for a chemical leak inside a beverage warehouse was sounded. The dispatch consisted of what we call a hazardous materials box: four engine companies, one truck company, a rescue squad, basic life support unit, hazardous materials company, and command officer. I was working at the hazardous materials company that day. While en route, a radio transmission by the first arriving company advised the Emergency Communications Center of a major ammonia leak inside the warehouse storage area and that they were taking protective actions. This area contained multiple storage of beverages and boxes within an enclosed and secured area that also contained valves and piping for the anhydrous ammonia refrigeration system.

Upon our arrival, the warehouse had been evacuated and a strong odor of ammonia had already consumed the entire area surrounding the warehouse. Once we had performed a hazard risk assessment and ensured that the first responders had taken appropriate protective actions, we then selected our level of protection. Three hazardous materials technicians and I entered the release area to shut off the valve to the leaking pipe. After locating the release area we located the valve and made an attempt to close the valve. While closing the valve a sudden release of gaseous and liquid ammonia covered the personnel working at and around the valve. Visibility was taken from us almost instantaneously because of the gaseous release and communications were lost between all four technicians. I was able to find my way out and noticed that my personnel were still inside the release area. Prior to making another entry to locate my personnel, I noticed a white smoke coming from my chemical boots.

After further investigation, I realized that the oil-based paint from the concrete floor was causing a chemical reaction under the soles of my boots. I reentered the release area, located my personnel, and immediately withdrew from the release area to the decontamination area. Once we were refreshed, a second entry attempt was made into the release area, where we were able to locate another sectional valve and stop the leak. The hardest part of the second attempt was removing our SCBA inside our suits to squeeze past piping and valves to get to the right one.

Prior to leaving the scene we finally determined that prior to our arrival a firefighter had entered the release area and closed the valve without notifying command and/or hazardous materials personnel. When hazardous materials personnel entered the release area thinking that the valve was not closed, they actually reopened it, which caused the valve to freeze in the open position. In this event, a number of factors affected our response. First-arriving crews needed to address isolation and evacuation issues, and the type of release, but one firefighter was endangered by not taking appropriate protective actions. No personnel were injured or exposed to the ammonia, but the incident proved to be very dangerous as a result of personnel freelancing and the lack of training present at an emergency scene.

—*Street Story by Gregory L. Socks, Captain,
Montgomery County, MD, HAZMAT Team*

LEARNING OBJECTIVES

After completing this chapter, the reader should be able to:

5-1. Discuss the various incident management systems.

5-2. Explain the four methods of vapor cloud movement.

5-3. Explain the methods used to make isolation and evacuation decisions.

5-4. Describe the determination and use of hot, warm, and cold zones.

5-5. Discuss the use of incident levels to describe the severity of the incident.

5-6. Describe common incidents within each hazard class.

5-7. Describe the four types of decontamination.

INTRODUCTION

This chapter provides a myriad of topics for the responder and focuses on some general tactics that should be followed at a hazardous materials incident. The tactical considerations provided here are for general situations and may not apply to specific situations because each chemical spill is different, and for each spill there may be another way of handling that release.

A lot of the information in this section may not apply to specific cases. For firefighters beginning their training, it is unlikely they will be making community evacuation decisions for the next couple of years, but the material in this chapter should be kept in mind for the time when they will be making these types of decisions.

NOTE

The basic concept for first responders is one of isolation—first responders should not allow other people to become part of the incident, and they should protect those involved with the incident or those who may become part of the incident in a short time.

HAZARDOUS MATERIALS MANAGEMENT PROCESSES

Several different management processes exist that can be used for hazardous materials incidents, many of which have been in use for many years and offer well-proven methods of organizing an incident. All of the systems have been adapted from fire service systems to fit the needs of a chemical release. The cores of all of these systems are basically the same, but they do differ in some areas. The core to all systems is the protection of life, property, and the environment.

One of the systems developed is the **8-Step Process,** which was devised by Mike Hildebrand, Greg Noll, and Jim Yvorra. Dave Lesak developed another system called the **GEDAPER process** of hazardous materials management. Another system developed early on by Ludwig Benner, Jr., is the **DECIDE process,** which is listed along with the other systems in **Table 5-1.** Regardless of which system a department chooses, it is important to choose a system that every-

one understands and can use. OSHA's HAZWOPER regulation requires the use of an IMS but does not state which type of system is required. In 2004, the Department of Homeland Security (DHS) mandated the use of the National Incident Management System (NIMS). The NIMS system was developed to merge the federal, state, and local response efforts to a disaster or terrorist attack. When developing NIMS, the various incident management systems were combined into one. For use in this text we will use Incident Management System (IMS). The reality is that for the overall incident an incident management team will be in place, and the group of responders responsible for hazardous material will fit into that IMS using one of the processes mentioned or a combination, depending on the situation.

NOTE

When arriving at a suspected chemical release, it is important to isolate the area from other people who may inadvertently wander into a hazardous environment.

TABLE 5-1   Hazardous Materials Management Systems

DECIDE	8-Step Process	GEDAPER	Hazmat Strategic Goals
Detect the presence of the hazardous materials	Site management and control	Gather information	Isolation
Estimate the likely harm	Identify the problem	Estimate potential course and harm	Evacuation
Choose a response objective	Hazard and risk identification	Determine strategic goals	Notification
Identify the action	Select personal protective clothing and equipment	Assess tactical options and resources	Product identification
Do the best possible	Information management and resource coordination	Plan and implement chosen actions	Determination of appropriate personal protective equipment
Evaluate your progress	Implement response objectives	Evaluate	Decontamination
	Decontamination	Review	Spill and leak control
	Terminate the incident		Termination

Sources: The information on the 8-Step Process® is from *Managing the Incident,* Gregory Noll, Michael Hildebrand, Jim Yvorra, Fire Protection Publications, Oklahoma University, 1995. The information on the GEDAPER® is from *Hazardous Materials Strategies and Tactics,* David Lesak, Prentice Hall, 1998. Both systems reproduced with permission of the authors.

Isolation and Protection

One of the most important tasks a responder trained to the awareness level can do is to isolate the area so that others do not become part of the problem. Methods of isolation can be as simple as barrier tape, **Figure 5-1,** to the use of law enforcement at traffic control points. Other methods such as traffic barriers, or even the use of emergency vehicles to block access, can be used. With a chemical hazards incident, it is important to control the incident quickly. The more people entering the suspected hazard area, the more people who may later need to be rescued or, depending on the situation, may need decontamination.

The protection of the people in a hazard area can best be accomplished by evacuation of the immediate area. This does not imply that everyone is simply told to leave, because they may need to be decontaminated or at least medically evaluated depending on the situation. A plan must be established for the holding of these people until a determination can be made as to their status. The hazardous materials team is usually the only group that can make this determination. When dealing with people in the suspected hazard area, frequent communication with the hazardous materials team is important. The other use of protection at a chemical spill is usually associated with the adjacent community evacuation or sheltering in place. Both of these issues are further discussed later in this chapter.

If the incident is suspected of involving criminal or terrorist activity, isolation is important not only for the chemical hazards but for evidence preservation as well. First responders at the awareness level should make efforts to keep people from leaving the area and keep note of persons or vehicles that are leaving the scene. They should isolate the area and ensure that others do not enter the hazard or crime scene area. First responders should establish a hazard area and restrict entry into that area. They should also notify law enforcement and the hazardous materials team of the conditions that they have found.

Rescue

When discussing isolation and protection, rescue is a topic that naturally follows those two important issues. The rescue of victims from a suspected hazard area can be extremely controversial. The decision to make a rescue is a personal one, because it may

FIGURE 5-1 One of the first priorities should be to isolate the area so as to prevent other people from becoming involved with the incident.

involve substantial risk to the rescuer. Local protocol and SOPs must be considered.

SAFETY

Firefighting is inherently risky. As much as firefighters would like to eliminate all risk from their occupation, it is impossible to do so. Their best hope is to safely manage that risk and use methods to identify that risk.

In reality, if responders arrive at the incident safe and sound, their chance for survival dramatically increases. One of the most dangerous parts of firefighting is responding to the incident—each year 20 to 30 percent of firefighter deaths occur while going to and from an incident.

With specific regard to Firefighters Protective Clothing (FFPC), scientific data published in August 2003 by the Soldiers Biological and Chemical Command (SBCCOM) provide Incident Commanders with some information as to the ability to make rescues in hazardous situations. The SBCCOM located at the Aberdeen Proving Grounds in Maryland was redesignated into several other organizations. Formerly SBCCOM, the now Research, Development, and Engineering Command (RDECOM) performed several studies involving firefighters' protective clothing, chemical protective clothing, and detection devices. The lead study, Risk Assessment of Using Firefighter Protective Ensemble (FFPE) with Self-Contained Breathing Apparatus (SCBA) for Rescue Operations During a Terrorist Chemical Agent Incident, is known as the 3/30 Rule, and researchers found that firefight-

ers' protective clothing can offer protection where significant hazards exist to someone without protective clothing. They tested military chemical nerve and blister agent vapor, both of which present toxicity hazards; the nerve agent is extremely toxic. In situations where a nerve agent has been released and there are both live and dead victims, firefighters in full PPE and SCBA can enter this environment and make rescues with no or minimal effects. In a situation where all the victims are dead, firefighters can enter this severely toxic environment for three minutes with no or minimal effects. There are a number of factors that go into entering this type of toxic environment. The study outlines the best way for firefighters to protect themselves with FFPC. There are methods discussed that use tape to add additional protection, as well as some other unique suggestions. The discussion on entering a terrorism event is controversial, and this study only offers some science behind the decision-making process. It goes without saying that in these cases emergency decontamination is a must and the gear should probably be destroyed afterward.

SAFETY

In order to understand the risks associated with the rescue of victims in a chemical warfare agent environment, responders should read the full report and understand the risks of operating in a dangerous environment. Within the report, there are several caveats to several scenarios. The report can be downloaded at http://www.ecbc.army.mil/hld/cwirp/ffpe_scba_rescue_ops_download.htm.

Once on the scene, firefighters need to evaluate the incident. Is it a rescue situation? This information should be confirmed as best it can; in some cases this information cannot be verified. What are conditions at the incident? Is there a confirmed chemical spill? If there is a release, can a person wearing no protection survive? In many cases the fire department is called to suspected chemical releases, and, after investigation, it turns out there really was not a chemical release. If the people who need to be rescued are alive, then the actual risk to the firefighters making the rescue attempt wearing full PPE is minimal. If, on the other hand, the first responders arrive at a local mall and are told that twenty people are unconscious and appear to be dead in the hardware store, the situation is different. When the responders approach the store entrance and see the twenty people lying on the floor, not moving, the responders may not want to enter that environment.

Other considerations need to be taken into account when making a rescue decision, such as response and notification time. The fact that the people in the hazardous environment have been there for a period of five to fifteen minutes, depending on the response time of the department, is critical to the decision-making process. They are not wearing any PPE and have been exposed to the material for a considerable amount of time. If they are alive when responders arrive, the risk to the rescuers is minimal. A method of emergency decontamination should be set up, and a backup crew should be standing by. The responding hazardous materials team should be consulted prior to entry so as to verify the chemical information. When rescuing the victims there should be no delay in their evacuation—a swoop-and-scoop technique is the order of the day. Stokes baskets or other methods of quick evacuation should be employed. Firefighters should not take vital signs, ask medical histories, or perform medical procedures; instead, they should quickly move the victims to a safe area. Responders should have established methods to notify the Incident Commander and other response personnel about critical conditions at the incident. Radios are the most common method, but hand signals may be needed. Some form of relay system may need to be set up to ensure communication between the response crews and the Incident Commander, so that critical information can flow both ways.

Once out of the area, decontamination should be performed if required on both victims and rescuers, and they should be isolated from the remainder of the responders. Once decontaminated, EMS personnel can begin to work on the patients using appropriate levels of PPE. Depending on the condition of the patients, decontamination may be hurried and therefore not perfect. Secondary contamination is of concern for EMS personnel. If a patient ingested a hazardous material, vomiting may result, which exposes personnel to potential contaminants in addition to the blood-borne (etiological) hazard. If the person has been contaminated with a material that is skin absorbable, then it is possible that the victim may off-gas the material or may release odors. After decontamination, the rescue team should remove their PPE and bag them. After evaluation by EMS providers and consultation with the hazardous materials team the rescuers should be sent to rehabilitation. Their turnout gear and SCBA will need to be evaluated and possibly sent for cleaning or replacement, depending on the possible contamination and the material in question.

SAFETY

The rescue of victims is made using simple risk/benefit analysis: a lot of risk is taken when a lot is at stake. No risk is taken when the benefit is little.

When handling hazardous materials incidents, encountering trapped victims is an unusual incident, but procedures should be in place to cover this contingency. The most likely scenario is a traffic accident, in which persons are trapped and chemicals are involved. This type of scenario is common if gasoline is introduced into the picture. Firefighters are used to extricating trapped victims with gasoline leaking from one or more vehicles. Protection lines are established, and a higher level of PPE is typically used by the rescue crews. The rescue crews limit the number of people in the hazard area, and a quick extrication is usually performed. If other chemicals are involved the scenario may change slightly. As hazardous materials companies arrive, personnel can be replaced by those with better chemical protection, and the hazardous materials company can work to control or eliminate the chemical hazard.

Making the decision to enter a hazardous environment takes training and experience. This decision should be made in direct consultation with the hazardous materials company. To make that decision, some hazardous materials teams use an approach known as *risk-based response*. By the heavy use of air monitors, an unknown chemical can be placed into one of four risk categories: fire, corrosive, toxic or radiation hazards. In examining chemical and physical properties, a chemical can be placed into one or more of these categories. The major factor of concern in a potentially hazardous situation is a chemical's vapor pressure. If the chemical does not have a high vapor pressure, the risk of entering an environment containing this spilled material is very low unless a responder

RISK-BASED RESPONSE

A risk-based response philosophy is based on the use of air monitors to guide responders safely. The monitors guide the responders as to which situations are safe to proceed into and when the use of PPE is appropriate. Unfortunately, many first responders do not have all of the required air monitors for a high level of protection. When confronted with rescue situations in potentially hazardous environments, first responders should use full protective equipment, including SCBA, as well as air monitors. They should also consult with the hazardous materials team, who can help guide them to the appropriate level of action. All emergency response involves some risk; steps need to be taken to manage (and thereby minimize) that risk.

In one situation, a truck carrying food rear-ended another tractor-trailer stopped in the middle lane of the highway as shown in **Figures 5-2, 5-3,** and **5-4.** It is estimated that the food truck was traveling at 65 mph when it hit the stopped truck. The truck it hit was carrying a mixed load consisting of mostly 55-gallon drums. The driver of the stopped truck was not hurt and was able to remove his shipping papers. The driver of the food truck was still alive but was extremely entangled in the wreckage. The driver was conscious and alert, and initial vital signs were stable. The trucks were tangled together, and the door to the truck carrying the drums was torn away from the truck. Drums had shifted onto the food truck and were laid over the front of that truck.

The first responders saw the placards on the first truck and requested a hazardous materials assignment. When they approached the food truck to evaluate the driver, they saw the drums in a precarious position in the back

FIGURE 5-2   The truck to the right rear-ended the front truck trapping the driver in the truck on the right. In the trailer to the left are 55-gallon drums, some of which are leaking. Using PPE, air monitoring devices, and protection lines, crews continued with the rescue. *(Courtesy of the Baltimore County Fire Department)*

of the truck. They already had turnout gear on but had also donned SCBA. They obtained the shipping papers from the driver and consulted with the hazardous materials company, which was about twenty minutes away. The contents of the drums were mostly flammable liquids and some combustible liquids. The first responders were advised to use a combustible gas indicator, continue with full PPE, establish foam lines, and begin the rescue.

Upon arrival, the hazardous materials company team members met with the incident commander (IC) to evaluate the scene. They confirmed the monitoring being done by the first responders and began to evaluate the other parts of the load. It was determined that whenever

touches or falls into the material. One of the other items that a hazardous materials team can use to its advantage is the fact that the majority of incidents they respond to involve flammable and combustible liquids or gases. Although not designed as such, firefighter turnout gear offers ample protection for a quick rescue situation. Hazardous materials teams and now many first responders have very capable detection devices for these types of materials and can determine the true risk during an operation. Included here is a listing of the top ten chemicals spilled in this country every year. This is a hazardous materials team's bread and butter, and they should be comfortable with the handling of these materials. First responders should know the locations in which these materials are stored and used. The response to an incident involving these

materials should be no different than a response to a bedroom fire. The top ten chemicals spilled are:[1]

1. Sulfuric acid
2. Hydrochloric acid
3. Chlorine
4. Ammonia
5. Sodium hydroxide
6. Gasoline
7. Propane
8. Combustible liquids
9. Flammable liquids
10. Natural gas

FIGURE 5-3 Rescue crews are extricating the driver of a second vehicle involved in a rear-end collision. The driver remained conscious throughout the extrication and was severely entangled. Complicating the rescue was the presence of a truck full of hazardous materials containers. *(Courtesy of the Baltimore County Fire Department)*

FIGURE 5-4 Leaking drums were found during this rescue operation. Rescue crews had a foam line standing by and were air monitoring for flammable levels. When hazardous materials crews arrived, they took over air monitoring, quickly removed the leaking drums, and secured the other drums. When the dirver was extricated, hazardous materials crews took care of the remaining drums. *(Courtesy of the Baltimore County Fire Department)*

the rescue companies moved a part of the dash of the truck, one of the leaking drums would increase its flow. The hazardous materials companies secured the leak, moved the drum, and secured the remainder of the drums. They examined the rest of the load to make sure there were no other problems and continued to monitor the atmosphere. Once the victim was removed and the rescue companies had moved away, the hazardous materials team **overpacked** and pumped the contents of the other drums into other containers.

The risk category was fire, and the PPE chosen was appropriate for the risk category. At no time were any

flammable readings indicated during the rescue, although some were encountered during the transfer operation. This was an example of the various disciplines working together to rescue a victim in a hazardous situation.

Site Management

The management of a hazardous materials incident can be very difficult even for a seasoned incident commander. A number of incident management strategies are available to the IC, some of which were outlined in previous chapters. The hazardous materials specific positions are outlined in **Table 5-2.** A fire department IC on a normal fire-type incident deals predominantly with resources within the IC's own agency. Occasions may arise when the IC will discuss items with the police department, and depending on the location a police officer may be in the command area. The IC may also deal with other municipal agencies such as the health department or social services. Other outside agencies such as the Red Cross

may also be involved and work under the direction of the IC. Depending on the size of the community, the media may play a factor in the management of the incident. But in the overall scheme of the incident the IC deals mostly with the IC's own agency and coworkers.

When involved in a chemical release, especially one of major proportions, many more agencies get involved. In some communities, the hazardous materials team may not be associated with the local fire department; instead it may be from a county mutual aid group or from an adjoining community. In many incidents, major road closures are necessary, which increases the presence of police officers, and depending on the size of the road may bring command level officers to the scene. If the incident involves a state road

TABLE 5-2   Hazardous Materials Branch Positions

- Backup: Personnel assigned to rescue the entry team if necessary. They are dressed in the same suits as the entry team and are fully prepared to make an entry, with the exception of being on air. The number of backup personnel is a minimum of two but can be expanded to more personnel depending on the incident. The backup team must be trained to the minimum of the Technician level.

- Decontamination: A person is assigned to oversee the setup and operation of the decontamination area, sometimes referred to as the decon area or the contamination reduction corridor. Other personnel will also be operating within the area performing the decon or PPE removal. People operating in this area are usually trained to the Technician level, but people trained to the Operations level may also be operating in this area. If Operations-level personnel are used, they should have received specific training in this job function.

- Entry: A minimum of a two-person team will enter the hazard area or hot zone. The type of chemical will determine the type of PPE that will be used. Prior to entry, the IC must brief the personnel who will be working in the hazard area. To be on the entry team, responders must be a Technician or a Specialist.

- Hazardous Materials Supervisor (HAZMAT Officer): This person is responsible for handling the tactical objectives, as assigned by the Incident Commander. This person is in charge of the HAZMAT team and therefore coordinates its efforts. May be referred to as HAZMAT operations.

- HAZMAT Safety Officer: This person assists the overall safety officer but is concerned mostly with HAZMAT-specific issues such as PPE selection and use. This HAZMAT safety position is usually a Technician or above. The overall safety officer is concerned with all personnel at the incident and usually focuses on other hazards not related to the chemical (e.g., slip, trip, and fall hazards).

- Information/research: This position provides information regarding the chemical and physical properties to the HAZMAT branch officer and the IC. After completion of these duties, these personnel may then assume responsibility for documenting the incident.

- Reconnaissance: These personnel use binoculars or may even enter the hazard area to determine incident severity, attempt identification of the hazardous material, and gather any additional information.

- Resources: These personnel are responsible for gathering and maintaining supplies required for the incident. May be referred to as logistics.

or highway or has the potential to impact these roads, the state highway department may attend the incident. County or state environmental representatives or responders may also arrive at the incident to assist. In some states the Department of Natural Resources (DNR) has jurisdiction in chemical spills and, hence, may respond to an incident.

On a large spill the media will be a much larger group and may be from outside the immediate area, such as at the incident shown in **Figure 5-5**. A chemical release incident usually requires the services of a cleanup contractor, who will be arriving at the incident to assist in the cleanup effort. Depending on the incident, it is not uncommon to see insurance adjusters arriving within a few hours of the spill. On larger spills or spills that may endanger a waterway, the EPA or Coast Guard may arrive to assist with the spill.

Although this listing of agencies is nowhere near complete, it gives the reader an idea of the number of

FIGURE 5-5   The media will play a role in the handling of an incident, from evacuation instructions to minimizing any hysteria that may have occurred during the incident. A good flow of public information is essential to keeping the surrounding community at ease. (*Courtesy of Baltimore County Fire Department*)

different agencies that may assist with the mitigation of the incident. In some cases the IC may have several alarms worth of equipment of his own to manage, not to mention these other agencies. Until the incident is moved from the emergency phase to the nonemergency cleanup phase, the fire department IC is usually still in charge.[2] In some cases the IC will have a difficult time and can end up handling the "assistance" rather than actually managing the incident. In this situation the department's public information officer (PIO) and liaison officer will be of great assistance.

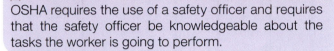

FIREFIGHTER FACT

At a hazardous materials incident, the incident commander is responsible for a number of tasks. Regardless of the local SOP, the IC has many legal obligations under the HAZWOPER regulation. The IC is responsible for the overall actions at the incident, regardless of who is performing the tasks. In most cases, the hazardous materials team performs the mitigation of the incident, and, depending on the interaction of the IC and the hazardous materials team, there may not be much verbal communication.

In the perfect world of IMS the IC makes all of the decisions, but the reality is that the technical group provides the suggested options. In some cases there may be only one option. If the incident goes to court, it will be the IC who has to answer for the hazardous materials team's actions, so the IC must stay informed of the incident action plan and be given the opportunity to choose an appropriate response.

At a chemical release the IC is also responsible for the pre-entry briefing, which is when the entry crew is informed of the hazards that exist and what actions are expected of them. If there are no predesignated emergency signals, these must be decided on prior to entry. The IC must use a system of monitoring the progress of the incident and make changes to the system as needed. The mitigation of an incident takes the cooperation of many persons and agencies, all working toward a common goal under the guidance of the IC.

The management scenario painted did not mention victims, hospitals, or an evacuation—all of which further complicate the incident.

NOTE

The use of an incident management system is not only mandated by OSHA but is a good idea so that personnel can be tracked and the outside agencies can be managed effectively.

Regardless of the IMS, only a small component of the overall system is typically used during routine fire operations, such as a finance branch being established during a house fire. It is likely, however, that in a major chemical release a finance branch would be established and have several persons assigned to assist in this function.

NOTE

OSHA requires the use of a safety officer and requires that the safety officer be knowledgeable about the tasks the worker is going to perform.

In some fashion a liaison must be established with all of the responding agencies, and their specific roles also have to function within the IMS. A hazardous materials incident usually requires a minimum of two safety officers, one for overall safety and one assigned to hazardous materials specific issues.

A person trained to the operations level is not adequate to be the safety officer for the hazardous materials operation. A hazardous materials safety officer should be trained to the Technician or Specialist level to make sure that the mitigation efforts are carried out effectively and safely. Some specific hazardous materials branch functions that may be utilized during an incident are listed in **Table 5-2.**

Establishment of Zones

The term **zone** or **sector** is used to refer to areas that are established to identify the various isolation points, such as those shown in **Figure 5-6.** These zones are referred to as the hot, warm, and cold zones. The hot zone may also be referred to as the exclusion zone, **isolation area,** hazard area, or a similar term. The warm zone may be known as the contamination reduction area and the cold zone as the support area. In many cases these areas are identified through colored barriers, cones, or other markings. Hot is usually identified by the color red, warm by the color yellow, and the cold zone by the color green, **Figure 5-7.**

SAFETY

The most important zone that needs to be established is the isolation area, and this needs to be established by the first arriving responder.

This area normally becomes the hot zone after the arrival of the hazardous materials team. This isolation or hot zone is the area immediately around the release

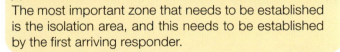

FIGURE 5-6 Proper isolation and work zones are needed to ensure the safety of responders.

and is an area that requires the use of the proper PPE. Entering the hazard area could expose personnel to the substance and possible contamination.

The minimum zone that should be established is this isolation area. The distances for this area can be determined by the use of the DOT ERG. The minimum distance established by this book is 330 feet (101 meters), which should be the absolute minimum for an unknown material. The distance for this area is obviously determined by the local situation, so it is difficult to make blanket statements about recommended distances. If it is convenient and easy to isolate a block then it makes sense to do so, but if isolating the block will create havoc in the community, then 330 feet (101 meters) may be recommended. This distance is only valid if there are no indicators of an actual release. Any indications of a flowing spill or a vapor cloud for an unknown substance will dramatically increase this distance. If time permits, and resources are available, then an attempt should be made to set up the warm and cold zones. These zones are usually established by the hazardous materials team after arrival and setup.

First responders should position themselves in an upwind and uphill position, as shown in **Figure 5-8.** If operating at a water-based spill, they should operate upwind and upstream. The first arriving apparatus should have set up in that position when it arrived, but if not, the vehicle should be moved to that type of position as quickly as possible. Although it cannot always be accomplished, being in an upwind position

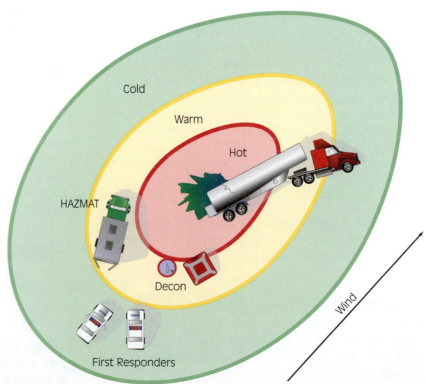

FIGURE 5-7 The establishment of zones is usually based on the types of hazards that may be present. For general chemical spills, the zones established are referred to as the hot, warm, and cold zones.

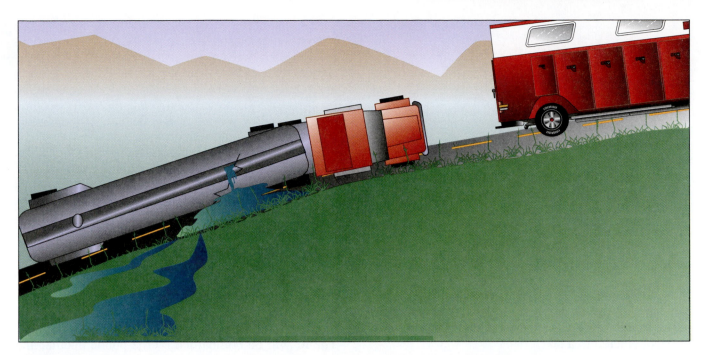

FIGURE 5-8 The best position for first responders is uphill and upwind from the release.

is the first priority, and uphill the second priority. If responders are upwind, but forced to set up in the downhill position, they must extend the distance away. One of the common mistakes made is that the first arriving apparatus does not communicate the on-scene wind directions and the best route of travel for other apparatus to follow to arrive and stage at the upwind and uphill position. All apparatus should be positioned so that they are pointed away from the incident so in the event of a catastrophic release or other emergency, the apparatus can be moved without turning around or backing.

Other items to consider for the hot zone are topography, accessibility for responding units and other resources, weather conditions, and bodies of water, including those for drinking water. Other items to consider are setup areas for the hazardous materials team including decontamination and dress-out locations. Other exposures must be examined, and part of the isolation process is the limiting of possible ignition sources.

Public exposure potential must be considered, and this can be broken down into specific time frames such as short term (minutes and hours), medium term (days, weeks, and months), and long term (years and generations). Emergency responders usually deal only with short-term and medium-term types of incidents, but the actions that first responders take can result in long-term exposures, for both the responders and the public. Sewer lines, both sanitary and storm sewers, must be taken into account as well as other utilities such as cable and phone, both aboveground and be-

lowground. Transportation corridors such as highways, rail lines, ports, and airports must be considered because the incident may well involve or affect these areas. For buildings the internal areas can be broken down into zones, but items that should be considered are floor drains, ventilation ducts, and air returns.

The warm zone is set up after the arrival of the hazardous materials team and is usually where the decontamination area is established. In some cases the warm zone is also extended to allow for the hazardous materials team setup. In unusual situations the warm zone is an area where some type of PPE may be required and depending on the circumstances may be an area that can be affected by wind shifts or a catastrophic failure. The establishment of these zones is arbitrary, although it is based on the best judgment of the IC or the hazardous materials team. There is no magic line of hot and warm zones, nor could one be established. The cold zone is the area where the incident command post is established and is where the first responding companies will be positioned. All support operations such as medical and rehabilitation are set up in this area, and movement between the zones is controlled at access control points. No PPE is required for the cold zone because there should be no chance for chemical exposure in this area. A person should be assigned to act as security for each of these zones to ensure that only authorized personnel enter these areas.

Some hazardous materials teams use monitoring devices to help establish the zones. The use of air monitors to establish these isolation points is crucial, because the distances provided in the DOT ERG

THE IMPORTANCE OF ISOLATING THE SCENE

The first arriving responders can make a hazardous materials release easier to manage if they begin the isolation process. When chemicals are released, there is great potential for liability and lawsuits are probable. Persons interested in joining these lawsuits may try and gain access to the perceived hazard area. In many cases, the insurance companies will try to settle with any potential victims prior to any lawsuits and will write checks to people who were involved, even if they were only in the proximity of the incident. One of the most difficult areas to attempt to control is a mall or a shopping center where hundreds of people may be involved. A simple discharge of pepper spray will result in a number of real victims and usually generate a few more suspect victims. Everyone must be treated and possibly transported, which can create a strain on the EMS system. Early isolation and evacuation can help reduce the number of potential victims.

are very conservative and are based on the worst-case scenario. Real-time, on-scene air monitoring cannot be replaced by plume projections, estimations, or models depicted in a text.

The zone distances are based on theories established by the risk-based response system. The materials that require the greatest isolation are those that have high vapor pressure or exist as gases in their natural state. A material such as sodium hydroxide (sometimes referred to as lye or caustic) requires an isolation distance of only a few feet. Although very corrosive and contact with skin would cause some irritation and burns if not washed off immediately, it has a very low vapor pressure. The vapor pressure of sodium hydroxide is 1 mm Hg at 1,390°F (754°C), which means that the spill would have to be heated to 1,390°F (754°C), in order to produce a very small amount of vapor, considerably less than the vapor pressure of water (25 mm Hg). With this material, it is unlikely that a large isolation distance would have to be established, nor would any evacuations probably be necessary.

A chlorine release from a railcar does present an extreme risk to the community because once in the atmosphere chlorine rapidly changes to a gas (see feature box, "Graniteville, South Carolina, Chlorine Case Study"). Chlorine is not only poisonous but also an oxidizer and a corrosive material with a very high vapor pressure of 5,168 mm Hg. A substantial evacuation area is required for chlorine. One advantage though is that chlorine is easily detected, so a true hazard area can be established using air monitoring.

NOTE

Making a decision to conduct an **evacuation** or to **shelter in place** can be one of the most difficult decisions for an emergency responder. Regardless of the decision, there are usually political ramifications, right, wrong, or indifferent.

Evacuations and Sheltering in Place

The IC will be bombarded by the public, media, and coworkers about the decision. To a first responder, unfortunately, not much assistance is readily available. The only resource is the DOT ERG, and that book is conservative and may not apply in a specific situation. The best way to determine whether to evacuate or shelter in place is to conduct real-time air monitoring that can determine the exact hazard area. Consultation with the hazardous materials team and the local conditions guided by the DOT ERG can establish a starting point. If the incident is at a SARA Title III facility, the jurisdiction's emergency plan should have some recommendations regarding evacuation. Some plans provide a checklist to follow to assist with the evacuation decision-making process. Close coordination with the local emergency management agency is essential to making the outcome successful.

If a decision is made to evacuate, a suitable location needs to be found, transportation may be required, and accommodations need to be established for the evacuees. This type of assistance is usually provided by the emergency management coordinator, who can be a good point of contact. The hazardous materials team can also run an ALOHA (aerial location of hazardous atmospheres) plume projection using the CAMEO (computer-aided management for emergency operations) computer program, but that is also conservative and may cause a larger evacuation than necessary. Chemical vapor plumes have characteristic shapes as determined by computer models, such as the ALOHA model. These standard plumes are to be used for worst-case scenarios and may not apply to a specific locality. Although the plumes are computed for a variety of topographies and types of weather, each local area differs, and only local weather conditions and air monitoring results can provide truly

GRANITEVILLE, SOUTH CAROLINA, CHLORINE CASE STUDY

On January 6, 2005, at 2:39 a.m., a train traveling north-bound ran into a parked train as the result of a misaligned switch. The northbound train had 42 freight cars—25 loaded and 17 empty, though chemical residue can be a factor in an otherwise empty car. Of these 42 cars, 14 held hazardous materials or residue from hazardous materials. When the trains collided, 16 cars derailed. Three of the derailed cars were tank cars carrying chlorine, one of which was leaking. The leaking tank car was carrying 13,830 gallons (52,352 liters) of chlorine, a lique-fied gas. Chlorine has an IDLH (immediate danger to life and health) value of 10 ppm and is very corrosive. Compounding the resulting chlorine release was the fact that Graniteville lies in a shallow valley, with a stream running parallel to the railroad tracks. Chlorine is heavier than air and remains low to the ground. The leaking chlorine was therefore accumulating over the streambed. Within the first few minutes, this chlorine cloud extended 2,500 feet (762 meters) to the north, 1,000 feet (305 meters) to the east, 900 feet (274 meters) to the south, and 1,000 feet (305 meters) to the west.

From the first 9-1-1 call, it was apparent that a chemi-cal release had occurred, and the Fire Chief requested additional resources, including a hazardous materials team. When the Fire Chief arrived on scene, he devel-oped difficulty breathing and had to abandon his imme-diate position. Several of the first-arriving law enforce-ment officers (deputy sheriffs) drove into the chlorine cloud and required medical treatment. Eleven minutes after the collision, the Fire Chief requested that the emer-gency notification system be activated and for citizens to shelter in place. Adjacent to the area of the derailment were a number of industrial facilities in which several per-sons were reported to be trapped or missing. Multiple decontamination areas were established and the local hospital was notified to expect patients. To effect rescue of persons in the hazard area, firefighters in FFPC would ride on pickup trucks, which were then used to transport victims. Later that morning, the Sheriff made the decision to evacuate people in a one-mile radius around the crash site, which resulted in 5,400 persons being evacuated from their homes.

About 21 hours after the crash, hazardous materials crews were able to place a polymer patch on the leak-ing chlorine car. Two days later, sodium hydroxide was pumped from another of the derailed cars. During the pumping operation, the patch on the chlorine car failed, releasing vapors in proximity to the sodium hydroxide car—a potentially dangerous and reactive combination. By reducing the vapor pressure in the leaking chlorine car, the leak was reduced and plans were developed for a more permanent patch. This new patch was applied the next day, on January 12. By midnight January 18, the off-loading of the leaking rail car was complete.

At least 554 people were taken to local hospitals and 75 of them were admitted for further treatment. Including the driver, who died four months after the incident, there were nine fatalities from exposure to chlorine gas: the engineer, six workers from the adjacent milling facility to the west and north of the accident, and a resident living south of the site. Apart from the driver, only the engineer lasted more than a few moments; all of the other fatalities occurred in the first few minutes of exposure. Six fire-fighters were treated and released and one was held for treatment for several days. Two sheriff's deputies were treated and released. As a result of chlorine exposure, two engines, one ambulance, and a service truck were lost to severe damage, costing $630,000.

Think about the impact of such an incident on this small community; lives were lost, injuries occured, and the economic impact was significant. Since the accident, a number of businesses and factories have closed. Al-though in this case, the chlorine exposure was the result of a violent train derailment, think about a perhaps more commonplace accidental release of a hazardous material from a faulty or loosened valve. The response to a chlo-rine (or comparable chemical) release must be quick, ef-ficient, and safe because the impact can be dramatic. This incident occurred in a relatively low-population area; think about the impact in a major metropolitan commu-nity. In the event of a hazardous chemical release, us-ing recognition and identification skills, wearing proper protective clothing, and making appropriate isolation and evacuation decisions are keys to survival for both the re-sponders and the community.

accurate results. The standard shapes for plumes are provided in **Figure 5-9.**

Some studies have shown that in most cases shel-tering in place is safer than evacuation. Just imagine evacuating the most populated area of a community; how would a responder notify all of the citizens? In most communities the police department may have this responsibility, but do they have the resources to accomplish this task? Is there an emergency alerting system for TV or radio in the community? In urban or metropolitan areas evacuating just a few blocks can affect thousands of people. The worst-case scenario using the DOT ERG is an evacuation distance of 4,000 feet (1,219 meters) by 7 miles (11 km), which in a city could be 25,000 to 100,000 evacuees. It would be nearly impossible to evacuate that many people,

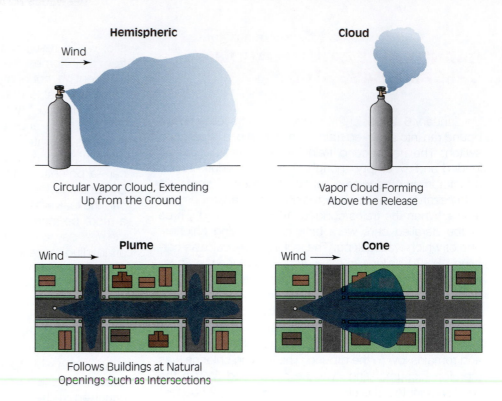

Hemispheric

Wind →

Circular Vapor Cloud, Extending
Up from the Ground

Cloud

Vapor Cloud Forming
Above the Release

Plume

Wind →

Follows Buildings at Natural
Openings Such as Intersections

Cone

Wind →

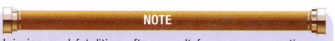

Stream

Stays Low to the Ground
Following Natural Barriers

Pool

Forms a Low-Lying Vapor
Cloud on the Ground

Irregular

Movement of the Material
by Responders or Other
Irregular Movement

FIGURE 5-9 Standard shapes for plumes or vapor clouds may form after a gas is released. The exact type varies with the topography and the buildings in the area.

not to mention the panic and chaos that such action would bring.

> **NOTE**
>
> Injuries and fatalities often result from an evacuation, even when the evacuation is announced days ahead of time.

There are certain times when evacuation is required such as when an explosion is probable, when explosives are involved, when a container may suffer a BLEVE or rupture with violent consequences, or when the release will continue for more than a few hours.

Sometimes an evacuation may not be recommended, such as for a hospital, nursing home, jail, or other facility in which rapid removal of the occupants is not practical. In these types of occupancies

air monitoring and control of HVAC are advisable, in addition to having a liaison remain at the facility who is in radio contact with the IC. The chemical properties of the released material will have an effect on the decision, because some materials will rise up in the air and dissipate quickly, whereas others may stay low to the ground causing evacuation problems. The type of leak must be considered: Can it be quickly and easily controlled or is there a probability that it cannot be controlled by the hazardous materials team? Will evacuating the citizens subject them to a higher level of the chemical than keeping them in place?

ACGIH has provided some planning levels for the chemicals that require planning under the Clean Air Act Amendments (CAAA). These planning levels are on three tiers and provide actual levels that can be used for emergency planning. Because the permissible exposure limits and threshold limit values were established for an eight-hour exposure during normal conditions, they do not have much applicability during an emergency release. These **emergency response planning (ERP)** levels are designed to assist with the emergency planners' preparation of the community emergency plan, and would be useful in determining the evacuation zone. When making the decision to evacuate or shelter in place, the flow of information to the public and the media is essential, because a lack of information can bring disastrous results.

When sheltering in place the citizens should shut all windows and doors, shut off air handling systems, and stay tuned to a TV or radio station. If a continual flow of information is not available, people will become frustrated and may make attempts to go find the information for themselves. When dealing with larger facilities such as high-rises, hospitals, nursing homes, schools, or jails, it may be best to station an emergency responder at that location who acts as a direct link to the incident, so that any questions can be immediately answered and fears alleviated. The emergency management office can also establish a rumor control hotline, which can be used to answer questions about the incident from concerned citizens and family members.

COMMON INCIDENTS

This section provides an overview of common incidents and the types of releases in each of the DOT hazard classes. This listing is far from complete; it merely represents some of the most common incidents or incidents that have a great potential to impact the community. The recommendations provided here are only suggestions; local policies and procedures should be followed.

Types of Releases

When talking about chemical incidents it is important to classify how the chemicals are released from their container. In many cases the manner in which the release occurred can help provide clues to the successful mitigation of the incident. The type of release can be classified as a breach in a container or as a release within a containment system. The NFPA provides two distinct categories for both breaches and releases.

There are several ways of looking at the potential release of a chemical. Either the chemical itself is stressed or the container is stressed. In an incident involving a gasoline tanker that is rolled over at 55 mph (88.5 kph) the container is stressed, and the contents are likely to come out of the container. In an incident in which a paint waste is reacting within a drum, the material is stressed, but if the drum is sealed tight, the container will be stressed due to the chemical reaction creating pressure. The three general types of stress are thermal stress, mechanical stress, and chemical stress. Thermal stress is the addition of heat or cold to a container, and on opposite sides cold stress can be as damaging as heat stress. Placing a metal drum into a pool of a liquefied gas could result in the drum becoming brittle and easily cracked. Dropping a drum is an example of mechanical stress, and putting a corrosive in a metal drum is an example of chemical stress.

Both pressurized and nonpressurized containers can breach in several different ways. **Figure 5-10** provides examples of each type of container breach. The most common container breaches are punctures and closures that open up. It is uncommon to find a container that has disintegrated or has runaway cracks, splits, or tears. Many factors are involved with container breaches, but the most common breaches are nail punctures in drums, forklift punctures, dropped drums, and in many cases closures not used or not in place. Many incidents have been quickly mitigated by the use of a drum lid. On tank trucks a common breach point is the frangible disk, which may rupture due to overfilling or a quick stop causing the liquid to slosh and rupture the disk, releasing some of the contents. The incident is quickly handled by the replacement of the disk.

The methods in which chemicals can be released through a containment system are detonation, violent rupture, rapid relief, and a spill or leak. When these methods result in a chemical release, the action can be violent and can have catastrophic consequences. When a container such as a propane tank, **Figure 5-11,** detonates, it can travel up to a mile. If a container ruptures violently, it means the container was under pressure, causing the violent release. Much like a BLEVE, a violent tank rupture can send the tank or portions thereof a considerable distance.

A container that has a rapid relief valve should be able to release enough pressure to bring a margin of safety to the incident. This is only true if the cause of the increased pressure is removed and the pressure does not continue to climb within the container. If a relief valve is operating, yet still not able to relieve building pressure, a true emergency condition exists and failure to bring the pressure under control can have catastrophic consequences. An operational relief valve is allowing the product to escape, and if the material is flammable it may find an ignition source and ignite. If the release ignites, this may cause the internal pressure to increase even more, causing more product to ignite. It is never recommended that a relief valve be stopped, nor that a fire coming from the relief valve be extinguished. The only time a fire

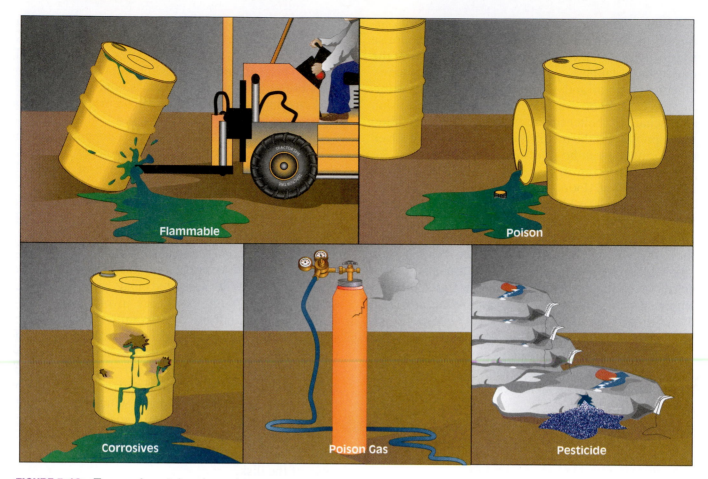

FIGURE 5-10   Types of container breaches.

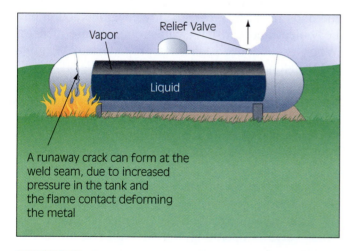

FIGURE 5-11   Propane tank detonation.

coming from any valve (relief or otherwise) should be extinguished is when the flow of the material can be stopped with 100 percent assurance that it will be successful. Extinguishing any other fire, including those that are impinging on a tank or container, should be a high priority. Hoselines should be directed such that

they cool the container, which will reduce the pressure and allow the relief valve to stop operating.

Spills and leaks both result in product being lost. A spill is defined as a loss of product from a naturally occurring opening such as a bung on a drum or a leaking valve. A leak is defined as a release from an unnatural opening such as a puncture in the side of a drum. The end result for a spill or a leak is that product is released and can create problems for the responders.

Explosives

All persons must be removed from the area, and a defensive operation should be established.

SAFETY

The general rule with explosives is that if the fire is near or is affecting the explosives, then emergency responders should isolate the area and back away. Life safety is the top priority.

Many other considerations come into play if the fire is not directly impacting the explosives. A brake

RESIST RUSHING IN !
APPROACH INCIDENT FROM UPWIND
STAY CLEAR OF ALL SPILLS, VAPORS, FUMES AND SMOKE

HOW TO USE THIS GUIDEBOOK DURING AN INCIDENT INVOLVING DANGEROUS GOODS

ONE IDENTIFY THE MATERIAL BY FINDING ANY **ONE** OF THE FOLLOWING:

THE 4-DIGIT ID NUMBER ON A PLACARD OR ORANGE PANEL

THE 4-DIGIT ID NUMBER (after UN/NA) ON A SHIPPING DOCUMENT OR PACKAGE

THE NAME OF THE MATERIAL ON A SHIPPING DOCUMENT, PLACARD OR PACKAGE

IF AN **ID NUMBER** OR THE **NAME OF THE MATERIAL** CANNOT BE FOUND, SKIP TO THE NOTES BELOW.

TWO LOOK UP THE MATERIAL'S 3-DIGIT GUIDE NUMBER IN EITHER:

THE ID NUMBER INDEX..(the yellow-bordered pages of the guidebook)

THE NAME OF MATERIAL INDEX..(the blue-bordered pages of the guidebook)

If the guide number is supplemented with the letter "P", it indicates that the material may undergo violent polymerization if subjected to heat or contamination.

If the index entry is highlighted (in either yellow or blue), it is a TIH (Toxic Inhalation Hazard) material, a chemical warfare agent or a Dangerous Water Reactive Material (produces toxic gas upon contact with water). **LOOK FOR THE ID NUMBER AND NAME OF THE MATERIAL** IN THE TABLE OF INITIAL ISOLATION AND PROTECTIVE ACTION DISTANCES (the green-bordered pages). Then, if necessary, **BEGIN PROTECTIVE ACTIONS IMMEDIATELY** (see Protective Actions on page 298). If protective action is not required, use the information jointly with the 3-digit guide.

USE GUIDE 112 FOR ALL EXPLOSIVES EXCEPT FOR EXPLOSIVES 1.4 (EXPLOSIVES C) WHERE GUIDE 114 IS TO BE CONSULTED.

THREE TURN TO THE NUMBERED GUIDE (the orange-bordered pages) AND READ CAREFULLY.

NOTES IF A NUMBERED GUIDE CANNOT BE OBTAINED BY FOLLOWING THE ABOVE STEPS, AND A PLACARD CAN BE SEEN, LOCATE THE PLACARD IN THE TABLE OF PLACARDS (pages 16-17), THEN GO TO THE 3-DIGIT GUIDE SHOWN NEXT TO THE SAMPLE PLACARD.

IF A REFERENCE TO A GUIDE CANNOT BE FOUND AND THIS INCIDENT IS BELIEVED TO INVOLVE DANGEROUS GOODS, TURN TO GUIDE 111 NOW, AND USE IT UNTIL ADDITIONAL INFORMATION BECOMES AVAILABLE. If the shipping document lists an emergency response telephone number, call that number. If the shipping document is not available, or no emergency response telephone number is listed, IMMEDIATELY CALL the appropriate **emergency response agency listed on the inside back cover of this guidebook.** Provide as much information as possible, such as the name of the carrier (trucking company or railroad) and vehicle number. AS A LAST RESORT, CONSULT THE TABLE OF RAIL CAR AND ROAD TRAILER IDENTIFICATION CHART (pages 18-19). IF THE CONTAINER CAN BE IDENTIFIED, REMEMBER THAT THE INFORMATION ASSOCIATED WITH THESE CONTAINERS IS FOR THE WORST CASE POSSIBLE.

Page 1

FIGURE 5-12 The DOT ERG provides some basic explosives information that provides a good margin of safety when dealing with these types of incidents.

fire on a truck carrying explosives is one example; as long as the fire has not reached the cargo area, firefighters may be able to extinguish it without incident. When making the decision to fight a fire in this situation, water must be applied quickly and in large quantities to cool the cargo and then extinguish the fire. Crews should be limited, and adjoining areas evacuated. The recommendations in the DOT ERG,

Figure 5-12, are a good starting point for isolation and evacuation.

Another incident that involves explosives arises when the fire department assists a bomb squad with a suspected explosive device. Close coordination with the bomb squad is required to make sure that the operation is conducted safely. Some example standoff distances are provided in **Table 5-3.** Although the distance varies with some jurisdictions, radios, cell phones, or other electronic devices should not be used within 500 to 1,000 feet (152 to 305 meters) of a suspected device. Many bomb squads detonate suspected devices in place, which may start a fire or cause a potential structural collapse. Other techniques are to "disrupt" the device or render it safe by the use of a water cannon. These techniques could cause the device to detonate. The time to discuss these scenarios and plans of action is prior to an incident, not during it. At incidents where there is a suspected device, firefighters should be aware of the potential for a secondary device, one designed to injure the responders. When operating at the scene of an explosion, it is important to have the bomb squad search the scene for secondary devices, **Figure 5-13.**

Other incidents may involve a shipment of explosives that has been involved in an accident. In most cases these incidents present little risk to the responder or the community. In this case the bomb squad is a great technical resource and should be used to evaluate the scene prior to moving any explosives. In some cases an explosives truck, such as the one shown in **Figure 5-14,** carrying ammonium nitrate may be involved in an accident and if it has fuel oil on board, the mixture known as **ANFO** could be created. Without an initiation charge or other substantial energy source, the material is unlikely to explode, but the ammonium nitrate does present a toxicity hazard and care should be taken around spills of this material. In incidents of this type the bomb squad should also be consulted.

TABLE 5-3   Bomb Threat Standoff Distances

Threat Description	Explosive Capacity	Lethal Air Blast Range	Mandatory Evacuation Distance	Desired Evacuation Distance
Pipe bomb	5 lbs. (2.2 kg)	25 ft. (8 m)	70 ft. (21 m)	850 ft. (259 m)
Briefcase or suitcase	50 lbs. (23 kg)	40 ft. (15 m)	150 ft. (46 m)	1,850 ft. (564 m)
Compact sedan	220 lbs. (100 kg)	60 ft. (18 m)	240 ft. (73 m)	915 ft. (279 m)
Sedan	500 lbs. (227 kg)	100 ft. (30 m)	320 ft. (98 m)	1,050 ft. (320 m)
Van	1,000 lbs. (454 kg)	125 ft. (38 m)	400 ft. (122 m)	1,200 ft. (366 m)
Moving van or delivery truck	4,000 lbs. (1,814 kg)	200 ft. (61 m)	640 ft. (195 m)	1,750 ft. (533 m)
Semi-trailer	40,000 lbs. (18,143 kg)	450 ft. (137 m)	1,400 ft. (427 m)	3,500 ft. (1,067 m)

Explosive capacity—Based on the maximum volume or weight of explosives (TNT equivalent) that could reasonably be hidden in the package or vehicle.

Lethal air blast range—The minimum distance personnel in the open are expected to survive from blast effects. It is based on severe lung damage or fatal impact injury from body translation.

Mandatory evacuation distance—The range to which all buildings must be evacuated. From this range to the desired evacuation distance, personnel may remain in the building (with some risk) but should move to a safe area in the interior of the building away from windows and exterior walls. Evacuated personnel must move to the desired evacuation distance.

Desired evacuation distance—The range to which personnel in the open must be evacuated and the preferred range for building evacuation. This is the maximum range of the threat from flying shrapnel/debris or flying glass from window breakage.

Source: Developed by the ATF, with technical assistance from the U.S. Corps of Engineers. Supported by the Technical Support Working Group (TSWG), a research and development arm of the National Security Council Interagency working group.

FIGURE 5-13 A bomb technician wearing explosives protective clothing removes a potentially explosive device from a vehicle.

FIGURE 5-14 An explosives truck, used to carry explosive materials such as ammonium nitrate. These trucks are commonly used at rock quarries and in other mining situations to deliver the explosive materials. *(Courtesy of Maryland Department of the Environment Emergency Response Division)*

Everything from distress flares and ammunition to hand grenades, pipe bombs, dynamite, and blasting caps is commonly brought to fire stations. Firefighters should not be handling these types of devices; the bomb squad should handle these types of situations. Depending on the type of device and the size, isolation and some evacuation may be necessary.

Citizens may also bring in old chemicals, some of which may be explosive and shock sensitive. With picric acid, just the simple removal of the cap or vibration of the container is enough to initiate the explosion of the container, with enough force to cause fatal injuries to the person who attempts to open or set down the container. Another common item that may be brought in is containers of old ether (ethyl ether), which is outdated after a year in storage. The opening of this container may also bring fatal consequences. The job of handling these containers usually falls to the bomb squad with a hazardous materials team interface.

SAFETY

It is a good general practice not to bring unknown materials into the fire station.

In some cases toxic materials may be brought in, which can contaminate the person who accepts the package and cause severe health problems for the firefighters in the station. Depending on state regulations, if fire department personnel accept a package from a citizen, the department may end up owning the package and may be required to pay for the proper disposal of the item—sometimes a costly good deed.

SAFETY

It is not uncommon for well-meaning citizens to bring explosives to the fire station. If the explosives are outside or in the citizen's vehicle, the firefighter should leave them there and call for the bomb squad.

Gases

Incidents involving gases include both flammable and nonflammable gases, with the most commonly released gases being flammable. Also in this category are poisonous gases. The poisons section provides more information on poisonous gases.

Luckily for many first responders, many departments carry gas detection devices that will warn of potentially explosive atmospheres involving flammable gases. Depending on the setup, the detector may warn of oxygen-deficient atmospheres, which may be caused by the release of a nonflammable gas. The two most commonly released flammable gases are natural gas and propane. Common propane bottles are shown in **Figure 5-15**. Although used for the same purposes, propane and natural gas do have differing characteristics that can affect a response to a gas leak. Both gases are odorless when they exist in their natural state, so when moved through a distribution system an odorant is added. In major interstate pipelines, however, the gas may be transported without an odorant added.

NOTE

The largest difference between natural gas and propane is the vapor density. Propane is heavier than air and will stay low to the ground trying to find an ignition source. Natural gas, on the other hand, will rise in air and should dissipate quickly.

This is weather dependent, and certain weather conditions can affect these characteristics. Under some conditions the gas may travel the length of a pipe, following the path of least resistance, usually coming up inside a building. The leak may not be detected for a few days or weeks until it surfaces. This is most common in areas of the country where the ground freezes and limits the upward movement of the gas. Many states offer training on natural gas and propane emergencies; this training is often provided by the gas companies themselves.

Incidents involving these two gases are commonplace, and for natural gas the most common incident involves a ruptured gas main. The gas system is set up on a grid, with both large distribution pipes and smaller delivery pipes. The pipes leading into a home are typically ½ to 1 inch (1.27 to 2.5 cm) diameter. Distribution pipes may be 2 to 36 inches (5 to 91 cm) depending on the region. First responder actions at these types of incidents generally involve isolation and protection, and then the team stands by until the line can be shut off. Gas detection devices should be employed to determine the true hazard area, and adjacent buildings should be checked for gas. When checking buildings, sampling should be continuous, and the highest point should be checked prior to declaring a building "safe."

Another significant issue with the release of flammable gases, which also applies to vapors coming from flammable liquids, is the presence of ignition sources. Pilot lights, electrical switches, electrical contacts, and some radio and cell phone transmissions are also possible sources of ignition. Static electricity is a concern when there is movement of persons or the chemical. Attempts should be made to eliminate these sources of ignition. The electricity to a building can be cut off at the pole, which is a safer alternative than shutting off the electricity inside the hazard building. At vehicle accidents, hot engines, vehicle batteries, and static electricity are potential ignition sources.

FIGURE 5-15 These 20-pound cylinders, found in most homes, can create large fireballs and can explode with considerable force if involved in fire.

CAUTION

At no point should first responders jump into a hole to attempt to shut off a leak, nor should underground valves be shut in an attempt to stop the leak.

SAFETY

If the tank becomes discolored, distended, or loses shape, or the sound from the relief valve increases, an immediate withdrawal is indicated.

Gas line valves may actually be keeping the amount of gas reduced, and moving the valve may increase the amount of gas escaping. If the leak is on the out-take site of the meter, it is acceptable for the valve on the meter to be shut off. The meter should be locked out and tagged out, and only the gas company should turn the flow of gas back on. Some departments carry special tags that mark the system as being out of service, and some actually lock the system in the off position. When the gas company arrives, they repair the leak and place the system back in service. When dealing with pipes and electrical systems, an OSHA regulation prescribes the procedure for shutting off a system and marking it so that the system is not accidentally turned back on prematurely. This regulation is known as the "lock out, tag out" regulation and applies to many fire service situations.

Propane releases generally involve cylinders, but leaks can also happen in a pipeline. A number of the pipelines running across many states carry propane as well as natural gas. Other than its flammability, the fact that propane sinks and stays low to the ground creates an additional hazard. If a propane cylinder is releasing liquid propane, the liquid will be very cold, sometimes as low as −90°F (−67°C). A common incident involving propane cylinders, especially those 20-pound (9 kg) cylinders designed for home barbecue grills, involves the overfilling of the cylinder. Propane cylinders are only supposed to be filled to 80 percent of the capacity of the cylinder to allow for expansion of the gas. If the cylinder is filled more than the 80 percent and the temperature increases, the gas may escape the relief valve or frangible disk.

For fires involving propane tanks, there is the potential for a BLEVE, an event that can be catastrophic to the responders. A BLEVE can occur with any flammable gas storage tank, but is usually associated with propane tanks. When pressurized tanks explode, they can travel for a considerable distance. Although no one can predict exactly when a tank will fail, some indicators that a BLEVE may be forthcoming are increased flame height, or the appearance that the flames coming from the relief valve are under high pressure. The sound of a relief valve is deafening, but when the pressure increases the pitch of the sound will get higher and may become louder.

When making the decision to fight a propane tank fire, a large quantity of water needs to be applied quickly and continuously. The vapor space of the tank should be concentrated on, and the fire should not be extinguished unless the responder is certain the flow of gas can be stopped. When preplanning, first responders should plan to establish a flow of water on the tank in excess of 500 gpm (1,893 lpm) within a few minutes of arrival. Each year a number of firefighters are killed by BLEVEs; in fact, in many cases whole alarm assignments have been killed and seriously injured during the BLEVE. More information in BLEVEs is found in Chapter 1 and Chapter 2.

Some cars now are powered by natural gas or propane; the most common are cars that are part of a fleet of vehicles, including government and utility company vehicles. To determine if a vehicle is powered by one of these gases, the firefighter should look for a sticker that reads CNG (compressed natural gas), LP (liquefied petroleum), propane powered, or natural gas powered. Sample stickers are shown in **Figure 5-16** and **Figure 5-17.** Other vehicles that may have compressed gas cylinders are recreational vehicles (motor homes) and work utility vehicles.

Although currently uncommon, there are cars powered by hydrogen. Hydrogen is also transported by tank truck, rail, and IMO (International Maritime

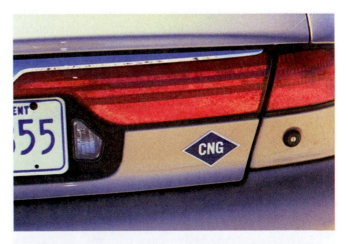

FIGURE 5-16 Vehicles that use alternative fuels such as natural gas, propane, or electric are marked to indicate the fuel source. Here a CNG sticker on the rear bumper, indicates that the car is fueled by compressed natural gas.

FIGURE 5-19 Note the severe damage to this high pressure hydrogen tube trailer. It was involved in a traffic accident and caught fire. *(Courtesy of Maryland Department of the Environment Emergency Response Division)*

FIGURE 5-17 This sticker is beside the front driver's quarter panel and indicates this vehicle is environmentally friendly alternatively fueled. Cars with alternative fuels typically still have a gasoline tank, and there may be gasoline present in emergencies.

Organization) style containers. This gas is extremely flammable and carries an unusual risk: it burns with an invisible flame and produces no smoke. As **Figure 5-18** and **Figure 5-19** show, this presents a major concern. Occasionally, the heat waves can be seen above a hydrogen flame, but in many cases a responder could not identify a hydrogen fire until they were too close.

When propane tank trucks overturn, as shown in **Figure 5-20,** the truck usually will not leak, but the situation is still quite dangerous. It is impossible to determine the exact damage to the tank, and just how much increased pressure might cause a violent release of the tank contents is another unknown. If the ambi-

ent temperature is increasing, so is the pressure in the tank. It is possible that the relief valve can become damaged during a rollover and may not function as designed. It would be dangerous to right any liquefied gas tank without the product being transferred or, in the case of propane, the contents flared, as shown in **Figure 5-21.**

Other common gas releases involve carbon dioxide, chlorine, and ammonia. Although carbon dioxide (CO_2) is a common fire extinguishing agent, it is also used for the distribution of beverages and is commonly found in restaurants, bars, and convenience stores. An increase of CO_2 in a building can make people sick, and it is becoming a common **sick building chemical.** Although not commonly looked for, when dealing with

FIGURE 5-20 This propane tank truck had a slight rollover in a toll booth. The tank was damaged and the propane was flared off. The situation is still dangerous because the extent of the damage to the tank is unknown, and righting the tank could create stresses that could cause a catastrophic release. *(Courtesy of Maryland Department of the Environment Emergency Response Division)*

FIGURE 5-18 Master streams are used to cool the hydrogen tanks on a tube trailer. *(Courtesy of Maryland Department of the Environment Emergency Response Division)*

FIGURE 5-21 A propane tank truck was involved in a collision and rolled over. Shown are two flares that are burning off the propane, which must be done prior to the tank being righted. Many hazardous materials response teams carry flaring equipment able to flare tanks from small, grill-sized propane tanks up to rail cars. *(Courtesy of Maryland Department of the Environment Emergency Response Division)*

emergency response to a **sick building,** CO_2 should be considered as a possible source of the problem. Both chlorine and ammonia have good warning properties. Their distinct odors are often easily identified by the people in the building. In most cases the odor is so irritating that people will self-evacuate the building prior to hazardous levels being built up.

SAFETY

Gasoline is the leading chemical when it comes to chemical accident fatalities, and transportation leads in those fatalities. Unfortunately, due to its familiarity, many emergency responders do not adequately protect themselves when responding to incidents of this type.

Flammable and Combustible Liquids

By far this is the leading category for the most common type of releases, because it includes gasoline and diesel fuel, **Figure 5-22.** Firefighters respond to these types of incidents thousands of times a day.

Gasoline has between 1 and 5 percent benzene, which is a confirmed human cancer-causing agent (carcinogen) to which firefighters are commonly exposed. A person who smells gasoline is receiving an exposure to benzene, an exposure that is repeated on a regular basis. When dealing with small spills firefighters may have a tendency to use little or no protective clothing, but their bodies are continually being exposed to this toxic material. In addition to its tox-

FIGURE 5-22 Responding to this emergency presented several challenges, due mainly to the final resting place of the overturned truck. Although hazardous materials response prefers an uphill position, the best place for the offload truck was downhill. Extra distance and additional protection lines were necessary to compensate for the disadvantaged position. *(Courtesy of Maryland Department of the Environment Emergency Response Division)*

icity, gasoline is also very flammable. It can even be ignited by the static electricity generated by clothing as a responder approaches an incident.

Although of less concern than gasoline, diesel fuel, fuel oil, and kerosene do have some toxicity concerns. This lower level of concern can catch a responder unaware. Although in normal circumstances it is difficult to ignite these materials, if their temperature increases they easily ignite. On a day with a temperature below 70°F (21°C) the potential for ignition is low, but a spill onto blacktop when it is 100°F (38°C) outside adds considerable risk for a fire because a larger quantity of vapors is going to be produced on the pavement, which may have temperatures in excess of 110°F (43°C).

One common incident involves overturned or burning tanker trucks. The product found most often in these trucks is gasoline, and when one of these trucks is involved in an accident, a fire usually results.

STREETSMART TIP

To Fight or Not to Fight? When first responding companies arrive they have a tendency to try to extinguish the fire. Without large quantities of firefighting foam it is very difficult to extinguish a tanker fire. If, on arrival, the truck is well involved and is not near any exposures and is not impacting an adjacent community, it is usually best to let the fire continue to burn.

When attempting to extinguish this type of fire there is considerable runoff, which usually contains both the gasoline and firefighting foam, both of which are best kept out of waterways. Also, if the fire is extinguished, a large amount of gasoline at an elevated temperature may remain so foam must be reapplied every few minutes to ensure that the fire does not reignite, **Figure 5-23.** It is then necessary to remove the hot gasoline from the truck and pump it into another truck, a dangerous proposition. If, on the other hand, the fire is allowed to burn, nothing would remain that could cause harm to responders or the environment. The resulting black smoke, although it looks horrible, is in reality less damaging to the environment than a liquid spill. The smoke is predominantly carbon, which provides the thick black smoke. However, if the truck is impacting a community or adjacent structure, then all attempts should be made to extinguish the fire. The truck may be wrecked under a bridge, which is usually considered a critical structure in a community. The cost of a bridge can be in the millions, so responders should try to calculate the financial impact in terms of hard dollars and inconvenience if the bridge were destroyed and it took a year to rebuild. If units arrive and are confronted with a large fire, the initial water should be directed toward the bridge. When sufficient foam and water have arrived, the truck fire should be extinguished, while continually applying foam.

The application of foam at a non-fire incident usually causes friction between firefighters and environmental agencies because the foam can be damaging to the environment. Even if an "environmentally safe" foam is used, the breakdown products of the fuel are not environmentally safe, so environmental agencies prefer the use of minimal foam.

Another concern arises when a truck is overturned and it must be drilled and pumped out prior to righting the truck. The application of foam makes the situation very slippery, hence creating additional hazards. The use of air monitors to determine when the foam blanket is breaking down is recommended, and the use of as little foam as possible is recommended.

In years past, spills of flammable or combustible liquids were commonly flushed down storm drains. This practice for the most part has been discontinued because it is severely environmentally damaging, not to mention that it only moves the problem to another location. Many areas of the country collect the spilled material and dispose of it in an environmen-

FIGURE 5-23 A diesel tank truck cab caught fire, impinging on the cargo tank. A quick and aggressive response by the Washington, DC, and Prince Georges County Fire Departments was able to knock down the fire before the contents were ignited. If the tank had become compromised, firefighting would have been very challenging because the tank truck was on a significant incline. Burning fuel would have traveled down the highway, possibly into storm drains. *(Courtesy of Maryland Department of the Environment Emergency Response Division)*

tally sound manner. The discharging of oil (includes fuels) is a violation of the federal Clean Water Act, and emergency responders could face serious fines if the material were flushed into a waterway.[3]

Flammable Solids, Water Reactives, and Spontaneously Combustible Materials

When dealing with materials in these categories, a specific identity and emergency response information are crucial. Consultation with the hazardous materials team is important, because using the wrong tactic can be devastating to the community. Most emergency responders have some experience with flammable solids, because road flares are classified as flammable solids. In most cases flammable solids are difficult to ignite, but once ignited they burn vigorously and are difficult to extinguish.

The water-reactive group can be defined in two ways. When a water-reactive material gets wet, a violent reaction, such as a fire or explosion, can result, as shown in **Figure 5-24, Figure 5-25** and **Figure 5-26.**

This is what most people think about when they hear the term water reactive. The water-reactive category also includes reactive metals that, once ignited, can create a problem, such as a violent reaction, if water is applied. Magnesium is one such reactive metal. Once ignited, magnesium burns vigorously and when water is applied it may explode, as shown in **Figure 5-27.**

FIGURE 5-24 A sea box container of magnesium was involved in an accident, and the trailer was ignited by a welder trying to make some repairs to the trailer. *(Courtesy of Cambria County, Pennsylvania, Emergency Services)*

FIGURE 5-25 The magnesium caught fire and resulted in an aggressive, very hot fire. The smoke is toxic and corrosive. Adding water to this fire would have been very dangerous. *(Courtesy of Cambria County, Pennsylvania, Emergency Services)*

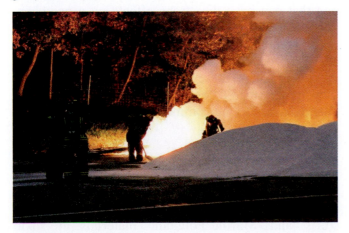

FIGURE 5-26 Sand, which should be very dry and stored indoors, can be used to attempt to extinguish smaller fires of flammable metals. *(Courtesy of Cambria County, Pennsylvania, Emergency Services)*

FIGURE 5-27 In this photo eight ounces of magnesium shavings were in a pool of burning diesel fuel. When the magnesium was heated, a slight water mist was sprayed over the fire. The white sparks are from the magnesium and the fireball is from the reaction as well. Relate the size of this violent reaction from a cup of magnesium to that of a truckload of magnesium.

Some materials such as calcium carbide are shipped as water-reactive materials. When water is applied to calcium carbide, a little bubbling occurs, which would not be considered very violent. The gas that is being released by the bubbles, however, is acetylene gas. When pure acetylene gas is produced in this manner it is unstable and reactive. If there is an ignition source nearby there will be a fire or explosion. So the actual application of water in itself does not produce a significant problem, but the action of the water may create additional concerns.

SAFETY

The use of water on flammable metals is not recommended and can cause severe injuries to the hose crew, because the metal has a tendency to explode, sending hot metal and other fragments flying.

Materials that are spontaneously combustible are usually transported in a manner designed to keep them stable. To remain stable, they may require storage at a specific temperature or have a material added to them. White phosphorus, which is usually covered with water, is an example of a spontaneously combustible material, because it ignites in the presence of air. Communities with facilities that use flammable metals should have a stockpile of metal extinguishing agents such as Metal-X or Lith-X, commonly referred to as Class D extinguishing powder. Sand that is known to be dry (e.g., it has been stored inside in a closed container) can also be used, but most sand has moisture contained within it.

The water will break down and release hydrogen gas, which will increase the amount of heat and flame being produced. Most hazardous materials teams have a limited supply of this type of extinguishing agent and usually rely on a local industrial resource for more.

Oxidizers and Organic Peroxides

Like the previous group, this is another class of materials for which help should be requested early. Oxidizers and organic peroxides can have explosive characteristics, and attempting to deal with them can lead to fatal mistakes. Organic peroxides can react explosively even if the responders have taken no action.

One of the best known oxidizers is ammonium nitrate. By itself it is relatively harmless, although it does have some toxicity associated with it. It cannot explode unless mixed with a fuel and an initiating charge added. Although all three items should not be transported together, on occasion ammonium nitrate, fuel, and an initiating charge might be on the same truck. In other situations the oxidizer does not require any oxygen to start a fire, because it will provide its own. Oxygen is placarded with a specific oxygen placard, or it can be shipped with an oxidizer placard. Oxygen itself cannot be ignited, but it will greatly intensify a fire when it is involved in amounts greater than the 20.9 percent that is found in air.

Liquefied oxygen (LOX), which is a cryogenic, presents even more hazards in addition to supporting combustion. On asphalt (or other hydrocarbon material) it is shock sensitive and can detonate if compressed, such as by a firefighter walking on it. Another concern is the absorption of oxygen from a LOX release by a firefighter's turnout gear, which can present a flammability problem. Although caution should be used, and the exposed gear should be given time to air out, this would be a very rare occurrence. Most hospitals have large upright cryogenic storage tanks of liquid oxygen, as do many nursing homes, **Figure 5-28.** Smaller in-home versions are available as well, presenting a large fire risk in a residential home, not to mention a freezing and contact hazard if knocked over.

Another commonplace situation involving oxidizers is encountered when dealing with pool chemicals.

FIGURE 5-28 A leak of liquid oxygen on asphalt can present a shock-sensitivity problem in addition to the increased risk of a fire.

Most pool chemicals are oxidizers or may react with oxidizers and can be involved in a chemical reaction or fire. If the materials do not ignite, a very dangerous situation can develop, because a large vapor cloud can be created from just a handful of pool chemicals—one that can have disastrous effects on a neighborhood. Specific advice from the hazardous materials team is required prior to handling or disposing of any of these types of chemicals. There is a lessened level of concern with these types of chemicals because they are considered "household," but they do present a large hazard to the community.

Poisons

Although Class 6 is poison liquids, included here are poisonous gases from Class 2 (gases). These types of materials, which include the "Stow Away from Foodstuffs" and "Marine Pollutant" materials, are toxic to both humans and the environment. Although they are toxic in varying degrees, for the first responder they are poisonous and should be treated as such. The materials labeled as poisonous are very much so, because the DOT does not include many of the materials that are toxic in this category. When choosing a hazard class for a material, the predominant hazard determines its placement into one or another category. Some improvements have been made in the subsidiary placarding and labeling category, but there is room for more. Poisonous gases may be accompanied by a label on a bulk container that reads "Poisonous by Inhalation" and would be listed on the shipped papers as PIH. By definition, these materials present a risk to humans and the community and require extra precautions.

The most common incidents with these types of materials result from pesticides and agricultural chemicals. Although most household materials are of lesser concentrations, on occasion higher strength materials have been used. There have been several incidents in which bug-spraying companies (or individuals) have used full-strength agricultural products in residential homes, requiring evacuations lasting a few weeks and removal of some homes due to the contamination.

SAFETY

Technical-grade pesticides are very dangerous and should be handled with care.

Incidents involving commercial home fertilizing trucks are common, but in most cases this material is diluted and ready for application. Fertilizers do not present much risk to the responders or the environment unless found in large quantities. Pesticides and insecticides do, however, present a risk to the responders and the environment, and responders should consult with a hazardous materials team prior to taking any action. Many home pest control vehicles carry a mixed load of premixed and undiluted materials. If such a vehicle is involved in an auto accident, the use of full gear is recommended, and anyone who comes in contact with any of the liquid or solid materials should be decontaminated.

Radioactive Materials

Although considerable emphasis was placed on radiation in the early days of hazardous materials, due to a low number of incidents involving these materials, some of the emphasis has been redirected to other areas. It is true that incidents involving these types of materials are rare, but they do have a significant impact on the well-being of the community, and with the potential now for terrorism, radiation is being emphasized again.

Radioactive materials are commonly used in the community, in smoke detectors, in ground imaging equipment, and in the medical community. Although many people express concern when around radioactive materials, most radioactive materials are not harmful even for a lengthy exposure.

Radiation is divided into two categories: ionizing and nonionizing radiation. Radioactive materials that emit alpha, beta, gamma, and neutron forms of radiation are ionizing types. The ionizing forms of radiation are able to make changes in atoms, which can lead to health problems with humans. Nonionizing radiation takes the form of sunlight (visible light), microwaves, and radio waves. These forms of energy are not as strong as their ionizing counterparts, but high amounts or repeated exposure (e.g., to sunlight) can cause health problems.

Some radioactive materials, such as those coming to and from a nuclear power–generating facility, can present some risk to the community. Radioactive materials that present a risk to the community are shipped in high-strength containers designed to withstand a substantial crash and to be involved in a fire situation without any release of radioactive materials. These shipments are usually well tracked and, when involved in an incident, specialized help is usually already on the way before local responders call for it. Knowing who is the local contact for radiation emergencies is key to an effective response when dealing with radioactive materials.

SAFETY

Keeping the material in the container that it is intended to be in is paramount to reducing the exposure levels of the responders.

When dealing with potential radiation hazards, the adage "time, distance, and shielding" should always be followed. This basic principle of radiation exposure should be applied to all chemical exposures. Firefighters must be careful to limit their time around a radiation source, keep their distance, and wear some type of shielding; then they can be protected against most forms of radiation. Most radiation exposure guidelines are based on a one-hour exposure time. By reducing their time, firefighters reduce their exposure. By doubling their distance, they can reduce their radiation exposure to one-fourth of the original exposure. Protective clothing and SCBA can provide some shielding for some forms of radiation, while more substantial forms of shielding such as lead or concrete may be needed for more dangerous types of radiation. Persons who have been exposed to a radiation source are not radioactive. They may suffer some significant health problems, but they do not present a risk to others. Persons who have radioactive material on them, such as a radioactive powder, are contaminated, may present a risk to others, and are considered to have an external contamination. Alpha and beta radiation sources are hazards and can contaminate persons through inhalation and ingestion, which would be considered internal contamination. Neutron and gamma radiation don't create contamination issues with regard to persons. Neutron radiation sources do have the ability to irradiate some materials, but the chances of encountering a neutron source are small, except in war or a terrorist attack. Simple decontamination with soap and water is needed to reduce the damage to the person. When dealing with radioactive materials it is best to try to contain the runoff so as not to further contaminate the incident scene. The best protection against radiation is to keep the radiation source in its protective container.

Corrosives

After the flammable and combustible categories, the next most likely hazardous materials incident will probably involve a corrosive. The most common incidents occur with sulfuric acid, hydrochloric (muriatic) acid, and sodium hydroxide. Both of the acids are transported as liquids, but the sodium hydroxide may be transported wet or dry. Both sulfuric acid and sodium hydroxide have little or no vapor pressure, so they present little risk outside of the immediate spill area. Hydrochloric acid, on the other hand, does have a high vapor pressure and may require some additional isolation and evacuation. Nitric acid is less common but is easy to detect as it has a characterisitic vapor cloud, as shown in **Figure 5-29.** When released, it forms a brown vapor cloud comparable to the release of bromine.

To handle any of these materials, chemical protective clothing is required, because after some time they can eat a hole in turnout gear. A responder who is splashed with any of these three materials should wash the material off as soon as possible. Note that the actual burns will take a couple of minutes to become evident, and will progress in severity if not quickly washed off. This applies to these three materials only, because some of the less common acids, such as oleum or sulfur trioxide, will cause burns immediately upon contact, but it is still a matter of time before disfiguring injuries occur.

Chemical neutralization may be the best choice for handling a corrosive spill. This type of operation needs to be performed by a hazardous materials technician who has donned appropriate levels of PPE. When neutralizing chemicals, heat and violent reactions may occur, so neutralization should take place only after consultation with a chemist.

Other Incidents

It is impossible to outline each specific action that first responders should take at a chemical release. The first rule if a responder suspects that hazardous ma-

FIGURE 5-29   The shipping papers did not indicate the presence of nitric acid. The brown vapor cloud is a result of a chemical reaction between bromine and red fuming nitric acid. When the hazardous materials team opened the back of the truck, they were greeted with these vapors. The team members in the photograph retreated when the vapors were released from the back of the truck and changed into chemical protective clothing. *(Courtesy of Maryland Department of the Environment Emergency Response Division)*

terials are involved is to isolate the area and evacuate other persons in the immediate area until further information is available. Request the assistance of the closest hazardous materials team because their technical expertise and equipment will be needed. First responders are not trained nor equipped to provide detection or characterization of unknown materials, and to use their senses, such as smell, is to play Russian roulette.

Many toxic materials are odorless and colorless and thus will provide little warning prior to causing lethal effects. It is not the job of first responders to seek the cause of a chemical release nor to provide an identification. If the information is presented to them or can be obtained without risk, then it is of benefit, but otherwise it is the job of a hazardous materials team to perform those tasks.

SAFETY

The absence of a vapor cloud or odor does not mean that a potentially deadly material is not present.

Common incidents that first responders may get involved with are sick buildings, or odor complaints that can occur in buildings such as the one shown in **Figure 5-30.** The use of air monitors is the key to survival with these types of incidents, but the sense of smell is not sufficient to determine immediately dangerous levels. When involved in sick buildings or odor complaints, the best course of action for first responders is to determine the validity of the complaint. Once it is determined that the call is valid and not related to a sunny Friday afternoon, the first responders should don SCBA and request the services of a hazardous materials team. The occupants of the building should be removed and the windows and doors shut. The HVAC system should be shut down so as not to remove the unknown material. These steps are crucial to the success of the hazardous materials team's attempts to find the source of the problem.

FIGURE 5-30 It is difficult to determine the origin of unusual odors in a high rise building. There are many occupants, offices, and potential sources.

First responders may also get called to gas leaks inside a building, and their ultimate safety is determined by a suitable air monitor. It is impossible to determine the level of a gas in a building based on smell. Once in an environment that has a gas odor, the body will desensitize itself to the odor and responders will not be able to smell the gas anymore, even though the amount has been increasing, moving closer to a potential explosion.

Gas grills are prone to propane leaks, and if the leak reaches an ignition source can cause severe damage to buildings. Many hazardous materials teams carry pipe flares that burn off the gas, preventing an explosion. First responder actions should be limited to isolation and evacuation and to preparing standby hoselines. If the vapors are traveling toward other homes, then the use of hoselines with fog nozzles to move the vapors is appropriate.

DECONTAMINATION

The task of decontamination may fall to a person trained at the first responder Operations level. If a first responder is expected to perform decontamination, then specific training in that area is required, as well as training in the use of chemical PPE. The definition of **decontamination,** or decon for short, is the physical removal of contaminants from people, equipment, and the environment. It is important to note that the concepts provided here may be for one or all three of these areas of concern. Just because one type of decontaminating solution is effective for a tool does not mean that it can be used on humans. Prior to performing any decontamination procedure on humans, first responders should consult with the hazardous materials team or a chemist. Some decon solutions use very dangerous materials, especially if they come in contact with skin. For equipment it may be acceptable to use large quantities of sodium hydroxide mixed with detergent, but this same solution would cause severe burns if placed on the skin. It is commonly recommended that no more than soap and water be used on victims.

The two ways to become contaminated are via direct contact or secondary contamination. If first responders go into a spill area and put their hands into a drum of blue paint, direct contamination has occurred. When they leave the area and shake hands with another responder, they have secondarily contaminated the other responder with the blue paint.

Types of Decontamination

With the revision to the NFPA 472 standard, there are three general types of decontamination levels: **emergency decontamination, technical de-**

contamination, and **mass decontamination.** One other type, **fine decontamination,** is added here because it is an important type of decontamination for victims. In a large, full-scale incident, all four types may be used, but in reality the majority of incidents require only minimal decon. There are two large categories of decontamination, wet and dry decontamination. The two categories should be self-explanatory: One requires the use of water or other liquid solution, whereas the other does not involve a wet process.

The process of decontamination is chemical specific so there are no absolute rules. First responders should as a minimum have a good understanding of the emergency decon procedures and develop a local procedure to handle this type of problem. When confronted with a contaminated patient who is in need of decon, there will be no time to develop the procedure. With potential terrorism in mind, responders should think about the emergency decon of not one person, but thousands of people.

FIGURE 5-31   One of the simplest forms of emergency decon is the use of a hoseline.

Emergency Decontamination

When a victim or a responder is contaminated with a hazardous material, it is vitally important to remove the contamination as rapidly as possible. To delay the removal of the hazard could result in fatal consequences. No matter what type of emergency activity is occurring, responders and the IC should have some thought of emergency decontamination. Even responding to a medical call can result in the need for emergency decon; having a plan is essential for survival. Emergency decontamination can be as simple as taking a hoseline and washing someone off, or it can be complicated by an unconscious patient on a backboard or in a Stokes basket. For most chemicals, plain water washing is sufficient to decontaminate the person, as long as the person's clothes are removed. Removal of the patient's clothes removes 60 to 90 percent of the contamination. As toxic as nerve agents are, if the patient has been contaminated with an aerosol, then just having the person stay in fresh air for fifteen minutes can be an effective decon method. Between fresh air and clothing removal, effective emergency decon can be accomplished.

The common methods of emergency decon involve the use of a water line and some method of runoff control, such as that shown in **Figure 5-31.** Once through emergency decontamination, the victim should receive technical or fine decontamination, since emergency decontamination is not likely to remove all of the contaminant.

NOTE

In reality when people's lives are on the line, runoff is not a large concern, but when not under those types of conditions, attempts should be made to recover any runoff. When dealing with radioactive decontamination, efforts should be taken to contain runoff.

There are a number of manufacturers who can provide decontamination containment. In the absence of formal measures, there are some practical alternatives. An inflatable children's pool, which can be easily carried on a fire truck, can be used to contain the runoff from decontamination. Another method involves the use of a tarp or plastic that is spread between two ladders. If none of these methods is possible, then using a tarp or plastic to catch as much runoff as possible is advised. If all other methods are not available, then it is best to take advantage of natural areas that would collect the runoff.

Decon should not be performed over storm drains or other environmental concerns. An industrial plant may have safety showers, and it is acceptable to use these safety showers. Prior to using the showers it is important to ask the people from the facility where the runoff goes. In most cases the runoff will go to the facility's own water treatment area and can be taken care of by that system. Caution should be used when the safety shower drains straight to the sanitary system; prior to using such a shower, the local sewer authority should be contacted.

One of the other general rules with regard to emergency decon involves the removal of corrosives from a victim. Large quantities of water should be used, because if small amounts of water are mixed with the corrosive there is the potential for a heat increase.

Massive amounts of water will eliminate this potential for further injury. Besides using large quantities of water on corrosives, it is necessary to continue the flushing for a minimum of twenty minutes. No matter how bad the temptation is for EMS providers to remove the patient, the best treatment for corrosive burns is the water flush, especially for burns caused by a base, such as sodium hydroxide.

Technical Decontamination

The first step in technical decontamination is the removal of the majority of the contaminants, and this is usually addressed by the first rinsing station within a full decon setup, as shown in **Figure 5-32.** This step is known as *gross decon,* and in many cases gross decon is just part of the whole process, but it can be the only step in a minor setup. Some hazardous materials teams use shower setups for gross decon, whereas others may use hoselines or pump tanks. Because this step is the most likely location for contaminants to collect, attempts should be made to recover any of the

runoff. When part of a multistep process, the entry team usually performs this step on themselves without any assistance.

When dry decon is used, overgarments such as booties or extra gloves may be used, and these are placed in a designated container when leaving the hot zone. For many materials, dry decon is an acceptable use of PPE and resources.

When using any type of decon, the bottom line is to make it safe for the responder to get out of the PPE and safe for the responders assisting with the PPE removal. When using reusable PPE, decon needs to be more formal, but even extensive cleaning can be done to the PPE after the incident and removal of the responder.

Further in the technical decon process is a step in which some actual scrubbing and cleaning take place to remove any residual contaminants, **Figure 5-33.** It is usually done after the gross decon and usually has a couple of steps. Attempts to recover any runoff should be used, but this is not as critical as for the gross decon step.

FIGURE 5-32 Gross decontamination is used to remove the majority of the contamination.

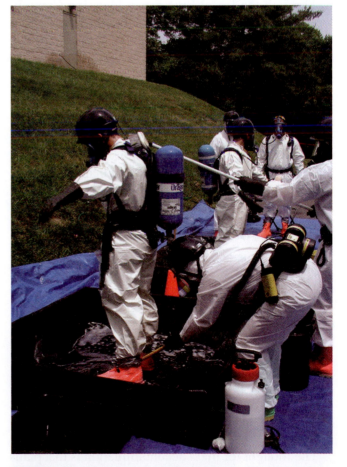

FIGURE 5-33 Formal decontamination is used to remove any further contamination that may remain after gross decontamination.

It is during the scrubbing step that decon solutions are typically added to the process. These solutions are chemical specific. This process may involve showers, hoselines, or pump sprayers, and one or two people dressed in a lower level of PPE may perform this activity. When performing decontamination, it is important to pay particular attention to the areas most likely to be contaminated, the hands and feet.

Fine Decontamination

This form of decon is not performed by first responders, but by hospital-based personnel. It is a cleaning that removes all of the contaminants from the body. It would be nearly impossible to clean eyes, ears, fingernails, and other areas of the body in the street; these tasks are better accomplished in a hospital. According to OSHA the staff must have proper training to perform fine decon, and they must be trained to the Operations level, although some training to the Technician level is preferred. Some hospitals have designated decon areas with a separate entrance, separate ventilation system, and a holding tank for runoff control.

Through the Local Emergency Planning Committee (LEPC) plan, hospitals are required to identify themselves as being able to accomplish decontamination of patients. Every hospital should have a policy to cover a walk-in contaminated patient, and emergency responders should test these plans on a regular basis. Regular training and interfacing are required to keep this type of system functioning well within a community. Responders should ensure that this capability is well run and that sufficient training has taken place, because they are likely candidates to become patients someday!

Mass Decontamination

Most hazardous materials decontamination setups are designed primarily to decontaminate the members of the hazardous materials team. They are designed to handle two to four people dressed in chemical protective clothing. Although most of these systems can handle civilians, they are designed to decontaminate responders' PPE. When decontaminating civilians, as shown in **Figure 5-34,** several issues arise: clothing, privacy, valuables, weather impact, and inconvenience. Civilians may not be willing to undergo decontamination or to separate from their relatives. They may not want to remove their clothes or valuables.

The thought for mass decontamination is that plans must be developed to handle decon of thousands of people. Communication will be difficult and bullhorns or loudspeakers should be employed to ad-

FIGURE 5-34 Mass decon should have the victims in water for as long as possible. If time permits, soap and water should be used to better the decontamination process.

dress large crowds. Keeping contaminated victims informed is one method to keep them calm. Some mass decon setups are provided in **Figure 5-35** and **Figure 5-36.** Most of the mass decontamination issues arise when terrorism events are discussed. It is best to have the victims decontaminated before they get to the hospital, because contaminated victims will create additional problems at the hospital. If victims are contaminated with military nerve or blister agents, speed is of the essence. If decontamination is not performed on symptomatic contaminated victims, their chance for survival may be slim. The hardest part of this process is identifying those who need physical decontamination and those who need **psychological decontamination.** If victims are symptomatic and there is a credible threat presented, then decontamination should be accomplished quickly. First arriving units need to establish this decontamination process. The best decontamination solution is water, followed by soap.

Studies from the military have shown that the use of bleach can actually cause more deaths of and injuries to contaminated victims. Decontamination is chemical specific, and unless first responders can identify the specific chemical compound they should use a simple soap and water solution. A hoseline can be used to accomplish emergency victim decontamination; as more assistance arrives, the system can become more elaborate, as one example in **Figure 5-37** demonstrates. When setting up the decontamination process, it is important to avoid victim backup and if there are delays, to make sure that the victims wait in

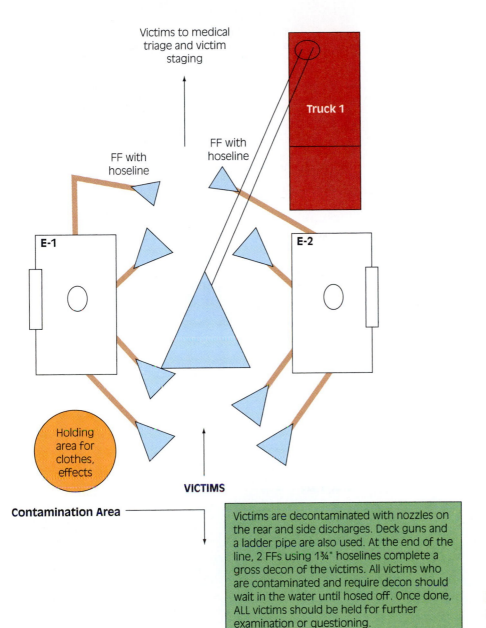

BASIC PLAN
Hazardous Materials Response Team
Mass Casualty Decontamination

Victims to medical triage and victim staging

Truck 1

FF with hoseline

FF with hoseline

FF with hoseline

E-1

E-2

Holding area for clothes, effects

VICTIMS

Contamination Area

Victims are decontaminated with nozzles on the rear and side discharges. Deck guns and a ladder pipe are also used. At the end of the line, 2 FFs using 1¾" hoselines complete a gross decon of the victims. All victims who are contaminated and require decon should wait in the water until hosed off. Once done, ALL victims should be held for further examination or questioning.

FIGURE 5-35A Mass decon set-ups. *(Courtesy of Baltimore County Fire Department)*

water. After the victims exit the decontamination area they should be guided to holding areas. It is important to keep symptomatic victims from nonsymptomatic victims. If these victims are mixed, the nonsymptomatic victims will tend to pick up the symptoms due to psychological reasons.

One method that is built into the engine company is shown in **Figure 5-38.** This engine company even carries robes and slippers to be distributed after decontamination. Another method of mass decontami-

nation is the use of mobile showers, shown in **Figure 5-39** and **Figure 5-40,** which can enable a number of victims to decontaminate themselves.

SAFETY

The use of a bleach and water solution in terrorism decontamination is no longer considered an acceptable practice.

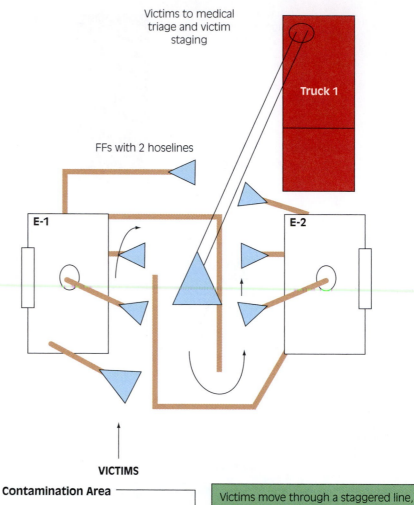

ADVANCED PLAN
Hazardous Materials Response Team
Mass Casualty Decontamination

Victims to medical triage and victim staging

Truck 1

FFs with 2 hoselines

E-1

E-2

VICTIMS

Contamination Area

Victims move through a staggered line, like that at an amusement park or bank. Gross decon is completed at the end of the line with two 1¾" hoselines. While waiting in line, nozzles from the rear, side, and deck pipe are washing the victims as well. Rope or barrier tape can be used for the line.

FIGURE 5-35B Once additional resources arrive then the advanced plan for mass decon can be implemented.

Decontamination Process

There are several variations to a decontamination process, but in the overall scheme of things they are basically similar.

Basic Decontamination Steps

1. Tool drop, **JPR 5-1A**

2. Gross decon, **JPR 5-1B**

3. Scrubbing and rinse, **JPR 5-1C**

4. PPE removal, **JPR 5-1D**

5. SCBA removal, **JPR 5-1E**

6. Clothing removal

7. Body wash and dry off, **JPR 5-1F**

8. Medical evaluation, including rehydration, **JPR 5-1G and JPR 5-1H**

One should consider a final step and that would be recordkeeping. While this may sound out of the ordinary, all exposure information (including the suit worn and any other pertinent information) should be logged for future use should it be needed. The varia-

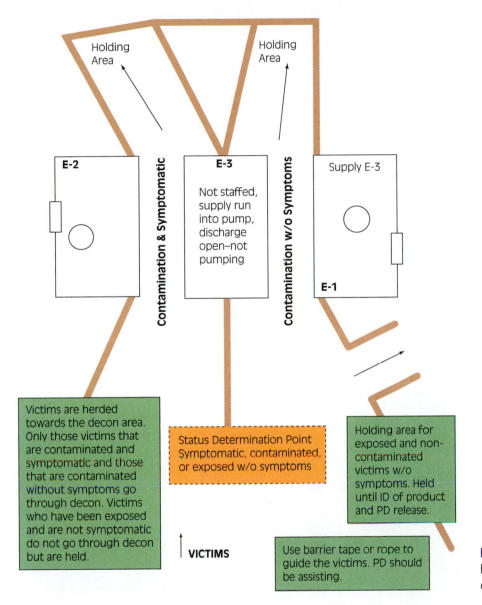

Herding Guide
Baltimore County Fire Department
Hazardous Materials Response Team

Holding Area

Holding Area

E-2

E-3
Not staffed, supply run into pump, discharge open–not pumping

Supply E-3

E-1

Contamination & Symptomatic

Contamination w/o Symptoms

Victims are herded towards the decon area. Only those victims that are contaminated and symptomatic and those that are contaminated without symptoms go through decon. Victims who have been exposed and are not symptomatic do not go through decon but are held.

Status Determination Point Symptomatic, contaminated, or exposed w/o symptoms

Holding area for exposed and non-contaminated victims w/o symptoms. Held until ID of product and PD release.

VICTIMS

Use barrier tape or rope to guide the victims. PD should be assisting.

FIGURE 5-36 Mass decon plan for herding mass numbers of potentially contaminated people.

tions include the use of shower systems, tents, number and location of steps, and personnel. The method of runoff collection varies team to team as well. For most situations that involve solids and liquids a minimum type of decon should be set up.

The HAZWOPER regulation requires that the IC have a plan for decontamination, which can vary from no setup through a multistep process as depicted in the decontamination photo series. For a material such as a technical-grade pesticide, a full setup with all stations is recommended. For gases, in reality no decon is required; however, a minimum setup should be established prior to entry in the event of an emergency and for psychological purposes. Regardless of the type or size of the setup used, it should be set up and ready prior to any entry into a hazard area.

METHODS OF DECONTAMINATION

There are general methods of decontamination that apply to humans, equipment, and the environment, **Table 5-4.** Each method may not apply directly to all three categories. It is important to consult with the hazardous materials team or a chemist prior to using any of these methods on a human, because they may be more dangerous than the contaminant.

Absorption

The spilled material is picked up by the absorbent material, which acts like a sponge. Common absorbent materials include ground-up newspaper, clay,

Water Flow for Herding Guide
Baltimore County Fire Department
Hazardous Materials Response Team

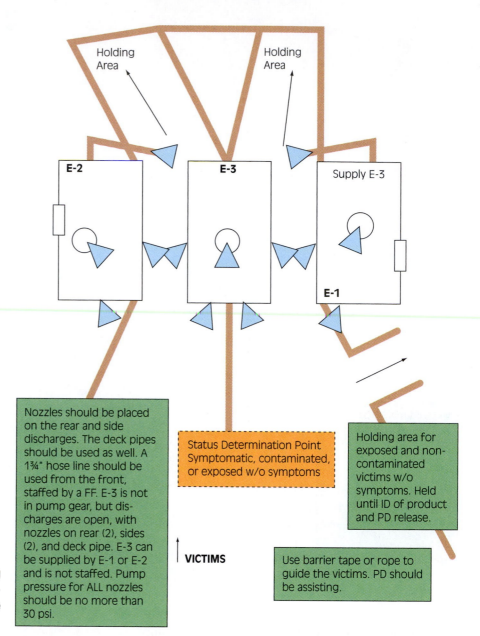

Nozzles should be placed on the rear and side discharges. The deck pipes should be used as well. A 1¾" hose line should be used from the front, staffed by a FF. E-3 is not in pump gear, but discharges are open, with nozzles on rear (2), sides (2), and deck pipe. E-3 can be supplied by E-1 or E-2 and is not staffed. Pump pressure for ALL nozzles should be no more than 30 psi.

Status Determination Point Symptomatic, contaminated, or exposed w/o symptoms

Holding area for exposed and non-contaminated victims w/o symptoms. Held until ID of product and PD release.

↑ **VICTIMS**

Use barrier tape or rope to guide the victims. PD should be assisting.

FIGURE 5-37 In addition to herding and organizing mass numbers of people, water for decontamination can be added to a plan.

kitty litter, sawdust, charcoal, and poly fiber. The absorbed material does not change, and the volume of material may increase. The whole mixture will have to be treated as waste and will not change any of the characteristics such as flammability. Compatibility needs to be researched prior to using these materials, because some materials could cause an adverse reaction.

Adsorption

The material to be picked up bonds to the outside of the adsorption medium. Activated carbon and sand are the most common adsorbents and should be

available locally at a chemical facility. Many chemical facilities have drums, bags, or totes of activated carbon set up to receive liquids within a process, and may have some available in the event of an emergency.

Chemical Degradation

The ability to degrade a chemical varies from compound to compound and is much like neutralization. To degrade a chemical, another chemical is added or the chemical is exposed to the elements, which breaks down the chemical into less hazardous components.

FIGURE 5-38 This engine has a canopy that unfolds to provide a decontamination corridor. The side of the engine is pre-piped with shower nozzles and garden hose extensions for additional water sprays. Sides can be added for privacy.

Dilution

The ability to dilute a contaminant is dependent on the chemical structure of the spilled material, and it does not work on all contaminants. Imagine getting

FIGURE 5-39 An example of a decontamination vehicle that has showers and can contain any runoff. *(Courtesy of Maryland Department of the Environment Emergency Response Division)*

coated with used motor oil; flushing with water will not remove all of the contamination. When dealing with corrosives, large quantities of water are required, sometimes into the hundreds of thousands of gallons. In recent studies, it has been determined that plain water is as effective as some of the available bleach and water solutions. But with some types of neutralization, dilution is a possible solution.

Disinfection

When dealing with humans a 0.5 percent bleach and water solution can be used for some etiological contaminants, although contact time is needed for success. It was initially thought that bleach solutions would also be effective for biological contaminants, but it has been determined that plain water is just as effective for humans. This is due to the contact time required for effectiveness: the exposure of skin to

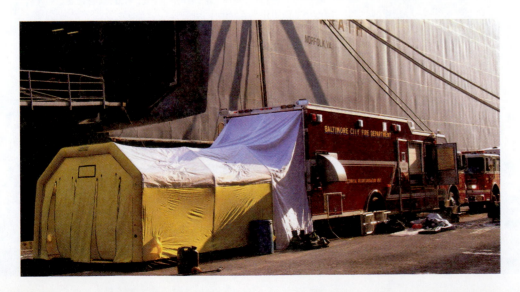

FIGURE 5-40 An example of the decontamination vehicle completely set up. The use of the tent adds an additional layer of privacy. *(Courtesy of Maryland Department of the Environment Emergency Response Division)*

JOB PERFORMANCE REQUIREMENT   5-1
Basic Decontamination Steps

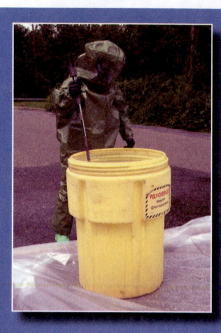

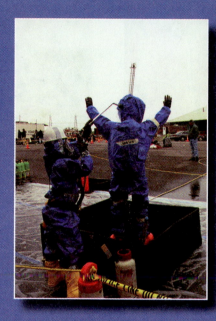

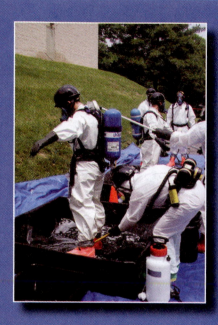

A   The tool drop is usually the first stage in the decon setup. The tools are placed for the use of another entry team or may be collected here for later decontamination.

B   In most systems gross decon is the next step. Some response teams use showers or hoselines to accomplish this task.

C   The next step is formal decon and may involve two stations in which the responder is rinsed, scrubbed, and rinsed again. The solution used is chemical dependent and varies depending on the contaminant. This is the first step in which other responders may assist.

D   After the wet portion of the decon process, the entry team removes their PPE, with the assistance of other responders.

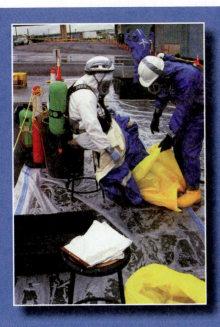

E After the removal of the PPE, the entry team will remove their SCBA. It is usually placed into some type of containment system for later cleaning.

F In some systems a fourth washing area is established for body washing. Some teams may set up a tent for this purpose.

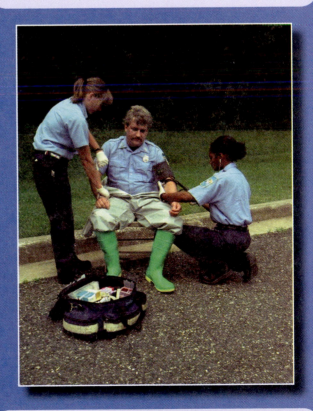

G The last steps are hydration and medical evaluation.

H Medical evaluation after entry is very important. Hazardous materials suit work is very stressful on the body.

TABLE 5-4   Methods of Decontamination

Type	Used on Humans	Used on Equipment	Used on Environment
Absorption	Yes	Yes	Yes
Adsorption	Yes	Yes	Yes
Chemical degradation	No	Yes	Yes
Dilution	Yes	Yes	Yes
Disinfection	Yes	Yes	Yes
Evaporization	Yes	Yes	Yes
Isolation and disposal	No	Yes	Yes
Neutralization	Under specific conditions	Yes	Yes
Sterilization	No	Yes	Yes
Solidification	No	Yes	Yes
Vacuuming	Yes	Yes	Yes
Washing	Yes	Yes	Yes

bleach solutions causes burns if left on for more than a short time. For equipment or the environment, disinfection can be accomplished using a higher percentage of bleach or other chemical.

Evaporization

Is allowing a chemical to evaporate changing its state of matter? If a solid or a liquid is allowed to remain in the open, depending on its vapor pressure, that solid or liquid will eventually change to a vapor. A chemical that approaches its boiling point will also evaporate, moving the material to the vapor state. The material does not disappear but merely changes its state of matter.

Isolation and Disposal

One of the easier forms of decontamination is to isolate the contaminant, collect it using appropriate protective clothing, and then dispose of the contaminant. The disposal should follow appropriate federal, state, and local regulations.

Neutralization

Neutralization is usually reserved for corrosive materials, but is also used here to describe the procedure that reduces the toxicity of a poisonous material. Responders are urged to consult with a chemist

prior to performing this type of activity. The chemicals used to reduce this risk are in themselves hazardous—but in a much lesser fashion than the contaminant. An emergency responder should avoid mixing a neutralization solution using sodium hydroxide because it is very corrosive, but with some chemicals it is extremely useful in small amounts in reducing the hazard of some types of contaminants. In general neutralization, it is exothermic and can produce violent reactions, although these are extremely rare.

Sterilization

There are two primary methods of sterilization, one through a combination of steam and high heat and the other chemical sterilization. The use of steam and high heat is useful for items that are contaminated with etiological materials. The use of chlorine dioxide has been used to decontaminate some buildings for anthrax contamination. Chlorine dioxide is very corrosive and would be considered a very concentrated form of bleach.

Solidification

Solidification is another method that, depending on the solidification agent used, may alter the suspect agent. In some cases, however, the solidification agent will not have any effect on the agent and will reduce the hazard and enable sampling to occur. Chemical compatibility is an item that needs to be confirmed prior to the use of a solidification material.

Vacuuming

For solid materials such as dusts or fibers, a vacuum would certainly reduce the hazard. An ordinary shop vac is not recommended for this task, however, because they are not filtered well enough to keep the exhaust from blowing the agent back into the air. Instead it is recommended to use a vacuum that is equipped with a high-efficiency particulate air (HEPA) filter. Special vacuums are made exclusively for picking up mercury without the mercury vapors being released into the air.

Washing

Typically washing is done with soap and water. This is one of the more effective decontamination solutions, but in reality the soap and water merely removes the contaminant. The contaminant may still present a risk but is removed from the person or item. Washing can be low pressure or high pressure, and can also use elevated-temperature liquids. Under the direction of a chemist, other solutions may be added to the washing solution.

To test the effectiveness of decontamination, responders can use air monitors, particularly a PID to determine if the contaminant has been removed. Other detection devices can be used as required. For the ultimate assurance, responders can take swipe samples and send the swipes off to a laboratory for analysis. As part of the termination procedures, the use of detection devices or other tests for the effectiveness of decontamination should be noted in the report. The type of decontamination and the specific type of decontamination solutions should also be recorded. The report should also include any potential exposures to hazardous materials and an activity log for the responders documenting their actions. These records should be kept for at least 30 years past the last date of employment for the responders. In addition to meeting NFPA requirements, these recordkeeping requirements are also part of the OSHA HAZWOPER regulation.

Some response teams and emergency plans outline incident levels to provide quick notifications. **Table 5-5** lists the incident levels and their associated concerns.

TABLE 5-5 Incident Levels	
Level	**Incident Scale**
1	Small-scale incident can usually be handled by the first responders but may also require minimal additional resources. Notifications are usually local and may only be internal to the fire department. The maximum level of PPE required is firefighter turnout gear and SCBA, and even that may not be necessary. The material is not toxic by skin absorption, although it may present an inhalation hazard. The size of the spill is usually small and will have minimal environmental impact. Incidents of this type include natural gas or propane leaks and small fuel spills.
2	Incidents at this level usually require additional assistance, such as from a HAZMAT team. The incident may require additional notifications to environmental or emergency management agencies on a local or state basis. The amount of material may be larger or more hazardous than in a Level 1 incident. The type of PPE will probably be chemical protective clothing or other clothing not carried by the first responders. The first responders may have switched from an active role to a support role. A Level 2 incident may require a small evacuation and possibly a large isolation. The incident will also probably impact the ability of the jurisdiction to respond to other emergencies, depending on the resources available. An example incident would be an overturned gasoline tanker or a leaking propane tanker. A leaking drum in the back of a tractor-trailer would be classed as a Level 2 incident.
3	A Level 3 incident requires substantial local resources and the assistance of other agencies, local and state. Resources on the federal level may also be required or, at a minimum, are to be notified. The incident will require the evacuation of the affected area and a substantial isolation area. The release is large or the material is extremely toxic. Examples of a Level 3 incident include a train derailment with leaking chlorine railcars. A substantial leak from an ammonia tank truck would also be an example of a Level 3 incident.

LESSONS LEARNED

The chlorine rail car accident in Graniteville offers an important lesson. Many of the topics in this chapter directly relate to that incident. Although the accident had a significant impact on that community and many lives were lost, the chlorine release was not as large as it might have been. Protective actions are used for a variety of purposes, the most important of which is to protect the public and, in many cases, the responder. Management of a chemical-release incident is not an easy task, because there are multi-agency jurisdictions and a variety of objectives may need to be accomplished at the same time. Unlike most fires or accidents, political pressure is usually associated with a chemical release, and there is the potential that thousands of people may require evacuation and relocation.

The determination for decontamination can also be a difficult decision, because most departments do not have the ability to decontaminate large numbers of victims. Choosing the method of decontamination can be difficult as well because of the many possibilities. The best options for protection are to provide prompt isolation and prevent further escalation of the exposed population. For those people who are contaminated, a water-based decontamination method is usually recommended.

KEY TERMS

8-Step Process   A management system used to organize the response to a chemical incident. The elements are site management and control, identifying the problem, hazard and risk evaluation, selecting PPE and equipment, information management and resource coordination, implementing response objectives, decon and cleanup operations, and terminating the incident.

ANFO   The acronym that is used for ammonium nitrate fuel oil mixture, which is a common explosive. ANFO was used in the Oklahoma City bombing incident.

DECIDE Process   A management system used to organize the response to a hazardous materials incident. The factors of DECIDE are detect, estimate, choose, identify, do the best, and evaluate.

Decontamination   The physical removal of contaminants (chemicals) from people, equipment, or the environment. Most often used to describe the process of cleaning to remove chemicals from a person. Often shortened to "decon."

Emergency Decontamination   The rapid removal of a material from a person when that person (or responder) has become contaminated and needs immediate cleaning. Most emergency decon setups use a single hoseline to perform a quick gross decon of a person with water.

Emergency Response Planning (ERP)   Levels that are used for planning purposes and are usually associated with the preplanning for evacuation zones.

Evacuation   The movement of people from an area, usually their homes, to another area that is considered to be safe. People are evacuated when they are no longer safe in their current area.

Fine Decontamination   The most detailed of the types of decontamination. Usually performed at a hospital that has trained staff and is equipped to perform fine-decon procedures.

GEDAPER Process   A management system used to organize the response to a chemical incident. The factors are gather information, estimate potential, determine goals, assess tactical options, plan, evaluate, and review.

Isolation Area   An area that is set up by responders and is intended to keep people, both citizens and responders, out. May later become the hot zone/sector as the incident evolves. Is the minimum area that should be established at any chemical spill.

Mass Decontamination   A process used to decontaminate large numbers of contaminated victims.

Overpacked   A response action that involves the placing of a leaking drum (or container) into another drum. There are drums made specifically to be used as overpack drums in that they are oversized to handle a normal-sized drum.

Psychological Decontamination   The process performed when persons who have been involved in a situation think they have been contaminated and want to be decontaminated. Responders who have identified that the persons have not been contaminated should still consider what can be done to make them feel better.

Sector   An area established and identified for a specific reason, typically because a hazard exists within the sector. The sectors are usually referred to as hot, warm, and cold sectors and provide an indication of the expected hazard in each sector. Sometimes referred to as a zone.

Shelter in Place   A form of isolation that provides a level of protection while leaving people in place, usually in their homes. People are usually sheltered in place when they may be placed in further danger by an evacuation.

Sick Building   A term that is associated with indoor air quality. A building that has an air quality problem is referred to as a sick building. In a sick

building, occupants become ill as a result of chemicals in and around the building.

Sick Building Chemical   When a building is referred to as a sick building, certain chemicals exist within that cause health problems for the occupants. These chemicals are referred to as sick building chemicals.

Technical Decontamination   The process used to perform decontamination of persons or equipment.

Zone   An area established and identified for a specific reason, typically because a hazard exists within the zone. The zones are usually referred to as hot, warm, and cold zones and provide an indication of the expected hazard in each zone. Sometimes referred to as a sector.

REVIEW QUESTIONS

1. What is the first priority of any incident management system?
2. What zone or sector should be established first?
3. When dealing with a flammable liquid spill, what could be used to establish the hot zone?
4. What level of incident would be used to describe a chemical release in which 2,000 people were evacuated?
5. What is the minimum amount of decontamination that should be set up for every incident?
6. What are the four types of decontamination?
7. Which decon step are hospitals involved in?
8. The decision to attempt a rescue in a hazardous materials situation depends on what information?
9. What causes the most common breach in a 55-gallon drum?
10. If fire has reached the cargo portion of a truck carrying DOT Class 1.1 materials, what tactics are used to fight the fire?
11. How can air monitors be used in the determination of hazard zones?

FOR FURTHER REVIEW

For additional review of the content covered in this chapter, including activities, games, and study materials to prepare for the certification exam, please refer to the following resources:

Firefighter's Handbook Online Companion
Click on our Web site at **http://www.delmarfire.cengage.com** for FREE access to games, quizzes, tips for studying for the certification exam, safety information, links to additional resources and more!

Firefighter's Handbook Study Guide
Order#: 978-1-4180-7322-0
An essential tool for review and exam preparation, this Study Guide combines various types of questions and exercises to evaluate your knowledge of the important concepts presented in each chapter of *Firefighter's Handbook*.

ENDNOTES

1. These top ten (which are in random order) are derived from the EPA, OSHA, and DOT by examining the chemical release records. Because each agency tracks releases in a different fashion, we calculated each agency's top ten and determined an average, the result of which is the list given.

2. In most states the fire department IC is in charge of the incident until command is relinquished to another agency. Check local and state regulations for more information. Although most fire departments will leave when the cleanup contractor begins to work, a problem exists if the contractor's crew is working in

a hazardous environment. As long as they are, the FD should maintain a presence because the FD is solely responsible for public safety, a responsibility that cannot be delegated.

3. Any location such as a storm drain, ditch, or culvert that eventually leads to a waterway is defined as a waterway.

ADDITIONAL RESOURCES

Bevelacqua, Armando, *Hazardous Materials Chemistry,* 2nd ed. Thomson Delmar Learning, Clifton Park, NY, 2006.

Bevelacqua, Armando and Richard Stilp, *Hazardous Materials Field Guide,* 2nd ed. Thomson Delmar Learning, Clifton Park, NY, 2006.

Hawley, Chris, *Hazardous Materials Air Monitoring and Detection Devices,* 2nd ed. Thomson Delmar Learning, Clifton Park, NY, 2007.

Hawley, Chris, *Hazardous Materials Incidents,* 3rd ed. Thomson Delmar Learning, Clifton Park, NY, 2007.

Noll, Gregory, Michael Hildebrand, and James Yvorra, *Hazardous Materials: Managing the Incident,* 3rd ed. Red Hat Publishing, Chester, MD, 2005.

Schnepp, Rob and Paul Gantt, *Hazardous Materials: Regulations, Response and Site Operations.* Delmar Publishers, Albany, NY, 1998.

Stilp, Richard and Armando Bevelacqua, *Emergency Medical Response to Hazardous Materials Incidents.* Delmar Publishers, Albany, NY, 1996.

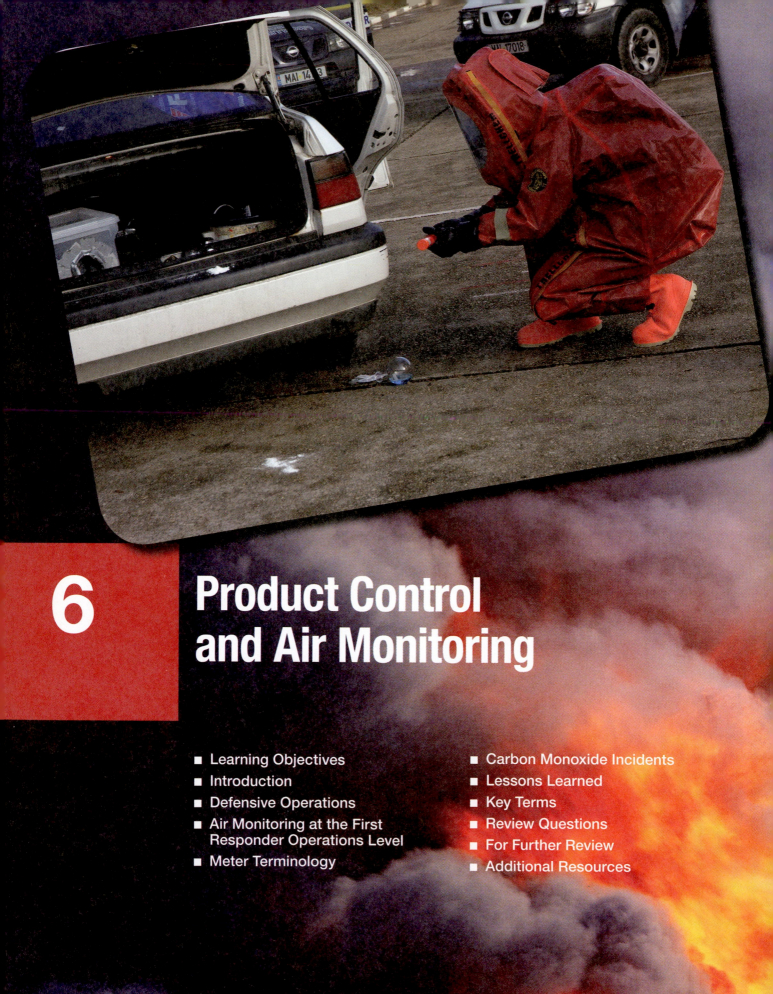

6

Product Control and Air Monitoring

On a weekday morning, our hazardous materials team was called to assist an engine company in the Clear Lake / NASA area of Houston, Texas. On arrival, the engine company officer advised us that a car had driven through a fence into the backyard of a one-story single-family residence and pushed a gas meter into the house. The vehicle was parked on top of the meter halfway inside the house. He did not want his crew to make entry into the house because they did not have any type of monitoring instrument to check for the level of gas in the house.

Because of the time of day and occupancy, we decided that it was necessary to do a quick primary search of the house to check for anyone stuck inside. Two of us suited up in structural firefighter protective clothing with SCBA and combustible gas indicators (CGI). The engine company was staged down the street due to the potential for explosion.

As we entered the house through a broken-down door near the car, my CGI went into alarm at 10 percent of the LEL and then into "over range" (OR). I was holding the monitor at arms length as high into the ceiling as I could reach because natural gas is lighter than air. As I looked at the monitor just past the end of my arm I saw a ceiling fan running in the high vaulted ceiling. I quickly processed the information and concluded that we were in an environment that was well above the upper explosive limit (UEL). Because we were already in the house, we did a quick primary search to ensure that no one was in the house, then quickly exited through the front door, which we left open for ventilation.

There was no one in the house. We had the gas shut off and it was safely ventilated.

It is important to carry monitoring equipment to aid in your decision making process with this very common type of call. Sometimes, firefighters have to make difficult and rapid decisions based on risk vs. benefit when dealing with natural gas emergencies. This was a very dangerous "common call" that ended with a positive outcome.

—*Street Story by Bill Hand, Houston Fire Department Hazardous Materials Response Team, Houston, Texas*

LEARNING OBJECTIVES

After completing this chapter, the reader should be able to:

6-1. Identify equipment that can determine hazardous environments and isolation areas.

6-2. Describe available defensive operations for a release.

6-3. Describe the various methods of damming, diking, diverting, and other defensive operations.

6-4. Explain what types of detection equipment are available to the first responder and equipment that a hazardous materials team might carry that could be used at a chemical release.

INTRODUCTION

The use of product control techniques can provide a quick reduction in the damage that is done to a community and the environment in the event of a spill. The reduction of the surface area over which a product has spread provides a direct reduction in the danger to responders and the community. By implementing these types of measures, first responders can provide an extreme advantage to the overall mitigation of the incident. The use of air monitoring devices is becoming more commonplace, and many first responders have some type of air monitoring device available to them. This chapter provides a brief overview of this complex subject and concentrates on the knowledge that the first responder should have. To ensure firefighter safety, basic air monitoring must be accomplished, even for basic firefighting, natural gas leaks, gasoline spills, or any other incident in which firefighters may be exposed to hazardous materials.

DEFENSIVE OPERATIONS

According to the NFPA, three large tasks fit into Operations-level tactical response: basic air monitoring, containment, and confinement. All of these can be combined into the task of product control, which is the one task that persons trained to the Operations level can perform. The tasks of containment and confinement should be performed away from the hazard area and to the smallest, safest area. To perform defensive operations, the responders do not really get close to the actual spill itself. The activities that fit into product control operations are absorption, diking, damming, diverting, retention, dilution, vapor dispersion, vapor suppression, and the use of remote shutoffs. The type and level of PPE required for each of these tasks will vary with the chemical involved, but a minimum would be Firefighting PPE and SCBA. If making a dam a considerable distance away from the chemical release—and no contact with the material is possible—then a lesser level could be used. The tasks discussed next are common tasks assigned to responders trained to the Operations level, and they are critical tasks that can protect the community and the environment. Although it is not necessary for first responders to carry all of this type of equipment, they should know its location in the event it is needed or know how to improvise, adapt, and overcome if the equipment is not available.

Absorption

First responders are often asked to clean up a spilled material, and on a daily basis, they probably use the **absorption** technique on a large quantity of fuel and oil products using a product similar to the one shown in **Figure 6-1.** First responders should know the type of absorbent materials carried on their apparatus. Some absorbent materials will not pick up water, while others absorb any liquid with which they come in contact. When trying to pick up fuels off water, having an absorbent that does not pick up water is important. Absorbents vary from clay kitty litter, ground-up newspaper, and corn husks to synthetic fibers. Their weight and absorbent capabilities vary, and not all can be used interchangeably. They may be available in loose bulk form, as pads, **Figure 6-2,** as large rolls, or in boom style.

To hold the absorbent material, construction of a filter fence such as the one shown in **Figure 6-3** is recommended. The filter fence can be made of "rat wire" or other small-weave fencing material. Wooden stakes can be used to hold the fence in place, unless the soil is rocky, in which case metal stakes are recommended.

FIGURE 6-1 A common defensive action is the application of absorbent to a spilled material.

Also available on the market are solidification agents that encapsulate the spill, as well as microbiological oil-eating bugs. Both of these agents are expensive and although sold as absorption materials, are not generally used by emergency services due to their cost and the quantity required. It is best to carry a variety of absorbent products, as well as several different styles.

Diking and Damming

The quantity of material spilled will determine to what extent **diking** and **damming** operations are performed. The actions required at a 50-gallon spill differ from those of a 500,000-gallon release. Diking, damming, and diverting are techniques to consider using when a spill occurs on the ground or on a waterway. A spill on the ground is much easier to control than one on the water, but the principles are the same. When creating a dike or a dam, the first responder is either stopping the flow of the material or keeping it from a specific area. Except for small streams, it is nearly impossible to stop a body of water from flowing. If the spilled material is soluble with water, and the waterway has a good flow, it will be impossible to collect the spilled material, unless heavy equipment and a large area are available that could be used to create a very large dam.

First responders often use earth, sand, or rocks for dikes or dams depending on what is available. Having local contacts available around the clock that can supply these items is important, and some jurisdictions have emergency contacts for dump truck loads of sand or dirt. Local or state highway departments are a good place to start when these items are required. Heavy equipment such as front-end loaders or track hoes can also be used to create dams or dikes. When constructing a dam or a dike, it is best to construct three setups. Responders should start at the farthest point away from the spill and work back toward the spill. It may be necessary to establish the first containment miles away from the spill. If the barrier is started near the spill, responders may be overtaken by the spill and be forced to play catch-up, not to mention the need for additional PPE.

FIGURE 6-2 Absorbent pads will pick up oil from the water, but will not absorb the water, allowing them to float.

FIGURE 6-3 A screen fence can be used to collect the absorbent material floating on the water after the absorbent has collected the spilled material. The water flows through the fence below the absorbent.

Two basic types of dams can be created: overflow and underflow dams, **Figure 6-4.** The specific gravity of the spilled material dictates whether the material will float on top of the water or travel on the bottom, which determines what type of dam needs to be constructed. An overflow dam allows the water to flow over the dam and contain the spilled material at the base of the dam. An underflow dam allows the water to flow under the dam and collect the material on top of the water. Most spilled materials float on top of the water and hence require the use of an underflow dam.

The materials required to construct either type of dam are shovels, dirt or sand, and pipes. Pipes should be at least 4 inches in diameter, but larger is better. The higher the flow, the bigger the pipe and the greater the number of pipes required. A good rule of thumb to use is to have enough pipes to cover two-thirds of the width of the waterway, which would be the minimum amount required. Standard PVC pipe is adequate and, for emergency situations, the hard sleeves (suction) on the engine company are a good substitute. **Figure 6-5, Figure 6-6,** and **Figure 6-7** show both styles of dams and the use of hard sleeves, respectively. For an overflow dam, first responders tend to establish a dam, place the pipes on top, and consider the dam finished. It is best, however, to set the pipes on top of the dam, and then continue with a layer of dirt or sand on top of the pipes, because there is a tendency for the water to erode the dam, and the containment will collapse without this extra layer of dirt or sand.

Diverting

Almost anything can be used, such as dirt, sand, absorbent material, tarps, and hoselines, to divert a running spill, **Figure 6-8.** The most common use of the

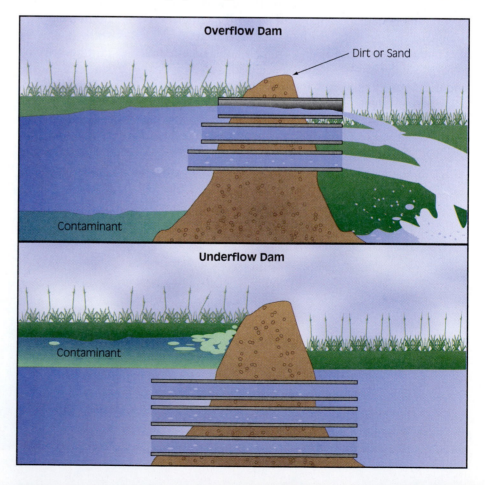

FIGURE 6-4 The overflow dam allows the clean water to flow over the dam and collects the spilled material at the base of the dam. The underflow dam collects the spilled material on top and allows the water to flow through the bottom of the dam.

FIGURE 6-5   Overflow dam.

FIGURE 6-6   Underflow dam.

FIGURE 6-7   First responders have equipment such as hard sleeves to establish overflow and underflow dams.

FIGURE 6-8   If a spill cannot be controlled, it may be better to divert the spill around items such as sewers and allow the spill to continue down the street.

diverting technique is to keep a running spill from entering a storm drain. A ring of dirt around the storm drain will keep the spilled material out, but does not stop the flow of the product. There are times when, due to the large quantity of spilled material, stopping the flow would be impossible, but it would be possible to keep it from the storm drain.

The diversion tactic can also be used for spills on water. By using a solid boom, such as a harbor boom, **Figure 6-9,** material can be diverted into another area or even contained. Another method of diversion is to dig a trench alongside the waterway to collect the spilled material off the top of the water. This is effective for large spills if there is a collection vehicle such as a vacuum truck standing by to collect the spilled material as it runs into the diversion. These types of vehicles are available from hazardous waste cleanup contractors. The most common method is to use a floating absorbent boom that absorbs the material off the water as shown in **Figure 6-10.**

Retention

The most common method of **retention** is the digging of a hole, as shown in **Figure 6-11,** either by hand or by machine, such as with a track hoe. To catch a running spill, the best method is to dig a hole large enough to collect the spill and then divert the spill into it.

Having the ability to create a large enough containment area is paramount to success. If unable to build a large enough containment area, then first responders should concentrate on building several in a row to collect the material. As with dikes and dams, it is best to start construction of the containment area farthest from the spill and then create several other areas closer to the spill.

STREETSMART TIP

When digging retention areas, it is important to be aware of other underground utilities such as water, electric, sewer, cable TV, phone, or natural gas lines.

Dilution

Much like **dilution,** when conducting decontamination processes, this technique is not always the solution to pollution. If a water-soluble material is in a waterway, then for all intents and purposes it is being diluted. If the waterway is small, the fire department

FIGURE 6-9 Shown is an example of using a harbor boom. The boom should have been angled so that the material being collected would be near a shoreline. In this case, the boom was stretched straight across the waterway causing the material to be collected in the middle.

FIGURE 6-10 Floating boom collects the spilled material that is floating on top of the water. Like the pads, the boom does not absorb water and should stay afloat even when saturated with the spilled material.

FIGURE 6-11 Building a retention area is a good method for holding released products. Retention areas can be dug from existing earth or created with truckloads of sand or dirt. Small holding areas can be hand dug, but for larger spills heavy equipment is required.

may need to add some water to the spilled materials. Simple flushing of a material into a waterway is not an acceptable tactic anymore and should not be considered. Items such as fuels, oils, or other hydrocarbons are not usually water soluble and cannot be diluted by water. For some specific chemicals, dilution is possible only when combined with some of the other containment tactics.

Vapor Dispersion

The topic of **vapor dispersion** can create some confusion for responders, because by their nature firefighters like to use water, usually in large quantities, and on some occasions the application of water will make spill situations safer. For instance, if a severe life threat exists because an adjacent nursing home cannot be evacuated, then using a water spray to disperse vapors would be a good idea. If, on the other hand, the release is occurring in a remote area away from any population, then a water spray to disperse vapors such as the one shown in **Figure 6-12** is not necessary. To be effective, the material that is being released must be water soluble or the vapor cloud must be able to be moved by the water streams. Although the DOT ERG states for many types of spills "use water spray to reduce vapors or divert vapor cloud," water is not recommended unless a life threat scenario exists.

Some departments use water sprays at natural gas leaks, which is not necessary and further complicates repair of the leak. Many of the leaks occur in a hole that was created by a backhoe. Water can fill the hole and make any attempted repair very dangerous and complicated. Also the use of water may actually knock down vapors that normally rise up and dissipate quickly.

FIGURE 6-12 The use of a water spray is not necessary unless the vapor cloud is impacting the public. In some cases it creates more hazards than it eliminates.

Use of a water spray can also result in the creation of another substance. When anhydrous ammonia is released into the air, and a water spray is applied, the resulting runoff is ammonium hydroxide, a very corrosive liquid that itself would require containment and cleanup. Both natural gas and propane are not water soluble and the use of water vapor will just relocate the vapors to another area.

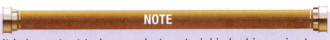
NOTE

It is important to know what material is leaking prior to the application of water, because the water spray may cause more problems. It is very important to consult with the hazardous materials team prior to using this tactic.

Vapor Suppression

With the use of firefighting foams, vapor suppression is another tactic that first responders can use. The type of material spilled dictates the type of foam to be used, because not all foams will work on all products. The use of foam to suppress vapors is generally limited to flammable liquid spills, but some chemical foams are available for other types of materials.

Before the application of foam, the responders should ensure that the material is contained, the application of foam will not cause any further problems, and the foam is compatible with the spilled material. As when using foam to extinguish a fire, responders need to make sure they have enough foam stockpiled to keep a layer of foam on the spilled material.

Remote Shutoffs

The shutting of valves is not usually a tactic used by first responders, but there are some exceptions to this rule. On most tank trucks and at some fixed facilities, there are **remote shutoffs** that could be operated by first responders. Each type of truck is different but in general an emergency shutoff is located behind the driver's side of the cab and another near the control valves. The shutoffs are both mechanically operated and in most cases are self-operating in the event of a fire. They are usually well marked as emergency shutoffs and in an easy-to-find location. The two most common locations are behind the driver's door at the tank or at the valve controls. At some facilities, typically those with loading racks, a remote shutoff may be located near the entrance so that in the event of an emergency the first responders can shut off the flow of product.

Implementing Defensive Operations

There are a number of considerations for the first responder that can help determine the overall outcome of the incident. JPR 6-1 walks you through the inherent clues found in various incidents critical for developing effective defensive operations as a first-arriving fire department responder.

1. **Scenario 1 (JPR 6-1A)**—The *visual clues* to potential hazards include:
   - There has been a traffic accident.
   - A double tank truck with the rear tank overturned is involved.
   - From this vantage point, there are no placards visible.
   - Although difficult to see, the dome on the rear tank is open, so if the tank was full, the product has been released.

Some *response priorities* include (order is dictated by on-scene conditions):
   - A high priority is to determine what the contents of the tank are (or were).
   - Update the other response units, such as the hazardous materials team, and obtain guidance from them.
   - Determine what type of PPE would be required.
   - Determine the status of the driver and determine the contents of the tank.
   - Determine where the police officer is and the officer's status.
   - Isolate the scene to remove bystanders and keep new bystanders from becoming part of the problem.
   - Perform decontamination on any victims or responders who may be contaminated.
   - Determine where the contents of the tank went and if any control measures can be implemented.
   - If appropriate, use your air monitors to determine the presence of hazardous gases.

2. **Scenario 2 (JPR 6-1B, JPR 6-1C)**—The *visual clues* to potential hazards include:
   - There has been a traffic accident with a number of vehicles involved.
   - A tank truck is involved.
   - There is a Class 2 non-flammable gas placard with "2187" in the middle indicating a bulk shipment. The product name "Carbon Dioxide Refrigerated Liquid" is stenciled below the placard.
   - Although the piping was involved in the accident, from this vantage point there does not appear to be a release.

   - In the upper right corner of JPR 6-1B, there is some damage to the outer shell of the insulated tank. In JPR 6-1C, think about the incident priorities if the occupants of this vehicle were trapped.

Some *response priorities* include (order is dictated by on-scene conditions):
   - A high priority is to determine what the hazards of carbon dioxide refrigerated liquid would be, and what impact this would have on any potential rescue situations.
   - Update the other response units, such as the hazardous materials team, and obtain guidance from them.
   - Determine what type of PPE would be required.
   - Determine the status of the other drivers and occupants.
   - Determine where the highway crew is and their status.
   - Isolate the scene to remove bystanders and keep new bystanders from becoming part of the problem.
   - If appropriate, use your air monitors to determine the oxygen content. Carbon dioxide is a colorless and odorless gas that is a simple asphyxiant.

3. **Scenario 3 (JPR 6-1D)**—The *visual clues* to potential hazards include:
   - There is a visible vapor cloud coming from the valve box on the bottom of the rail car. The box seems to be very cold.
   - From this vantage point there are no placards visible, but the rail car is placarded with a Class 2 "Flammable Gas" placard. The material is ethylene and liquefied flammable gas.
   - Although it is not known where the release is coming from, all of the transfer piping and valving is inside the box.

Some *response priorities* include (order is dictated by on-scene conditions):
   - A high priority is to confirm the contents of the tank.
   - Update the other response units, such as the hazardous materials team, and obtain guidance from them.
   - Determine the status of the train crew and determine the contents of the tank.
   - Isolate the scene to remove bystanders and keep new bystanders from becoming part of the problem.
   - Determine what downwind exposures may be at risk.

A Scenario 1: Visual clues

B Scenario 2: Visual clue *(Photo courtesy of Maryland Department of the Environment Emergency Response Division)*

C Scenario 2: Visual clue *(Photo courtesy of Maryland Department of the Environment Emergency Response Division)*

D Scenario 3: Visual clues *(Photo courtesy of Mike Austin)*

E Scenario 4: Visual clues *(Photo courtesy of Mark E. Brady, PGFD)*

■ Determine where the vapors will go and determine if any control measures can be implemented, such as water spray. Since ethylene is a flammable gas and not water soluble, this tactic would have minimal impact.

■ If appropriate, use your air monitors to determine the presence and location of flammable and hazardous gases. Ethylene is one of the few gases that will rise, and it has a vapor density of 0.98, which is very close to the vapor density of air. What would change if the humidity was 85 percent or higher?

4. **Scenario 4 (JPR 6-1E)**—The *visual clues* to potential hazards include:

■ There is a fire involving a tank truck and some sort of traffic accident.

■ From this vantage point, there are no placards visible, but there are Class 3 "Flammable Liquid" placards on the truck.

To fight the fire or not to fight the fire?

■ If the truck contained gasoline, overall it would be more hazardous, but the decision to fight the fire or retreat is the same for both gasoline or diesel; the difference between the gasoline and diesel fuel are minimal. Once either is burning, comparable resources are required to extinguish the fire.

■ Do you have enough resources to extinguish the fire?

    ■ If the cargo were involved, foam would be required.

    ■ If it is gasoline, do you have the appropriate foam for ethanol gasoline?

    ■ Do you have enough foam?

    ■ Can you maintain a foam blanket every five to ten minutes?

■ Are there exposures or persons at risk?

■ What if you did nothing?

■ What is the impact on the community?

Some *response priorities* include (order is dictated by on-scene conditions):

■ A high priority is to determine what the contents of the tank are (or were). Given the placarding and markings on this truck, the most likely products are gasoline or diesel fuel.

■ Update the other response units, such as the hazardous materials team, and obtain guidance from them.

■ Determine what type of PPE would be required.

■ Determine the status of the driver and determine the contents of the tank.

■ Isolate the scene to remove bystanders and keep new bystanders from becoming part of the problem.

■ Perform decontamination on any victims or responders who may be contaminated.

■ Determine where the contents of the tank went and if any control measures can be implemented.

    ■ The truck is sitting on a downhill incline.

    ■ There are storm drains on the downhill side.

    ■ The storm drains feed to a creek.

■ If appropriate, use your air monitors to determine the presence of hazardous gases.

AIR MONITORING AT THE FIRST RESPONDER OPERATIONS LEVEL

The use of air monitoring is one of the most important tasks to accomplish when responding to chemical incidents. Air monitoring can keep responders alive. The NFPA 472 Committee recognized this and incorporated a number of air monitoring compentencies at the Operations level. The new objectives are added as Mission Specific Competencies. These are additional competencies that are optional. If firefighters are going to use air monitors at the Operations level, then the mission-specific competencies should be used for training. With many departments purchasing detectors to assist with carbon monoxide alarms, first responders are experiencing associated benefits. Depending on the type of alarm purchased, it may also be used to detect flammable gases, oxygen levels, and one or two toxic gases, **Figure 6-13**. Many departments use these instruments for flammable gas leaks and for confined space entries.

When responding to situations that involve unknown

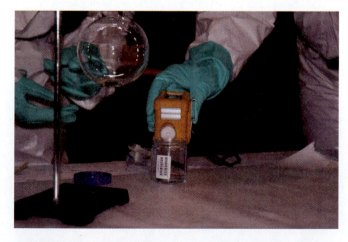

FIGURE 6-13 First responders commonly use devices that detect the presence of flammable gases, carbon monoxide, and hydrogen sulfide and determine the oxygen content. This detector also detects the presence of common toxic vapors.

materials, hazardous materials responders need pH detection for corrosives, a combustible gas detector for the fire risks, a **photoionization detector (PID)** for the toxic risks, and a radiation monitor for radiation risks. Unfortunately, many responders only rely on one or two of these detectors, and the most common detector is a three-, four-, or five-gas instrument. Other instruments are not available or not used. By not using a method of detecting toxic materials, responders (and the public) can be lulled into a false sense of security. Combustible gas detectors are not made or designed to measure toxic gases in air; they only read those types of gases when they become fire hazards. Many flammable gases are very toxic at considerably lower levels than their lower explosive limit. Many first responders have purchased air monitors, but may not have a full understanding of how they work.

NOTE

The use of air monitoring and sampling equipment can be complicated and requires practice.

Most fire service responders have a reasonable understanding of combustible gas indicators but even that knowledge can be limited. Society is becoming much more sophisticated and the citizens (the fire service's customers) expect more. This section presents some concepts on air monitoring, monitoring strategies, and information on how the monitors work and their uses. The field of air monitoring technology is ever changing and new technology emerges each year, although the basic principles of safe decision making remain the same. When purchasing air monitors it is important to understand the basic features of the instruments so that the purchase benefits the organization making the purchase.

NOTE

The general rule when purchasing an air monitor is to figure one-quarter of the purchase price into the department's annual budget for upkeep and repairs. Even though the purchaser may have been told differently, it is necessary to keep the instrument calibrated and maintained on a regular basis.

Regulations and Standards

The Occupational Safety and Health Administration (OSHA) HAZWOPER regulation (29 CFR 1910.120) does not provide a lot of specific requirements for air monitoring, even for the hazardous materials technician, but air monitoring is the principal safety element throughout the document. OSHA wants the incident commander (IC) to identify and classify the hazards that are present on a site. The use of air monitoring

is the primary key to fulfilling this obligation, but in order to fulfill this obligation a hazardous materials technician must be present along with other detection devices.

The NFPA 472 document also has some requirements for air monitoring but, like the OSHA regulation, they are fairly generic. The 2008 edition of NFPA 472 includes some new Mission Specific Competencies that provide training objectives for the use of air monitors. The Technician level was also revised with some additional WMD monitoring competencies. There always have been air monitoring competencies at the Technician level, but as technology has improved, NFPA 472 needed to address these new technologies. The WMD-specific monitoring competencies were added as mission-specific, much like the air monitoring competencies at the Operations level.

It is important for first responders to understand how monitors work and what their deficiencies are because they may be placing their lives on the line depending on the readings received on an air monitor. First responders are more affected by the confined space regulation with regard to air monitors than they are by the hazardous materials regulation.

It is unfortunate, but it can be predicted that a lack of understanding and/or maintenance will play a factor in future firefighter fatalities and injuries based on the large number of departments using these instruments.

Air Monitor Configurations

Most departments purchase an instrument that is known as a three-, four-, or five-gas instrument, commonly referred to as a **multi-gas detector,** such as the one shown in **Figure 6-14.** In other words it samples for the presence of three to five different gases. Nor-

FIGURE 6-14 The miniature detector on the left detects flammable gases, carbon monoxide, hydrogen sulfide, and oxygen content. These types of instruments require calibration and regular maintenance.

mal units sample for oxygen levels, a lower explosive limit (LEL), and two or three toxic gases. The normal "toxics" are carbon monoxide (CO) and hydrogen sulfide (H_2S). If a third gas can be sampled for, most departments choose chlorine or sulfur dioxide. For most departments, though, sampling for CO and H_2S is more than sufficient.

Although these instruments are considered expensive, in the grand scheme of things they are inexpensive compared to other detection devices used by a hazardous materials team. Many departments may purchase a device with little or no training, which can be detrimental to the successful use of the instrument. When buying a detector it is essential to purchase several maintenance items designed to keep the instrument functioning. Typically these items are included in a special kit price and include a battery-powered sample pump, calibration gas, and hardware. The calibration gas usually has an expiration date and will need to be replaced in six months to a year. The sensors in the detector need regular calibration, and the instrument should be turned on and used on a regular basis for a period specified by the manufacturer. Depending on the charging system, rechargeable batteries may need to be replaced on occasion.

The sensors also need replacement. Although most manufacturers provide two-year warranties for the sensors, in general, the LEL sensor will last four to five years, the oxygen sensor eighteen to twenty-four months, and the CO and H_2S sensors eighteen to twenty-four months. Other toxic sensors may last six to eighteen months depending on the type.

Most of the perceived problems with instruments are created by a lack of maintenance and adequate training. Regardless of what a salesperson says, no device exists that can conclusively identify an unknown material for a first responder.

METER TERMINOLOGY

To comprehend fully the use of air monitors, the responder must understand the basic terminology that is generic to all monitors and applies across the board. More detailed information related to the specific monitors is provided in later sections.

Bump Test

Two terms that need further explanation are bump test and calibration. A **bump test,** also known as a field test, exposes a monitor to known gases, allowing the monitor to go into alarm mode, and then removes the gas. By exposing the monitor to a known quantity of gas, a person can verify the monitor's response to that gas. Most manufacturers provide bump gas cylinders, which contain the gases required to check their

instrument. When bump testing, firefighters should follow the manufacturer's recommendations.

In most cases, the bump test is used to ensure that the alarms function as intended and the instrument is reading. By regularly calibrating and bump testing, the user can determine how accurate the bump test will be in the field. Some instruments react very well to bump testing and will display the levels as provided on the bump gas cylinder, while others will be off slightly. When responding to confined spaces, responders are required to bump test the instrument prior to entry into the confined space.

Calibration

Calibration is used to determine if a monitor responds accurately to exposure to a known quantity of gas, **Figure 6-15.** When new sensors are installed, they will usually read higher than intended; calibration electronically changes the sensor to read the intended value. As the sensor gets older, it becomes less sensitive and calibration electronically raises the value that the sensor displays. When a sensor fails, it cannot be electronically brought up to the correct value. Regular calibration gives a picture of the expected life of a sensor, which will deteriorate over time.

The regularity of calibration is subject to great debate. Some departments calibrate daily, others every six months. The only item found in the regulations (for anything that requires air monitoring) is in the confined space regulation, which requires calibration according to the manufacturer's recommendations. Most of the written instruction guides from the manufacturers require calibration before each use. The International Safety Equipment Association (ISEA), an industrial association, has lobbied OSHA to add regulations governing calibration. The ISEA recom-

FIGURE 6-15 Calibration involves exposing the instrument on the right to a known quantity of gases to ensure that it is reading the gases correctly. This kit calibrates the monitor with a variety of gases.

mends bump testing prior to use and, depending on the conditions where the device may be used, recommends calibration every thirty days. The definition of a calibration at this point is also subject to debate, because the department can verify a monitor's accuracy by exposing it to a known quantity of gas, but not perform a "full" calibration. The ISEA calls this type of test a functional test. Most response teams establish a regular schedule of calibration (weekly/monthly) and then perform "bump" or "field" checks during an emergency response. It is essential to check with manufacturers as to what calibration/bump test policy they recommend because they are all different.

Reaction Time

All monitors have a lag time or, as it is better known, reaction time. The use of a pump or not will vary the reaction time. Monitors operating without a pump are in what is known as the diffusion mode and will generally have a fifteen- to thirty-second lag time. Monitors operating with a pump have a typical reaction time of seven to thirty seconds, depending on the sensor type. Some instruments, primarily WMD detection devices, may take up to ninety seconds to react. Hand-aspirated pumps usually require ten to fifteen pumps to draw in an appropriate sample. Hand-aspirated pumps are not recommended, because the readings they provide will be too varied depending on how well the user operates the hand pump. The goal is to provide a given amount of volume across the sensors. It is important to follow the manufacturer's recommended lengths of hose to ensure that the pump operates correctly.

NOTE

When using sampling tubing, one to two seconds of lag time should be added for each foot of hose.

Recovery Time

Monitors also have a recovery time. Recovery time is the amount of time it takes the monitor to clear itself of the air sample. This time is affected by the chemical and physical properties of the sample, the amount of sampling hose, and the amount absorbed by the monitor. Some monitors take an extended period of time to clear if they are exposed to a large quantity of a gas. In some cases the instrument must be taken out of the environment and shut off, then started again. The reaction time will affect the overall recovery time.

Relative Response

When a gas monitor is purchased, it is set to read a specific type of gas, such as methane. If any other gas is sampled, the instrument will usually read the gas but at different values. Flammable gases have varied lower explosive limits (LELs) and therefore react differently with the flammable gas detector. The term relative response is used to describe the way the monitor reacts to a gas other than the one it was calibrated for, and is a term that is not commonly used by emergency responders.

To maintain a high level of safety, each person operating an air monitor must have a basic understanding of relative response so as not to let the monitor lead them into dangerous situations. Each detector has what is known as a relative response curve that compensates for different types of gases, **Table 6-1.** The monitor's manufacturer has tested the monitor against other gases and has provided a factor, commonly known as a correction factor, that is a relative response factor that can be used to determine the amount of gas present when sampling. The user multiplies the displayed reading by the factor to arrive at the reading for the gas that is actually present. For instance, consider a responder at a xylene spill who is using an ISC TMX-412 calibrated for pentane. The detector is reading 50 percent of the LEL. According to Table 6-1, the response curve factor for xylene is 1.3, which is multiplied by the LEL reading:

$$\text{Detector reading} \times \text{Response curve factor} = \text{Actual LEL reading}$$

$$50 \times 1.3 = 65$$

So the actual LEL is 65 percent, a number higher than what the instrument was providing. On the other hand, if the responder used the same instrument for a spill of propane, the response curve factor is 0.8. Using the same scenario—that the instrument was reading 50 percent of the LEL—the actual reading is 40 percent ($50 \times 0.8 = 40$), a safer situation than reported by the detector.

Oxygen Monitors

Oxygen is one of the most important things to sample for. Humans need it to survive, and the other instruments need it to function correctly. Normal air contains 20.9 percent oxygen; below 19.5 percent is considered oxygen deficient and is considered to be a health risk, and above 23.5 percent is considered a fire risk. If an oxygen drop is noted on the monitor, one or possibly more than one contaminants are present, causing the reduced oxygen levels, and another material (i.e., toxic, flammable, corrosive, or inert) is causing the oxygen-deficient atmosphere. When in oxygen-deficient atmospheres, any combustible gas readings will also be deficient and cannot be relied on.

TABLE 6-1   Correction Factors (Calibrated to Pentane)[a]

Gas Being Sampled	ISC TMX 412 Factors[b]	MSA 261 Factors[c]	RAE System Factors[d]	Molecular Weight
Methane	0.5	0.6	1	16.04
Propane	0.8	0.9	1.88	44.1
Ethanol	0.8	N/A	1.69	46.1
Pentane	1	1	1	72.2
Benzene	1	1.1	2.51	78.1
Toluene	1.1	1.2	2.47	92.1

[a]Always use the correction factors suplied by the manufacturer, keeping in mind these are laboratory estimations.

[b]Factors for Industrial Scientific Corporation LEL Sensor 1704A1856-200 calibrated with pentane.

[c]Factors for the Mine Safety Appliances MSA 360/361 calibrated with pentane.

[d]Factors are for the RAE sytem catalytic bead LEL sensor calibrated with pentane.

If in an oxygen-enriched atmosphere, then the combustible gas readings will be increased and not accurate. Many oxygen-enriched atmospheres result from a chemical reaction involving oxidizers and typically present a dangerous situation. Monitoring for oxygen is important and the alarm values of 19.5 percent and 23.5 percent are set by OSHA regulation, but the reality is that responders can be in danger prior to these values. As an example, an oxygen drop to 20.8 percent can be a very dangerous situation depending on the contaminant. A drop to 19.9 percent would not result in an alarm, but considerable contaminants may be present in the air at this value nevertheless. **Figure 6-16** provides additional information regarding this topic.

Oxygen Monitor Limitations

There are three types of oxygen sensors: two versions of lead wool technology and a solid polymer electrolyte (SPE) sensor. In all of the sensors, there is a chemical reaction and the calibrated sensors are able to determine the level of oxygen in the air. As long as the sensor is exposed to O_2, it will cause a reaction within the sensor. This is the most commonly replaced sensor and usually needs to be replaced every eighteen months, although some may last longer. Chemicals that hurt O_2 sensors are those with lots of oxygen in their molecular structure such as carbon dioxide (CO_2) and strong oxidizing materials such as chlorine and ozone. The problem with CO_2 is that it is always present in the air, and the higher the percentage the faster the sensor will deteriorate.

The optimal temperature for operation is between 32°F (0°C) and 120°F (49°C). Between 0°F (−18°C) and 32°F (0°C) the sensor slows down (electrolyte is like a slushy), and temperatures below 0°F (−18°C)

can permanently damage the sensor. Operation depends on absolute atmospheric pressure, and calibration is required at the atmospheric pressure at which the user will be sampling. The sensor should also be calibrated for the temperature and weather conditions for the area being sampled.

Flammable Gas Indicators

NOTE

Flammable gas indicator (FGI) is a new term. The old terminology for these detectors described them as combustible gas indicators (CGIs), which is a misapplied term. These devices only detect the presence of flammable gases, not combustible gases. The term *combustible gases* is derived from their history and development in mining safety. Another term that could be used to describe these units is *LEL sensors*, as they are used to determine the percent of the lower explosive limit (LEL) of the material present. This text uses FGI, but LEL sensor would also apply.

All of these types of sensors work, some better than others in different situations. It is important to understand how each of these sensors works, because budgetary considerations may dictate the purchase of only one type. (Note, however, that to be able to detect a wide variety of chemicals effectively, it may be necessary to have more than one type of sensor.)

Most of the new FGIs are used to measure the LEL of the gas for which they are calibrated. The majority of FGIs are calibrated for methane (natural gas) using pentane gas. When calibrated for methane, the FGI sensor will read up to the LEL and, with some of the new units, will in fact shut off the sensor when the atmosphere exceeds the LEL. This is an important consideration, because the longer the sensor is exposed to an atmosphere above the LEL, the quicker it loses its life.

The FGI reads up to 100 percent of the LEL, so that if a FGI is calibrated for methane and the user is sampling methane, the FGI will read 100 percent, but the actual concentration in that area will be 5 percent for methane (the LEL for methane is 5 percent). If the FGI displays a reading of 50 percent, then the concentration for methane is 2.5 percent. Any flammable gas sample that passes over the sensor will cause a reaction; how much of a reaction depends on the gas. To further complicate this issue, the reading on the FGI can either be higher or lower depending on the gas. If the reading is below the actual percentage the user is safe, but if the actual percentage is above what the FGI is reading then the user may no longer be in a safe atmosphere because the LEL may have been exceeded.

The basic principle of most FGIs is that a stream of sampled air passes through the sensor housing, causing a heat increase and, conversely, creating an electric charge and causing a reading on the instrument. The three combustible gas sensor types are shown in **Figure 6-16.** When purchasing or using a monitor, responders should be aware of the different sensor types. Readings can and do vary among the three, and the safety of responders is in the hands of that instrument so they must understand how it works.

Catalytic Bead

The **catalytic bead** sensor is the most common sensor in use today. A catalytic bead sensor is wire between two poles with a solid bead of metal in the center. There are two wires and beads in the sensor: one burns the gas and the other is used to compare the heat increase or decrease. The beads are coated with

a catalytic material that facilitates the burn off of the gas. These sensors replaced the wheatstone bridge sensor, which was a coiled piece of platinum wire between two poles.

Metal Oxide Sensor

The **metal oxide sensor** is commonly referred to as an **MOS** and is sometimes referred to as a tin dioxide sensor. Its very nature makes it a very sensitive sensor, which causes all of its "perceived" problems. If used correctly (and interpreted correctly) this sensor can provide many clues and answers at chemical incidents.

The MOS is a semiconductor in a sealed unit that has a wheatstone bridge sensor in it surrounded by a coating of a metal oxide. Heater coils provide a constant temperature. When the sample gas passes over the heated bridge, it combines with a pocket of oxygen created from the metal oxide. This reaction causes an electrical change, which causes the FGI to provide a reading. This sensor is very sensitive and will pick up almost anything that crosses it, which can be confusing to the user. It requires regular maintenance and calibration, more so than the other types of sensors. Most MOS FGIs do not provide a readout of the percentage; most provide only an audible warning or provide a number from within a range. The range extends from 0 to 50,000 units and is a relative scale. If a tablespoon of baby oil is spilled on a table and an MOS sensor passed over it, the reading will be in the range of 5 to 30. If an MOS sensor is taken into a room with 5 percent methane, the reading will be within the range for a flammable gas, probably near 45,000 to 50,000 units. Some monitors allow the MOS to read in percentages of the LEL, in addition to a general sensing range of 0 to 50,000 units. The MOS reacts to tiny amounts, which is an outstanding feature of the monitor—most monitors are not nearly as sensitive. Note, however, that if a department can afford only one combustible gas sensor, a catalytic bead sensor is preferred because the MOS can be erratic. The MOS sensor does have great applicability for a hazardous materials team or for a department that already has another type of LEL sensor.

Infrared Sensors

The **infrared sensor** is common in industry but not all that common in the emergency response field. It uses a hot wire to produce a broad range of wavelengths, a filter to obtain the desired wavelength, and a detection device on the other side of the sensor housing. The light that is emitted from the hot wire is split, one wavelength going through the filter, the other to the detection device to be used as a reference source. When a gas is

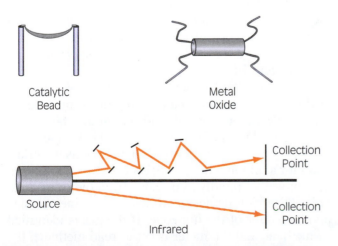

FIGURE 6-16 Examples of flammable gas indicators: catalytic bead, metal oxide, and infrared.

sent into the sample chamber, the gas molecules will absorb some of the infrared light and will not reach the detection device, which will read the amount of light reaching the detector. The amount of light reaching the detection device is compared and a reading is provided. The big advantage of infrared is that it does not require oxygen in which to function, that is, it can take readings in oxygen-deficient atmospheres. The device also is not affected by temperature, nor is it easily poisoned by high exposures. The disadvantages include cost and its many cross sensitivities. This is another sensor that should not be purchased as a stand-alone unit, but should be considered after the purchase of a catalytic bead or wheatstone bridge sensor.

Toxic Gas Monitors

This section describes the sensors that are commonly used in three-, four-, and five-gas units (LEL, O_2, toxic 1, toxic 2, toxic 3) and are usually used to measure carbon monoxide (CO) and hydrogen sulfide (H_2S). Toxic sensors are available for a variety of gases, however, including chlorine, sulfur dioxide, hydrogen chloride, hydrogen cyanide, and nitrogen dioxide. The most common unit sold today by far is a four-gas unit that measures LEL, O_2, CO, and H_2S. This combination is a direct result of the confined space regulation issued by OSHA. Many people understand why the first three sensors were chosen, but do not understand why H_2S was chosen because OSHA did not specify H_2S. Most confined space entries (and consequently the most deaths and injuries) take place in sewers and manholes, and H_2S is commonly found in sewers and anyplace else things are rotting. So when choosing one of the optional gases to sample for, it is best to choose H_2S unless the department has a specialized need for some of the other gases.

Many response teams consider choosing one of the other toxic gases mentioned in the preceding paragraph as their fourth or fifth gas, but they must weigh the cost of doing so. More cost-effective methods of detecting toxic gases such as chlorine or sulfur dioxide are available. The average cost for the sensor is $400 to $500 and is usually guaranteed for one year. The calibration gas is $300 to $400 and is only good for six months before it expires. The other problem is that if the instrument is taken into an environment with more than 20 ppm chlorine, it will be ruined.

Most toxic sensors are electrochemical sensors that have electrodes (two or more) and a chemical mixture sealed in a sensor housing. The gases pass over the sensor, causing a chemical reaction within it, and an electrical charge is created, which causes a readout to be displayed. All toxic sensors display in parts per million. Some toxic sensors use metal oxide technology and react in the same fashion as FGI metal oxide sensors.

Tactical Use of Multi-Gas Detection Devices

It is important to understand the types of sensors that are in the multi-gas detection device so that responders can make effective tactical decisions. Among the three flammable gas sensors, there are some notable differences. The metal oxide sensors (MOS) will react very quickly and will detect low levels of flammable gases, whereas the other flammable sensors have a seven to ten second delay in reading the gas levels and a detection threshold that is not as low. The MOS does not report readings that would be considered accurate, and they fluctuate considerably. In high humidity situations, the MOS will detect water vapor and give erroneous readings. MOS sensors offer an advantage in some situations and disadvantages in others. They are useful when determining if there is a contaminant in the air but are not accurate for determining LEL levels. In contrast, the catalytic bead and infrared flammable gas sensors are both accurate for determining LEL levels.

The other consideration when using flammable gas detectors lies in identifying the gas or vapor that is present. Most flammable gas sensors are calibrated for methane and will read only methane accurately. They will detect all other flammable gases, but the readings will not be accurate. Using correction factors will make the devices accurate. For instance, any reading on the flammable gas sensor indicates that there is a flammable gas or vapor present. Most detection devices alarm at 10 percent of the LEL for methane. If the device is calibrated for methane, this reading is accurate only for methane. There are correction factors that exceed a factor of 3, which means your monitor readings are off by a factor of 3. This would not be dangerous if the meter read 10 percent, since the actual reading would then be 30 percent of the LEL. If the meter reading is 40 percent, then multiplied by a factor of 3, the actual reading is 120 percent and well into the flammable range, and in some cases above the upper explosive limit (UEL) for some materials.

When an LEL sensor alarms at the 10 percent level for an unidentified material, first responders should evaluate why they are in a building or in this level of flammable vapor. They should retreat and take steps to reduce the level of flammable gas. They should immediately consult with their local hazardous materials team for additional tactical decisions. Citizens should be removed as rapidly as possible from the immediate area. Detection devices can help determine the size and boundary of the hot zone. If the gas is identified as methane and the meter is set to read methane, the reading of 10 percent is of concern and the source must be identified as rapidly as possible. If the reading at the ground floor door of a three-story apartment building is 10 percent of the LEL, no matter where the source is, the levels on the top floor will be extremely high, and probably well into the flammable range. Any spark could ignite the vapors causing a massive explosion of the building and putting responders at great risk. The electrical company should be immediately contacted to shut off the electricity to the building as rapidly as possible. The electricity should be shut off away from the building, preferably at the closest pole or transformer.

As for the toxic sensors, they are calibrated to their specific gases and are accurate for only those gases. Both CO and H_2S react to some other gases and will give readings for about thirty-five other materials. There is no way for a first responder to determine which gas may be present. Victims may be able to provide some clues; for example, CO is odorless, so if victims or bystanders reported an odor, then CO is not a likely culprit. The exception to this would be if the victims reported a vehicle exhaust odor; carbon monoxide makes up approximately 7 percent of vehicle exhaust. If H_2S is present, victims will report a rotten egg odor; if they do not report this odor, then H_2S is not the culprit. Both CO and H_2S are toxic and any exposure should be avoided. Citizens should be removed from the area if these gases are found. Carbon monoxide is only produced from incomplete combustion, so this source must be located to ensure the occupants safety.

Other Detectors

In most cases the detectors discussed in the following subsections will be used by a hazardous materials team, but the first responder should be aware of the devices and their capabilities. In areas without a hazardous materials team or where the hazardous materials team response times are unusually long, the first responder may want to consider the purchase of these items. As is true of the monitoring devices already discussed, these instruments are complicated, and this is not by far a complete listing of the instruments that should be carried by a hazardous materials team. First responders are capable of beginning the detection process, but in almost all cases further detection capability is required.

Photoionization Detectors

The photoionization detector (PID) is is shown in **Figure 6-17.** Because of their ability to detect a wide variety of gases in small amounts, PIDs are becoming essential tools of hazardous materials response teams. The PID does not indicate what materials are present, much the same as the FGI will not identify the specific material that is present, but when used as a general survey instrument, the user can identify potential

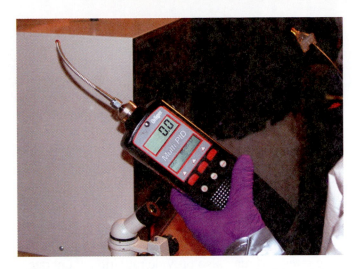

FIGURE 6-17 This instrument is a photoionizing detection device, which has the ability to detect a wide variety of potentially toxic gases or vapors. It doesn't identify the gases, it only indicates their presence. Some instruments combine this detection capability with other sensors for flammable gases, carbon monoxide, hydrogen sulfide, and oxygen.

areas of concern and possible leaks/contamination. These are the instruments that look for toxic materials in the air, and they are essential to determining exposures to many toxic materials. The flammable gas detector will start reading toxic materials in a range of 50 to 500 ppm, which for some gases is extremely high and could have immediate effects on a responder. The PID starts to read at levels near 0.1 ppm, which is very sensitive. There are PIDs that can measure in parts per billion, a very sensitive device. Because of their sensitive nature, they can detect small amounts of hydrocarbons in the soil. Sick building calls are on the increase, and the PID is a valuable tool in identifying possible hot spots within the building.

SAFETY

The dose makes the poison, and so responders should have some understanding of dose levels, as discussed in Chapter 4. Most toxic materials report dose levels in parts per million (ppm) or parts per billion (ppb) levels. These designations typically are indicated for airborne inhalation hazards. Most first-responder toxic sensors report their readings in ppm. Some alarms are for instantaneous readings, some for a fifteen-minute exposure, and others for an eight-hour exposure. An alarm with a carbon monoxide sensor in an industrial application is different than one found in a home. Workplace exposures may be legal at certain levels, and some sensors alarm at levels lower than those required by law. Responders should consult with their local hazardous materials team when confronted with exposure levels that cause their meter to alarm.

Radiation Detection

Many first responders are carrying some form of radiation detection, as shown in **Figure 6-18.** There are two primary types: pagers/dosimeters and detection devices. Many agencies are using radiation pagers or comparable dosimeters. Radiation pagers such as the one shown in **Figure 6-19** provide an indication that a radiation level above background has been detected. It alerts the user that a radiation source is nearby. The dosimeter shown in **Figure 6-20** also alerts when levels above background are indicated, but it provides a continual reading of radiation dose as well. The ability to know the dose level is crucial to responder safety. Detection devices can read radiation dose rates and may also provide radiation count rates. Some devices may also indicate the type of radiation source that may be present.

Colorimetric Sampling

Colorimetric tubes are used for the detection of known and unknown vapors. With sick building calls becoming more and more frequent, the use of colorimetric tubes by hazardous materials teams has greatly increased. Colorimetric sampling consists of taking a

FIGURE 6-19   This radiation pager will alert the user to radiation levels above background.

FIGURE 6-20   This radiation pager will alert when the radiation level increases, and it also provides the radiation dose rate.

glass tube filled with a reagent (usually a powder or crystal). The reagent is placed into a pumping mechanism that causes air to pass over the reagent. A colorimetric sampling device is shown in **Figure 6-21.** If the gas reacts with the reagent, then a color change should occur, indicating a response to the gas sample.

FIGURE 6-18   This detection device monitors the radiation dose and will also identify the radioactive isotope.

FIGURE 6-21 This is a single colorimetric tube, which samples for a specific gas or a chemical family of gases.

Detection tubes are made for a wide variety of gases and generally follow the chemical family lines (i.e., hydrocarbons, halogenated hydrocarbons, acid gases, amines, etc.). Although the tubes may be marked for a specific gas, they usually have cross sensitivities (react to other materials), which at times is the most valuable aspect of colorimetric sampling.

One system that has been developed involves the use of bar-coded sampling chips. This system, called the **chip measurement system (CMS),** involves the insertion of a sampling chip into a pump, **Figure 6-22.** The pump recognizes the chip in use and provides the correct amount of sample through the reagent. The pump, by means of optics and light transfer system reflective measurement, provides an accurate reading of the gas that may be present.

Technology related to the detection of hazardous materials is growing at a rapid rate. There are many detection devices that are available for hazardous materials teams; some are simple indicators and others are laboratory-style devices. Some of the other instruments used by a hazardous materials team include pH detection, flame-ionization detectors, a variety of warfare agent detection devices, and characterization sample kits. The detection of unknown materials is complicated even for some hazardous materials teams. Accordingly, when first responders are confronted with a possible chemical release, no matter how insignificant they think it is, they should request a hazardous materials team or at a minimum consult with the closest hazardous materials group available.

Using Air Monitoring Equipment

Air monitoring is a mission-specific skill at the NFPA Operations level. To use air monitors, responders must understand the response actions based on air monitor readings.

FIGURE 6-22 The chip measurement system uses a chip to analyze a gas and provides a digital readout. Standard colorimetric sampling requires an interpretation of the color change and the length of the change. The CMS takes the guesswork out of the system by providing the reading in ppm.

1. Assure battery level, and bump test the instrument.
   - If the instrument does not respond appropriately to the bump test, then it should be calibrated.
2. Understand what the flammable gas sensor detects and what it does not detect.
   - Most instruments detect flammable gases and vapors.
   - They do not identify the gas, they only indicate the presence of the gas.
   - This sensor takes seven to ten seconds to indicate.
   - The levels indicated are only accurate for the gas the sensor is calibrated for, such as methane (natural gas).
     - If other gases are detected, the readings will need to be corrected.
     - Correction factors are used to correct the readings.
     - Some instruments can be changed to read various gases.

- Readings are usually based on the lower explosive limit for a given flammable gas based on the calibration gas.
    - Most detectors are factory set for methane, based on pentane gas calibration.
- The lower explosive limit (LEL) for methane is 5 percent.
- The flammable gas sensor indicates with a 0 to 100 percent scale. When the detection device reads 100 percent, there is 5 percent methane present.
    - When the device indicates 50 percent, there is 2.5 percent methane present.
3. The oxygen sensor indicates the level of oxygen.
    - Normal oxygen is 20.9 percent.
    - This sensor takes twenty to thirty seconds to indicate.
4. Alarm levels:
    - The flammable gas sensor typically alarms at 10 percent of the LEL for methane.
        - If you do not know the type of gas or vapor released, a reading of 30 percent is extremely dangerous—you do not know the correction factor.
        - Once you identify the gas or vapor, then the correction can be made.
        - An indication of 1 percent means there is a flammable gas or vapor present.
        - A reading of 10 percent or more indicates the need to isolate and evacuate the area.
    - Oxygen alarms at 19.5 percent (low) and 23.5 percent (high).
        - A reading of 20.8 percent means that there is a contaminant in the air causing an oxygen drop.
        - A reading of 19.9 percent means that there is a significant contaminant in the air causing the drop, and the meter has not yet alarmed.

CARBON MONOXIDE INCIDENTS

The CDC (Centers for Disease Control) estimates that more than 40,000 people are treated for carbon monoxide (CO) poisoning each year. In the United States, accidental CO poisoning causes the death of about five hundred people each year and is responsible for 50 percent of the fatal poisonings. More than 2,000 people die each year from intentional CO poisoning. Many communities as part of the fire protection code also require the installation of carbon monoxide (CO) detectors, as shown in **Figure 6-23.** Since CO detectors first became available to homeowners in the mid-90s, the fire service is responding to an increasing

number of calls involving CO detector alarms. In 1995 the city of Chicago experienced several thousand carbon monoxide detector alarms in one day, due to an inversion that kept the smog, pollution, and carbon monoxide at a low elevation within the city. Because carbon monoxide is colorless, odorless, and very toxic it is important that first responders understand the characteristics of carbon monoxide and how the detectors in the home work. As with other chemicals, carbon monoxide (CO) can be an acute or chronic toxicity hazard. It is only acutely toxic at high levels, typically at levels in excess of 100 ppm. At levels of less than 100 ppm, the hazard comes from a chronic exposure, which can be hazardous. The EPA (Environmental Protection Agency) has established an outdoor air quality level of 9 ppm for an eight-hour exposure, and 25 ppm for a one-hour exposure. The Consumer Product Safety Commission recommends that a long-term (eight hours) exposure indoors be less than 15 ppm,

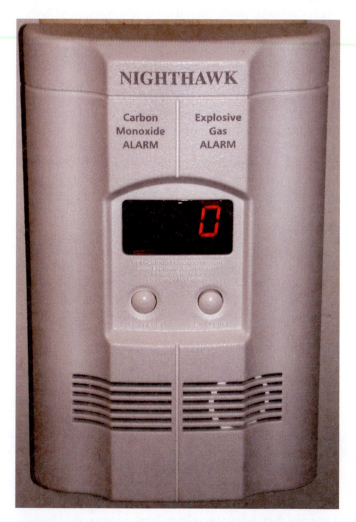

FIGURE 6-23 This carbon monoxide detector for home use not only monitors the CO levels but can also detect flammable gases such as propane and methane. Responders should be aware that this detection device can activate due to a number of causes, not just CO.

and 25 ppm for an hour. Women who are pregnant and are exposed to CO should be encouraged to seek medical attention. Fetal hemoglobin has a higher affinity for CO, and it is possible that the mother may be asymptomatic while the unborn child suffers its ill effects. CO is also a suspected teratogen (agent capable of inducing developmental malformations) and possible **abortifacient** (abortion-inducing agent), and chronic exposure can cause low birth weights as well. The fetus can not clear the resulting carboxyhemoglobin as well as the mother can. First responders' CO detectors typically alarm at 35 ppm. Because of carbon monoxide's properties, it only be detected by a CO detector. In extremely high concentrations, CO can be explosive. Exposure to CO causes flulike symptoms, headache, nausea, dizziness, confusion, and irritability. Exposure to high levels can cause vomiting, chest pain, shortness of breath, loss of consciousness, brain damage, and death.

NOTE

If CO is found in a home, its residents should seek medical attention because signs and symptoms of CO poisoning may be delayed for twenty-four to seventy-two hours.

Although confronted with levels of CO in a house, the residents may not be exhibiting any signs or symptoms. Levels over 100 ppm are extremely dangerous and the residents should be medically evaluated. Monitoring with a CO monitor is essential to determine the possible exposure to CO. Persons who may only show minor effects of CO poisoning and who would normally be transported to the closest hospital need to have the residence monitored. If high levels are found using a monitor, then the exposed residents will need alternative treatment such as the **hyperbaric chamber** and such treatment should not be delayed. An oxygen saturation monitor will often be used incorrectly to determine the O_2 level in a patient. Patients who have been exposed to CO will cause an oxygen saturation monitor to read 100 percent because the monitor reads the oxygen molecule in CO as being O_2. The elderly, children, or women who are pregnant are especially susceptible to CO and may have had a serious exposure without showing any effects.

If the first arriving units do not have a CO monitor and victims may still be in the residence, personnel are to have SCBA on and functioning when searching the residence. After determining no victims are present, crews are to ensure that the house is closed up and then wait outside for a CO monitor. Crews should not enter an area with a CO detector that is activated without the use of SCBA. When using a monitor, if

crews find levels that exceed 35 ppm, crews should use SCBA to continue the investigation. Crews should be suspicious when responding to reports of an unconscious person or reports of "several people down" and should not enter an area without SCBA if it is possible that CO (or other toxic gases) may be present. When several persons are reported to be ill in the same area, the cause is usually chemically related, and first responders should use extreme caution. An air monitor will ensure responder safety with regard to the gases for which it samples. Failure to use an air monitor to check the atmosphere for contaminants is a dangerous practice.

Occasionally there are reports that people may be found unconscious due to a natural gas leak, but this is very unlikely. Natural gas has a distinctive odor, is a non-toxic gas, and will only asphyxiate a person by pushing oxygen out of an area. The only sign of this exposure is unconsciousness or death; any of the flu-like symptoms are due to CO poisoning, not natural gas. If the level of natural gas is high enough to cause unconsciousness, then a very severe explosion hazard is present; in fact, an explosion would be imminent. Air monitoring is critical to responder safety.

When home CO detectors are activated, it is possible that a standard fire department air monitor will not pick up any CO when first responders arrive. This is because the CO detectors purchased for the homes are made to detect small amounts of CO over a long period of time, but fire service detectors provide "instant" readings and only pick up 1 ppm or more. The fact that firefighters may not pick up any readings, however, does not mean the residence's detector is defective. Many factors may cause fire service monitors not to get any readings, including if low amounts of CO are present or a momentary high level of CO activated the alarm but then dissipated prior to fire department arrival. The amount of time the residence is open will also dramatically affect fire service readings. Crews are reminded to keep the residence closed up so that the air monitor has a chance to monitor the level of CO. As a reminder, any time units respond to unknown odors or sick building calls, responders should remove any people from the building and keep it closed up. Because the amounts of toxic gases in sick building incidents are usually small, keeping them contained is very important. A patient cannot be treated for toxic gas exposure unless the source and type of exposure are known. For the patient's long-term health as well as the responder's, quick, reliable gas samples are a necessity.

The brand and type of CO detector will determine how well the device actually performs. The three basic sensing technologies are **biomimetic,** metal oxide, and electrochemical, each having advantages and disadvantages. Location, weather conditions, and the type of sensor will determine the types of readings that can be expected from a particular brand of detector.

Biomimetic

A biomimetic is a gel-like material that is designed to operate in the same fashion as the human body does when exposed to CO. This type of sensor is prone to false alarms, because it cannot reset itself unless it is placed in an environment free of CO, which in most homes is impossible. The sensor may need twenty-four to forty-eight hours to clear itself after an exposure to CO. The actual concentration of CO at the time the detector sounds may be low, but the exposure may have been enough to send the detector over the alarming threshold. If responding to an incident in which one of these detectors has activated, it will need to be placed in a CO-free environment for twenty-four to forty-eight hours to allow itself to clear. It is important for some type of detection device to be left in place for the residents until their detector clears, because it is not advisable to leave them unprotected.

Metal Oxide

This is the same type of sensor that is used in the combustible gas detector, but it is designed to read carbon monoxide. This detector—although superior to the biomimetic sensor—does have some cross sensitivities and will react to other gases.

Although it is hoped that responders would be using a three- to five-gas detection device to check a home,

it is possible for this type of sensor to alarm for natural gas, propane, and other flammable gases. It will even react to the flammable vapors from nail polish remover. Responders using only a CO instrument may find themselves walking into a flammable atmosphere. This type of detector can usually be identified by the use of a power cord because the sensor requires a lot of energy and, in most cases, provides a digital readout. Once activated this sensor needs some time to clear itself, but this is usually less than twenty-four hours.

Electrochemical

An electrochemical sensor is also referred to as an instant detection and response (IDR) sensor, and it is the same type of electrochemical sensor found in three- to five-gas instruments. It has a sensor housing with two charged poles in a chemical slurry. When CO goes across the sensor it causes a chemical reaction, which changes the resistance within the housing. If the amount is high enough, it will cause an alarm. It provides an instant reading of CO and does not require a buildup of CO to activate. It has an internal mechanism that checks the sensor to make sure it is functioning, which is a unique feature. Out of the three types of residential detectors, based on sensor technology, the electrochemical sensor would provide the best sensing capability.

Common sources of CO include furnaces (oil and gas), hot water heaters (oil and gas), fireplaces (wood, coal, and gas), kerosene heaters or other fueled heaters, gasoline engines running inside (basements or garages), barbecue grills burning near the residence (garage or porch), and faulty flues or exhaust pipes.

LESSONS LEARNED

The use of defensive product control methods is a key component for the protection of a community and the environment. In most cases, first responders have the equipment necessary to handle these tasks. With some modification or adaptation, first responders can accomplish the control of many spills. The limiting of spills will mitigate the incident sooner and prevent its spread. If first responders cannot stockpile the necessary equipment, then contacts should be made with those facilities that may have the materials. (Under the requirements of the Oil Pollution Act (OPA) of 1990, certain facilities are required to maintain stockpiles of emergency equipment.)

First responders are also becoming more involved with air monitoring and more aware of the hazards

chemicals present. Incidents involving flammable gases such as methane or propane and response to carbon monoxide incidents are the most common calls that first responders are involved in. When using air monitors, first responders are reminded that they are not all-encompassing and their use requires training and experience. Understanding the action levels are an important consideration for safety—when is safe really safe? Detection devices require testing and regular maintenance and can not be expected to function properly without this upkeep. To determine true levels, first responders must use a range of instruments, which may be above their level. When in doubt, first responders should consult with the local hazardous materials team.

KEY TERMS

Abortifacient   A chemical or material that can cause abortions.

Absorption   A defensive method of controlling a spill by applying a material that absorbs the spilled chemical.

Biomimetic   A form of gas sensor that is used to determine levels of carbon monoxide. It is of the type of sensors used in home CO detectors. It closely recreates the body's reaction to CO and activates an alarm.

Bump Test   Used to determine if an air monitor is working. It will alarm if a toxic gas is present. It is a quick check to make sure the instrument responds to a sample of gas.

Calibration   Used to set the air monitor and to ensure that it reads correctly. When calibrating a monitor, it is exposed to a known quantity of gas to make sure it reads the values correctly.

Catalytic Bead   The most common type of flammable gas sensor that uses two heated beads of metal to determine the presence of flammable gases.

Chip Measurement System (CMS)   A form of colorimetric air sampling in which the gas sample passes through a tube. If the correct color change occurs, the monitor interprets the amount of change and indicates a level of the gas on an LCD screen.

Colorimetric Tubes   Crystal-filled tubes that change colors in the presence of the intended gases. These tubes are made for the detection of known and unknown gases.

Damming   The stopping of a body of water, which at the same time stops the spread of the spilled material.

Diking   A defensive method of stopping a spill. A common dike is constructed of dirt or sand and is used to hold a spilled product. In some facilities, a dike may be preconstructed, such as around a tank farm.

Dilution   The addition of a material to the spilled material to make it less hazardous. In most cases water is used to dilute a spilled material, although other chemicals could be used.

Diverting   Using materials to divert a spill around an item. For instance, several shovels full of dirt can be used to divert a running spill around a storm drain.

Hyperbaric Chamber   A chamber that is usually used to treat scuba divers who ascended too quickly and need extra oxygen to survive. The chamber recreates the high-pressure atmosphere of diving and forces oxygen into the body. It is also successful in the treating of carbon monoxide poisoning and smoke inhalation, because both of these problems require high amounts of oxygen to assist with the patient's recovery.

Infrared Sensor   A sensor that uses infrared light to determine the presence of flammable gases. The light is emitted in the sensor housing and the gas passes through the light. If the gas is flammable the sensor will indicate the presence of the gas.

Metal Oxide Sensor (MOS)   A coiled piece of wire that is heated to determine the presence of flammable gases. Also called tin dioxide sensor.

Multi-Gas Detector   A term used to describe an air monitor that measures oxygen levels, explosive (flammable) levels, and one or two toxic gases such as carbon monoxide or hydrogen sulfide.

Photoionization Detector (PID)   An air monitoring device used by hazardous materials teams to determine the amount of toxic materials in the air.

Remote Shutoffs   Valves that can be used to shut off the flow of a chemical. The term remote is used to denote valves that are located away from the spill.

Retention   The digging of a hole in which to collect a spill. Can be used to contain a running spill or collect a spill from the water.

Vapor Dispersion   The intentional movement of vapors to another area, usually by the use of master streams or hoselines.

REVIEW QUESTIONS

1. What type of dam would be required for a fuel spill in which the fuel has a specific gravity of less than 1?

2. What type of dam is required for a chemical spill in which the material has a specific gravity of greater than 1?

3. Describe the two key items needed to construct an underflow or overflow dam.

4. Who should be consulted prior to using vapor dispersion techniques?

5. Describe the normal configuration for a multi-gas detector.

6. If a detector is calibrated for methane and a person responds to a propane release, describe whether the instrument will detect propane and, if it will, how it does so.

7. Explain why and how often detectors should be calibrated.

8. With regard to safety and the ability to detect various toxic gases, how would you rate the use of a multi-gas detector with this combination: LEL, O_2, CO, H_2S?

9. When responding to situations involving unknown materials, which four detection devices are needed?

10. Does NFPA 472 address hazardous materials Operations level responders' use of detection devices?

11. When do most manufacturers recommend detection devices be calibrated?

12. How long do some WMD detection devices take to indicate?

FOR FURTHER REVIEW

For additional review of the content covered in this chapter, including activities, games, and study materials to prepare for the certification exam, please refer to the following resources:

Firefighter's Handbook Online Companion
Click on our Web site at **http://www.delmarfire.cengage.com** for FREE access to games, quizzes, tips for studying for the certification exam, safety information, links to additional resources and more!

Firefighter's Handbook Study Guide
Order#: 978-1-4180-7322-0
An essential tool for review and exam preparation, this Study Guide combines various types of questions and exercises to evaluate your knowledge of the important concepts presented in each chapter of *Firefighter's Handbook*.

ADDITIONAL RESOURCES

Bevelacqua, Armando, *Hazardous Materials Chemistry,* 2nd ed. Thomson Delmar Learning, Clifton Park, NY, 2006.

Bevelacqua, Armando and Richard Stilp, *Hazardous Materials Field Guide,* 2nd ed. Thomson Delmar Learning, Clifton Park, NY, 2006.

Bevelacqua, Armando and Richard Stilp, *Terrorism Handbook for Operational Responders,* 3rd ed. Delmar, Cengage Learning, Clifton Park, NY, 2003.

Greingor, J. L., J. M. Tosi, S. Ruhlman, and M. Aussedat, "Acute Carbon Monoxide Intoxication During Pregnancy." *Emergency Medicine Journal,* http://emj.bmj.com, 2000.

Hawley, Chris, *Hazardous Materials Air Monitoring and Detection Devices,* 2nd ed. Thomson Delmar Learning, Clifton Park, NY, 2006.

Hawley, Chris, *Hazardous Materials Incidents,* 3rd ed. Thomson Delmar Learning, Clifton Park, NY, 2007.

Henry, Timothy V., *Decontamination for Hazardous Materials Emergencies.* Delmar Publishers, Albany, NY, 1998.

Noll, Gregory, Michael Hildebrand, and James Yvorra, *Hazardous Materials: Managing the Incident,* 3rd ed. Red Hat Publishing, Chester, MD, 2006.

Schnepp, Rob and Paul Gantt, *Hazardous Materials: Regulations, Response and Site Operations.* Delmar Publishers, Albany, NY, 1998.

Stilp, Richard and Armando Bevelacqua, *Emergency Medical Response to Hazardous Materials Incidents.* Delmar Publishers, Albany, NY, 1996.

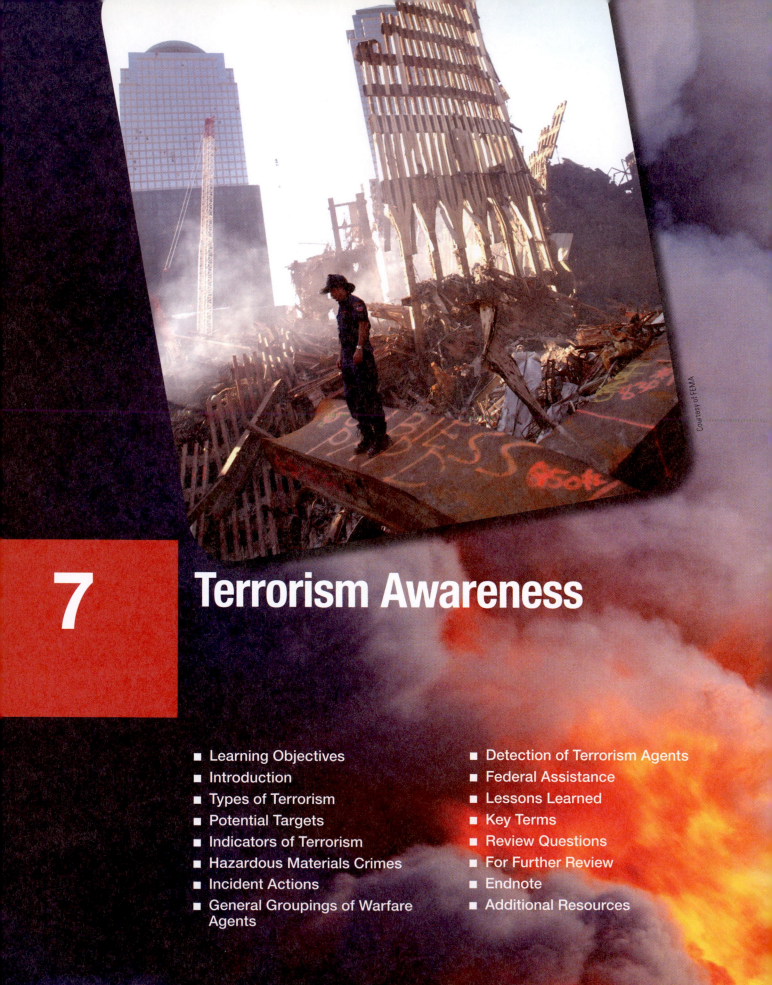

Courtesy of FEMA

7

Terrorism Awareness

The fire service is made up of a large, tight-knit family, and the special bond we have is carried over into the other emergency services.

On September 11, 2001, our family and our nation were attacked in a manner that was unprecedented. After the attacks in New York and Arlington, Virginia, and the plane crash in Shanksville, Pennsylvania, America was at war. Terrorists had struck, and in a few precious minutes we suffered a grievous loss. America changed that day, and the emergency services took a severe hit, but we persevered. During the first few minutes of the fire in the World Trade Center, my thoughts were that it was a large fire, and that it would be a tough battle, but one the FDNY could win. When the towers collapsed, my heart sank. I knew that I had just witnessed hundreds of my family members being struck down. My thoughts were immediately for my good friends who work in FDNY and for all of the other emergency responders at that scene.

The events of that day caused the deaths of many emergency responders and an enormous amount of emotional damage to the survivors, family members, and population of the United States. But on that day, when citizens across the country dialed 9-1-1, emergency responders showed up. . . . We proved that we were not defeated. We are emergency responders and the keystones of our communities. We are the first line of defense against terrorism in this country. We must remain focused on our mission, which is to protect the citizens in our community. With that said, we as emergency responders must look at the real threat that terrorism presents to our communities.

There are many events that occur in this country that present substantial risks to you as a responder and to the community. The events discussed in this chapter are events that can have catastrophic consequences for your community. You must prepare for these events and become tuned in to the potential for terrorism or other criminal behavior in your community. The tactics and motivation of terrorists have changed even since September 11. Suicide/homicide bombers (SHB) are attacking commuter systems throughout the world. SHB attack locations where large numbers of people congregate. In many cases, the profiles of these bombers indicate that they are sympathetic to al Qaeda but not directly associated with them. In Iraq, terrorists are now using explosive devices combined with chlorine and other chemicals. The idea of suicide/homicide bombers is disturbing, and the use of toxic industrial chemicals adds a dangerous twist for terrorism responders. Domestic terrorism is rearing its ugly head again, and although there have been only a few arrests, these individuals present a risk as they typically act alone. Every community has the potential to be attacked or to be a base of terrorist operations. Read the listing of the case histories, and you will learn that all communities are at risk. As you respond in your community, remember those who served on September 11, 2001, and dedicate your career to being the most knowledgeable and prepared firefighter that you can be. . . . You never know what the future may bring.

—*Street Story by Christopher Hawley, Baltimore County Fire Department (Ret.)*

LEARNING OBJECTIVES

After completing this chapter, the reader should be able to:

7-1. Discuss potential target locations.

7-2. Discuss indicators of potential terrorist activity.

7-3. Describe incident actions to be taken at a terrorist attack.

7-4. Describe additional hazards at a terrorist attack.

7-5. Describe other specialized resources to assist with a terrorist attack.

7-6. Describe methods of requesting federal assistance.

7-7. Identify common agents that may be used in a terrorist attack.

INTRODUCTION

It is unfortunate that a chapter on terrorism needs to be included in this firefighting text, and until recent times this would not have been necessary. Although everyone has seen terrorism on the evening news, it used to occur in places such as Northern Ireland, Beirut, Israel, or somewhere other than the United States. Until recently, the United States remained for the most part immune to the reign of terrorists. This changed in February 1993 when the World Trade Center was bombed in New York City, **Figure 7-1.** Even when this bombing occurred, the fire service did not pay much attention, because the bombing was looked at as the type of incident that happened only in big cities. In addition, the persons who were found to be responsible for the bombing were controlled by an influence from outside the United States, so the thought was that it was an isolated foreign attack.

However, when the Alfred P. Murrah Federal Building in Oklahoma City was devastated by a bomb on April 19, 1995, the United States fire service took notice. The attack on the federal building, **Figure 7-2,** was brought on by a person who did not have ties to another country and was a natural born citizen of the United States. It was perceived as an attack on America from one of its own, not from some unknown citizen from a foreign nation.

And perhaps the most devastating attack on the United States occurred on September 11, 2001, when the World Trade Center and the Pentagon were hit with a total of three airplanes. Then, in Shanksville, Pennsylvania, a hijacked plane crashed into a field. As a result of the impact of the planes and the resulting fire, the 110-story twin towers of the World Trade Center collapsed. The death toll from all three sites was more than 2,600 civilians and 346 emergency responders.

Shortly following three horrific crashes, in October 2001 a flood of noncredible anthrax letters hit the country. Mixed in with the thousands of noncredible letters were several letters containing real anthrax. These let-

ters were responsible for five deaths and up to twenty-two other illnesses. The letters were sent to news media outlets and to members of Congress.

Regardless of the origin, the potential for terrorism is here in this country and has to remain in firefighters' thoughts as they respond to any incident. Other incidents are occurring on a regular basis, and, although they do not fit the exact definition of terrorism, they involve the use of large-caliber weapons or a large number of weapons. This chapter looks at terrorism, **hazardous materials crimes,** and other potentially dangerous criminal situations that are sometimes just as deadly as terrorism. Firefighters have to think outside the box in regard to terrorism. There are many events

FIGURE 7-1   In 1993 a van was used to carry explosives to an underground parking garage in the World Trade Center. Six people were killed and more than 1,000 were injured. There were 50,000 people in the building, and the goal of the terrorist was to collapse the building into the adjacent tower.

FIGURE 7-2 A truck bomb caused the devastation in the Oklahoma City bombing in which 167 people were killed and 759 injured. The damage extended several blocks in each direction, and 300 buildings were damaged. Fatalities occurred in 14 separate buildings. *(Courtesy of John O'Connell)*

in Los Angeles and in St. Petersburg, Florida, usually after a sporting event. In Chicago, twenty-one people were killed and fifty were injured in a nightclub after a fight broke out. These deaths may have been caused by a lack of exits and the use of pepper spray by security personnel, which panicked the crowd. The workplace is also becoming an increasingly dangerous place to be. A firefighter in Jackson, Mississippi, entered the headquarters fire station intent on killing the fire chief. Several persons were killed, not to mention the emotional damage that occurred.

Crimes are becoming more violent, and it is perceived that this trend will continue to increase. Emergency responders are immediately placed into dangerous situations, and can get caught—literally and figuratively—in the cross fire.

The following sections provide some information about the type of terrorism agents that currently exist. Some are unique ideas and thoughts, and some represent possible scenarios. There exists some thought within the fire service that this type of information should not be published. All of this information is readily available in the public domain and in training programs and texts throughout the United States. Most of the exact recipes and "how-to" instructions for these and any other device are easily obtained through printed texts and the Internet. This text does not provide any information that cannot be easily obtained elsewhere. If an ordinary citizen is able to obtain it in an easy, normal, and legal fashion, it is certain that the terrorist already has it. Responders need to be aware of the various chemicals and devices that someone may design to kill them and others in their community. The best defenses are education and trying to stay one step ahead of the terrorist. By being informed as to how a terrorist may operate or what some of the devices may involve, responders will be alert to a potentially fatal situation.

that occur in this country that place responders at risk. Terrorism, hazardous materials crimes, murders, and other criminal events all present a risk to firefighters. Someone using ricin to kill another person is committing murder, but the weapon is as deadly to the responders as it is to the intended victim. A booby-trapped drug lab is not terrorism, but it presents significant risk to responders. Many criminals are protecting themselves with body armor and fortified vehicles. Many crimes such as bank robberies involve the use of explosives and sophisticated weaponry. Street violence that can be associated with gangs is increasingly violent, and when people are killed or injured the fire service is called into action. Crimes such as assaults and murders are increasing in the school system. In recent times there have been deadly riots

TYPES OF TERRORISM

The types of terrorism are divided into two distinct areas: foreign based and domestic. Until the Murrah building bombing, the fear of terrorism was aimed at a foreign source, and it was thought that any terrorist attack would be from a foreign country. To be foreign based, the motivation or supervision must come from a foreign country. Domestic terrorism originates from within the United States and is not influenced by any foreign party. From the list of terrorist acts provided earlier and from other statistics, it is clear that the largest percentage of terrorism is domestic in origin.

HISTORY OF MODERN TERRORISM AND HAZARDOUS MATERIALS CRIMES

As a result of the terrorist attacks on September 11, 2001, the previous bombing of the World Trade Center in 1993, the Alfred P. Murrah Federal Building bombing in Oklahoma City, and other incidents, the fire service's response to some types of incidents will be forever changed. In the late 1990s there were two other bombings that are fairly well known to the fire service: the Atlanta Olympic Park bombing in 1996 and the Atlanta abortion clinic bombing in 1997. That particular abortion clinic bombing is significant because a secondary device was used. It is thought that it was strategically placed with the sole intention of harming the responders. The device was placed near the location where the incident command post was set up. Luckily, only minor injuries were received by the responders who were near the blast. In 1998 there was another bombing at a Birmingham, Alabama, abortion clinic in which an off-duty police officer employed as a security guard was killed by a device that was more accurately placed to target the responders. This secondary device was activated by Eric Rudolph, the person responsible as he watched the victim approach the radio-controlled device. A brief overview of some of the recent acts of terrorism, hazardous materials crimes, use of terrorism materials, and criminal acts that presented risk to responders follows:

1980s A series of bombings targeted primarily at the Internal Revenue Service (IRS) included an attempt at a chemical bomb using ammonia and bleach. Another attempt included the use of a hot water heater as a very large pipe bomb, but the vehicle carrying the bomb caught fire while the terrorist was driving the vehicle.

1984 Dr. Michael Swango had a long history of suspicious deaths while he moved throughout the United States and abroad. He was stripped of his license to work as a doctor and got a job as a paramedic. Although his crimes do not fit the pattern of standard terrorism, he was finally arrested for attempting to poison his paramedic coworkers with chemicals.

1984 The Rajneesh Foundation was responsible for poisoning 715 people with Salmonella bacteria. They poisoned the salad bars in ten restaurants. The group had used other biological agents and had targeted a water supply for attack. They had previously used raw sewage and dead rodents in an attempt to poison the system. They used nursing home patients as their test targets for some of the biological attacks. The attacks were an attempt to effect a change in a local election.

1985 Members of a militant group developed a plot to poison a water supply. They were going to use 35 gallons of cyanide, which they had in their possession. Other members of the group were arrested for church arsons and attempting to blow up gas pipelines.

1993 The World Trade Center was damaged by a van bomb. Six people were killed.

1995 A man was arrested after manufacturing **ricin,** an extremely deadly toxin, with the intention of killing someone he was jealous of.

1995 The Alfred P. Murrah Federal Building in Oklahoma City was the target of a truck bomb. Several hundred people were injured and 186 people were killed in the blast.

1995 The members of a militia group known as the Patriots Council were arrested for the manufacture of ricin in the attempted assassination of a U.S. marshal.

1995 Two separate ricin incidents involved two doctors. Dr. Deborah Green killed two of her three children by burning her house, and she attempted to kill her husband three times with ricin. In Virginia, Dr. Ray Mettetal attempted to murder his boss with a syringe filled with boric acid and saline. He also possessed ricin and a number of other materials.

1995 (and throughout the 1990s) The Aum Shinrikyo are the cult group that is best known for a 1995 sarin nerve agent attack in the Tokyo subway. For a long period of time the group carried off chemical and biological weapons attacks and went to great lengths to develop their weapons program. An examination of this group yields many valuable lessons on chemical and biological attacks. The group had assets of more than $300 million. The group was also home to more than a hundred scientists whose sole function was to develop chemical and biological weapons. The group used, or explored the use of, Clostridium botulinum, Bacillus anthracis (anthrax), Q fever, Ebola virus, and viral hemorrhagic fever. They carried off ten chemical attacks and nine biological attacks, not to mention numerous small and large full-scale tests. They killed seven people and injured 200 in a sarin nerve agent attack in Matsumoto in 1994. The subway attack in 1995 killed thirteen but injured more than 5,000. The attack consisted of placing sarin in three subway lines, with a total of eleven sarin bags that were pierced with umbrellas. The deaths occurred to those who came in contact with liquid or were in small confined spaces. Over 4,000 people went on their own to 278 hospitals. The Tokyo Fire Department ambulances transported 688 patients. The breakdown of the injured was seventeen critical and thirty-seven severe. There were 948 patients suffering from miosis (pinpoint pupils). All of the moderately injured patients were treated and released within six hours after they arrived at the hospital. Although 85 percent of the patients did not require any treatment, they still flooded the hospitals.

1995 The Anaheim Fire Department in California was made aware of a potential sarin nerve agent attack at

Disneyland. The chief was notified at midnight to be at the command post four hours later. Police agencies, federal law enforcement, and the military had known about the threat for five days. Up until notifying the fire chief and assembling the resources prior to the event, no other planning had occurred. The fire department was placed in charge of the incident and quickly developed a plan of action. Disneyland did not close and 30,000 to 40,000 people visited the park during the threat period. Luckily, the threat never materialized.

1996 Members of a paramilitary group who had access to preplanning information obtained through the fire department were arrested when they attempted to blow up a Department of Justice complex in West Virginia.

1996 and 1997 The Atlanta area was besieged with bombings, including one at the Olympic Park that killed one person. At an abortion clinic in Birmingham, Alabama, a booby-trapped device killed an off-duty police officer and wounded a nurse.

1996 Theodore Kaczynski, known as the Unabomber, was responsible for a string of bombings that lasted eighteen years. He sent sixteen bombs, which killed three people and injured twenty-three. He wrote a manifesto that appeared in the New York Times and was recognized by his brother, who turned him in to the Federal Bureau of Investigation (FBI).

1996 A disgruntled firefighter in Jacksonville, Mississippi, went to his department's headquarters with the intent of killing the fire chief. The fire chief was not injured, but several other fire department personnel were killed.

1996 A lab worker brought in pastries that were poisoned with Shigella dysenterie to coworkers. She made twelve people ill and was found guilty on five felony assault charges. Police also learned that she had attempted on more than one occasion to poison a former boyfriend with Shigella and other biological agents.

1997 EMS and police responded to a shooting incident, and the suspect was found to have ricin, E. coli, and a mixture of nicotine and dimethyl sulfoxide (DMSO).

1997 Four members of the Ku Klux Klan were arrested for plotting to blow up a hydrogen sulfide tank in order to create a diversion for an armored car robbery.

1998 In Nairobi, Kenya and Dar-es-Salaam, Tanzania, bombs exploded in two U.S. embassies, killing 224 people.

1998 Kathryn Schoonover attempted to mail 100 envelopes with cyanide disguised as diet powder all over the United States.

1998 Two men were arrested on the suspicion of **anthrax** possession. They were later released, since it was determined that they only possessed a possible anthrax vaccine. Larry Wayne Harris had previously been arrested in 1995 for the possession of plague, a biological

toxin that he had ordered through the mail. He was a previous member of the Aryan Nation but was deemed too radical for the group.

1998 In Charlotte, North Carolina, a man with an explosive device held some hostages in a government building. The explosive device was thought to also contain some type of chemical agent. It was later determined that the filling agent was harmless, although the explosive was live.

1998 Abortion clinics in Florida, Louisiana, and Texas were affected by attacks using **butyric acid,** a material with a horrible, irritating odor.

1998 In Lafayette, Indiana, a pickup truck rammed into the courthouse. The bed of the pickup had flammable and combustible materials as well as several explosives.

1999 In Colorado, two students who were armed with an array of guns and explosives attacked their own high school. The suicide attack resulted in fifteen deaths and spurred a rash of bomb hoaxes throughout the country.

1999 The FBI investigated hundreds of anthrax hoaxes, none of which involved the actual use of the biological agent. Abortion clinics were the targets in most of the cases.

1999 Ahmed Ressam was arrested in Port Angeles when he crossed over the border into the United States. He was responsible for plotting the millennium "border bomb" and had the components to a large bomb in his car. His target was the Los Angeles Airport (LAX) or the Space Needle in Seattle on New Years Eve. This plot is the responsibility of an al Qaeda sympathetic group, who have worldwide ties to other terrorist cells.

2000 Dr. Larry Ford's house in Irvine, California, was searched after he attempted to murder his partner. Dr. Ford committed suicide, and his house was found to have held numerous chemicals and chemical agents. The search took several weeks, and chemicals, guns, and biological materials were found.

2000 During a genetics conference in Minneapolis, Minnesota, protestors attacked three restaurants with hydrogen cyanide.

2000 Members of a militia group in Sacramento, California, were arrested for plotting to blow up a 24-million-gallon propane tank.

2001 Over 2,600 civilians and 365 emergency responders lost their lives at the World Trade Center in New York City; the Pentagon in Arlington, Virginia; and a field in Shanksville, Pennsylvania. Four planes were used in the attack, two hitting the World Trade Center and one hitting the Pentagon. The fourth plane crashed into a field. Members of the al Qaeda group have been held responsible for the attack.

2001 In October there was one death as a result of an anthrax-laden letter. Later, there were a total of seven deaths from anthrax. The emergency services across the country were deluged with calls about white powder. A very small percentage of the calls was investigated by the FBI, but they opened more than 14,000 cases regarding the white powder events. Only five cases involved real anthrax, which was sent to members of Congress, to New York City, and to Boca Raton, Florida. In December, Clayton Lee Waagner was arrested and admitted to sending more than 500 noncredible anthrax letters to women's reproductive health centers (WRHCs). He was a member of the Army of God, a group known for pursuit of the right to life cause. As of 2007, no one has been arrested for the anthrax attacks. The FBI is still actively pursuing the perpetrator. They have traveled all over the world looking for clues and have interviewed more than 9,100 persons.

2001 The Animal Liberation Front (ALF) and the Earth Liberation Front (ELF) committed a number of terrorist acts throughout the year. The ALF admitted to 137 illegal actions at a variety of locations. Both groups target buildings and businesses that have any connection to the use of animals or perceived violation of the environment. The acts typically create inconveniences, such as glued locks, but on occasion do include arson and other violent destruction of property. For the most part these groups take great care to avoid any potential injuries to humans, but responders could be killed or injured while responding to or handling these acts.

2001 Two Jewish Defense League (JDL) members were arrested for plotting to blow up a mosque and a congressman's office. Other members of the JDL have been quoted as having a desire to continue the militant work of the two men arrested.

2001 Richard Reid was arrested on a Paris-to-Miami flight for attempting to detonate a PETN explosive that was located in his tennis shoes. PETN is one of the more powerful explosives. He was overpowered by other passengers and the flight crew while trying to light the fuse. He has ties to al Qaeda and is supportive of their cause.

2002 Two different bombers created fear in the United States, one in Philadelphia and the other in the Midwest. Preston Lit set pipe bombs off in U.S. Postal Service mailboxes in the Philadelphia area. Lucas Helder, a college student, set off pipe bombs in residential home mailboxes in five states. Helder's intention was to set off bombs so that the explosions when drawn on a map would form a smiley face. His bombings injured six people. He set a total of eighteen pipe bombs, six of which exploded.

2002 A fifteen-year-old stole a small plane and flew it into a high-rise building in Tampa, Florida, killing himself. His suicide note stated that he wanted to be just like Tim (McVeigh), Eric (Rudolph), and Osama (bin Laden).

2002 Joseph Konopka, a member of the Realm of Chaos group, had stashed potassium and hydrogen cyanide in a Chicago subway tunnel. The group is known for wanting to destroy public utility, water, sewage, and telecommunications systems.

2002 The leaders of two militia groups in Kentucky and Pennsylvania were arrested for possession of weapons and explosives.

2002 The FBI arrested Josě Padilla, also known as Abdullah al Muhajir, who is associated with al Qaeda, in Chicago. He was researching the use of, and looking for materials to detonate, a radiological dispersion device. He had traveled to Pakistan and had studied methods to pull off such an attack with al Qaeda operatives. In late 2002 the FBI was still looking for 100 al Qaeda members and investigating 150 persons and groups who may have al Qaeda ties. They made two large arrests in Detroit and Buffalo, as well as other arrests throughout the United States, apprehending a number of suspects who had al Qaeda ties.

2003 Mr. Iyman Faris was sentenced to twenty years in prison for providing assistance to al Qaeda and conspiring to commit a terrorist act. Mr. Faris was a U.S. citizen who was born in Kashmir, and last lived in Columbus, Ohio. He was accused of scouting targets, one of which included the Brooklyn Bridge. He had traveled to Afghanistan in 2000 to attend a training camp with Bin Laden. His experience as a truck driver and access to cargo planes was of interest to al Qaeda. He performed a number of research tasks for the al Qaeda leadership. In early 2003, he became a double agent for the FBI, providing information on al Qaeda. In 2004, Nuradin Abdi was charged with conspiracy to commit a terrorist act, by setting off explosives at a shopping center. When Adbi arrived in Columbus, it was Faris who picked him up.

2004 There was a series of bombings in Madrid, Spain. The attacks, which took place on four commuter trains, killed 191 persons and injured 2,050. Although it was suspected that the attacks were carried out by al Qaeda, the actual terrorists were local Islamics and several others who were not Islamic. The terrorists used thirteen backpack explosives, of which ten detonated. The attack was very well coordinated and each of the explosions took place within several minutes of each other in different parts if the city, overwhelming the emergency response system. The attacks did create some attention among other terrorist groups, including al Qaeda, as the attack changed the outcome of the national elections. The government immediately placed blame on the Euskadi Ta Askatasuna, which is commonly called the ETA. This group has committed a number of terrorist attacks in Spain and was a likely suspect group. There were a number of misstatements made by government officials, which did not make the citizens happy. On the days following the attack, thousands of citizens protested in the streets. The demonstrations increased in size and intensity each day new facts were learned that contradicted the government's position. As a result, the party that was in power prior to the attack lost the election and government control went to the opposition party.

2004 William Krar was sentenced to eleven years, along with Judith Bruey who was sentenced for just over four months in federal prison. They were charged with possession of sodium cyanide and other chemicals which could be used in weapons of mass destruction. The chemicals were primarily acids, which could be used to produce sodium cyanide gas if mixed with the sodium cyanide. In addition to the chemicals, the FBI found hundreds of thousands of ammunition rounds, a substantial number of pipe bombs, machine guns, and a number of remote-controlled explosive devices disguised as briefcases. Mr. Krar had sent false identification documents, but the package was delivered to the wrong address. The exact motive or intended target of the weapons is not known, but Mr. Krar had a history of being anti-government and had a number of previous weapons-related arrests.

2004 Demetrius Van Crocker was convicted in a domestic terrorism case. Working through intelligence, the FBI was able to work the case with an undercover FBI agent. The FBI was posing as a security guard at the U.S. Army Pine Bluff Arsenal, one of the locations that stored military chemical weapons. Van Crocker conspired with the FBI agent to purchase sarin nerve agent and C-4 explosives, which the agent purported that he could steal from the arsenal. Van Crocker was a white supremacist who disliked the government and supported Timothy McVeigh. Although Van Crocker had a low IQ, he had studied the process to manufacture military chemical weapons and other toxic industrial chemicals that could be used as effective weapons as well. Through discussions with the undercover agent, he was able to describe several scenarios involving chemical weapons, and he had more than a general interest in the topic.

2005 Much like the Madrid bombings, London's public subway system was the location for a series of coordinated attacks. There were two sets of bombings, a series on July 7 and July 21. Within one minute of each other, three TATP-based bombs exploded on the London Underground. An hour later a fourth backpack bomb exploded on a double-decker bus. The attack killed fifty-two, including four of the attackers, and injured more than seven hundred. The placement and timing of the explosions created hardships for the rescue crews, as the trains had left and were in between stations. The relationship with al Qaeda is not known, but it is suspected that the terrorists were sympathetic to their cause. The bombings on the 21st were similar, but the four explosive devices malfunctioned and did not detonate. Three trains and a bus were targeted, but only

the detonators exploded and the main charges failed to detonate. Like the earlier bombs, these were made of TATP, which is very unstable. Two days later, another bomb was found similar to the other devices. The explosives in both cases had been homemade and considered very unstable. It is possible that the explosives were made at the same time, and due to the length of time intervening, the explosives deteriorated and were no longer explosive. In both bombings, sixteen people were arrested and a number of others were held and released.

2005 Bali, Indonesia, and two adjacent communities were the targets of a number of explosions. The bombings occurred in markets and other areas where tourists are known to congregate. In one bombing, the SHB walked into a crowded café and set off a backpack bomb. In all, twenty-three people were killed, including the three SHB. Among the 129 injured, there were six Americans. The attacks came before Ramadan, which is a period of significance for Muslims. A number of Australian tourists were in Bali for vacation during this period, and the victims included four Australians who died and 19 who were injured. In 2004, the Australian Embassy in Jarkarta was bombed. In August 2003, a SHB detonated a car bomb outside Jakarta's Marriott hotel. This attack against a U.S. interest killed 12 and injured 150 persons. In 2002, there was another series of well-coordinated attacks in Bali, where 202 persons were killed and 209 were injured. The majority of those killed were from countries outside of Indonesia, including seven from the United States. Australia suffered the most casualties, as the area is popular with Australian tourists. The terrorists used diversionary attacks to move victims to a central location where a larger van bomb was exploded. The SHB in the van did not have to activate the bomb, as it could have been remotely controlled if the bomber had a change of mind. In all of the bombings, members of the Islamic group Jemaah Islamiyah have been arrested or suspected to be behind these years of terrorism.

2006 Police in Ontario, Canada, arrested seventeen persons thought to be part of an Islamic terrorist cell. Allegedly the group had made plans to use truck bombs to target government and other high-profile buildings and to attack events where there would be large numbers of civilians. The group had ordered 6,600 pounds of ammonium nitrate, a common explosive material. As the police had intelligence information and had infiltrated the group, they were able to make the arrests prior to any terrorist actions occurring. In addition to the truck bombs, the group had made plans to storm government buildings, take hostages, and execute high-level government officials.

2006 The FBI was alerted that Derrick Shareef, also known as Talib abu Salam Ibn Shareef, had a desire to commit a terrorist act in Rockford, Illinois, planning to attack government buildings and commit other crimes to raise money. During discussions with an informant, Shareef changed his plan of attack to that of the Rockford Mall. Shareef discussed with an undercover FBI agent the desire to purchase four hand grenades and two pistols. He traded two stereo speakers for the grenades and one pistol, and as the U.S. Attorney stated, he fit the profile of a "wannabe" terrorist. Shareef planned his attack on the mall with the thought to maximize casualties and, it being Christmastime, he hoped to panic the citizens across the United States.

2006 Police in London arrested twenty-four persons for conspiring to hijack and destroy ten airplanes. The planes in question were departing London and headed to the United States, and the explosives were going to be hidden in the carry-on luggage. As a result of these arrests, liquids were thereafter banned from carry-on luggage, as the explosives were allegedly liquid-based. The explosives most likely would have been triacetone triperoxide (TATP) and hexamethylene triperoxide diamine (HMTD). The use of TATP has attracted the attention of terrorists due to its ease of manufacture using common chemicals.

2007 The American Embassy in Greece was hit with a rocket-propelled grenade (RPG) in the second such attack of the American Embassy. The Marxist group Revolutionary Struggle is suspected in this attack. There were no injuries or deaths as a result of the attack, although the Embassy suffered some minor damage. The Greek terrorist group November 17 had previously attacked the embassy in 1975.

The FBI defines terrorism as a violent act or an act dangerous to human life in violation of the criminal laws of the United States or any segment to intimidate or coerce a government, the civilian population, or any segment thereof, in furtherance of political or social objectives. The key to this definition is the intimidation of the government or the civilian population. A militant group trying to influence the local political process sprayed Salmonella bacteria on a fast-food restaurant salad bar and was successful in making more than 600 people sick. The Tokyo subway sarin attack was an attempt to destroy a good portion of the police department in an effort to prevent a raid on the terrorist compound.

SAFETY

The fire service will be called to many incidents that will not fit the exact definition of terrorism, but the hazards from a pipe bomb are the same regardless of the motivation of the builder.

Many responses that would have been routinely handled in the past must now be treated much differently, and responders must always be on their guard for terrorist-style devices or potential acts of terrorism.

The terrorist's motivation is to produce fear that may be aimed at the general public or the government. Fear can be provoked by large-scale actions such as the acts on September 11, 2001, the original bombing of the World Trade Center in 1993, or the bombing of the Alfred P. Murrah Federal Building in Oklahoma City. The rail attacks in Madrid affected the outcome of Spain's national election. Even acts that are not terrorism can create terror in the community. The Chicago nightclub incident, which was caused by pepper spray, and the Rhode Island nightclub fire have sparked fear and concern about safety in nightclubs. A terrorist can also incite fear just by planting the thought of potential terrorism or by devising a hoax. The latter scenario is the more likely and can be very difficult to handle from an emergency service perspective.

The thought process for determining if a threat is credible or not has five elements. If the person known or thought to be responsible is determined to have several of these capabilities, it increases the credibility factor:

1. The first of the five qualifiers involves the potential terrorist's educational ability to make a device or agent that, unlike explosives or ricin, is very difficult to attain. To truly make a biological pathogen agent, in most cases, one needs an advanced knowledge of biology. Ricin, a biological toxin, does not require any advanced knowledge compared with anthrax. Some of the threats with letters or packages have misstated the origin of the material, such as calling anthrax a virus, or have misspelled the agent's name. If the terrorists do not know the true origin of the material or cannot spell the material correctly, they probably do not have the education necessary to make the material they state they produced. This does not take into account a person who may purchase the material.

2. The next qualifier is a person's ability to obtain the raw materials necessary to make the agents. Many of the materials necessary to make chemical warfare agents are banned for sale. Others appear on hot lists, which means they are only sold to legitimate businesses. This would not preclude someone from buying the raw materials on the black market or stealing them from a legitimate business.

3. The third qualifier is the ability to manufacture the devices or machinery required to make the agent. To manufacture chemical warfare agents requires the use of a reactor vessel, which requires about a 10-foot by 10-foot (3-meter by 3-meter) space to produce less than a gallon (3.7 liters). There are some agents that could be produced in a bathtub using backyard chemistry, but these are not the high-end agents that attract much attention. Many people who have attempted to make agents in less than ideal conditions have died during the production process. Many criminals do not take the time to follow standard industrial safety precautions.

4. One qualifier that is often overlooked is the ability to disseminate these agents. The military conducted many tests on chemical and biological warfare agents, and although they have some good methods of dissemination, even they lack a 100 percent effective method of dissemination. The Aum Shinrikyo cult in Japan is a perfect example, as they were a group with millions of dollars in assets and full chemical and biological lab and production facilities. They employed the services of 235 scientists to develop and manufacture chemical and biological agents. The Aum Shinrikyo abandoned their biological weapons program after a full-scale release of anthrax that failed. They used sarin nerve agent twice, the first time in Matsumoto, Japan, in which seven people were killed and 200 injured. The dissemination method used in the Matsumoto attack was much more effective than the one used in the Tokyo subway attack. If they had used the same dissemination method they would have greatly altered the course of events. They would have been limited by the amount of agent that could have been produced in a short period of time.

5. The last qualifier, which is the most important, is whether the person or group has the motivation to pull off the attack. The intentional killing of one person takes significant motivation, and the intentional killing of hundreds requires a whole lot of motivation. There is always the potential for an attack, but it takes considerable education, raw materials, manufacturing, and dissemination ability to pull off a chemical or biological terrorist attack. On the other hand, explosives are easy to manufacture and do not require much education, only simple tools, and the materials required are easy to assemble. It is for this reason that explosives are used in the majority of cases and are quite successful in completing an attack. In many cases the terrorist can be successful because of the hysteria associated with a potential terrorism incident. A balance must be struck between a cautious approach and one that does not allow the terrorist to win by crippling a community and causing hysteria.

Another consideration, as we have seen in England and Ireland, is the disturbing trend of viewing emergency service personnel as targets. One theory currently under examination is that the second explosion at one of the Atlanta abortion clinic bombings was aimed at the emergency responders.

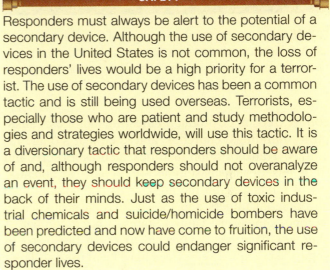

SAFETY

Responders must always be alert to the potential of a secondary device. Although the use of secondary devices in the United States is not common, the loss of responders' lives would be a high priority for a terrorist. The use of secondary devices has been a common tactic and is still being used overseas. Terrorists, especially those who are patient and study methodologies and strategies worldwide, will use this tactic. It is a diversionary tactic that responders should be aware of and, although responders should not overanalyze an event, they should keep secondary devices in the back of their minds. Just as the use of toxic industrial chemicals and suicide/homicide bombers have been predicted and now have come to fruition, the use of secondary devices could endanger significant responder lives.

POTENTIAL TARGETS

Potential targets exist throughout every community in the nation and can be commercial buildings, high-rise buildings, and even residential homes. Although some incidents do not fit the definition of terrorism, the materials used are the same as those a terrorist might use. Whether the objective is murder or terrorism, the danger to the responder is the same. When looking at terrorism, potential targets can be grouped into several categories: public assembly such as the area shown in **Figure 7-3;** federal, state, and local public buildings; mass transit systems; high economic impact areas; telecommunication facilities; and historical or symbolic locations.

While obviously not an exclusive list, buildings that could be targeted include the Federal Bureau of Investigation (FBI); the Bureau of Alcohol, Tobacco, and Firearms (ATF); the Internal Revenue Service (IRS); military installations, Social Security buildings; transportation areas; city or county buildings, including fire and police stations; abortion clinics and Planned Parenthood offices; fur stores; laboratories;

FIGURE 7-3 Any location is a potential target for a terrorist. Any location where large numbers of people are present, such as a mall or sports event, is a prime target.

colleges; cosmetic production/testing facilities; banks; utility buildings; churches; and chemical storage facilities. Transportation facilities such as airports and train, bus, or subway stations are high on the potential list of targets, given the number of people who may be potential targets and the relative ease of escape. In the southeastern United States, a large number of churches have been subjected to arson fires, and in some cases explosive devices have been used. A large number of abortion clinics have been the subject of bombings, attacks, and other threats. Any incident in or near one of these facilities should be approached with caution. Responders should know the location of these facilities in their jurisdiction. Preplans for these facilities should be thought out by the company of-

ficers, but it is not recommended that these plans be committed to paper. As the battle between pro-life and pro-choice groups continues to rage, incidents at these facilities can only be expected to rise, with emergency responders caught in the cross fire.

Many of the potential targets of terrorism have not been buildings at all but events where large numbers of people are present. The Atlanta Olympic Park bombing is an example. Other scenarios involve sports stadiums, such as the one pictured in **Figure 7-4,** public assembly locations, transportation hubs, and fairs and festivals. First responders should have some preplans for these types of locations. One possible scenario for a stadium, devised by Captain Richard Brooks (ret.) of the Baltimore County Fire Department, describes the first-in medic unit arriving at a stadium where in Section 300 there are 40 people projectile vomiting. After five minutes 200 people are projectile vomiting, and as time goes on the number increases. What happens to the responders when confronted with a situation of this nature? How many responders would be affected by this massive amount of people vomiting? How many responders would be needed to handle this incident? This act of terrorism could be accomplished by putting syrup of ipecac in the ketchup container beside one of the hot dog vendors, an easy task. Imagine the hysteria if a note was found stating that a biological agent was distributed in that section. What impact would that have on the remaining 50,000 people in the stadium if that information got out? Planners and responders involved on the national level in trying to develop response profiles to terrorism are grappling with how to plan for incidents involving 100 people, 1,000 people, 10,000

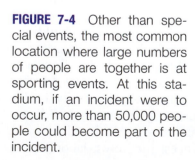

FIGURE 7-4 Other than special events, the most common location where large numbers of people are together is at sporting events. At this stadium, if an incident were to occur, more than 50,000 people could become part of the incident.

people, and 50,000 people. Terrorism incidents can very quickly overwhelm the responders and their whole emergency response system.

When dignitaries visit locations, a lengthy planning process typically takes place in which the fire department should be involved. When the Pope visited Baltimore in 1997 the planning process took more than eight months. Planning for such a large event takes the cooperation of local, state, and federal agencies. Even when dignitaries visit locations such as New York City or Chicago, advance planning occurs. Other events such as political conventions or other large political gatherings all bring the potential for an incident. When one of these events comes to a community, local responders need to be prepared for not only the people arriving to the event, but also the massive federal response that may be pre-positioned. For many special events, whole task forces of federal resources may be hidden away just in case of an incident.

Certain dates have significance to several militant groups. The date of April 19 is the anniversary of the Waco, Texas, incident in which the ATF stormed a compound that housed the Branch Davidians, a group thought to be a militia group. April 19 was also the date of the Oklahoma City bombing, and the date was chosen by the bomber as a way to retaliate for the Waco incident. Other dates provoke the potential for terrorist acts. For instance, the anniversary of Roe v. Wade,[1] January 22, could incite a strike by antiabortion groups. Within the United States, forty states are suspected to have members of militia, patriot, or con-stitutionalist groups. Membership counts vary from fifty people in some states to several thousand members in other states. Groups of concern include anarchist groups and white supremacy groups such as the Ku Klux Klan, Zionist Occupation Government, skinheads, and neo-Nazi groups, including the Aryan Nation. Other groups suspected of activity or thought to have terrorism potential include patriots, New World Order militia, constitutionalists, and tax protesters. To learn which groups are active, it is relatively simple to use a Web browser and search the Internet for many of these groups, because most have Web sites.

INDICATORS OF TERRORISM

An explosion or explosive device is the most common tool of the terrorist, and police across the country have made several arrests of persons for making or storing large quantities of explosives.

FIREFIGHTER FACT

According to an FBI source, more than 93 percent of terrorism incidents use explosives as the weapon of choice.

The most common device is a pipe bomb such as the one shown in **Figure 7-5.** Any incident where an explosion has occurred or first responders believe that an explosion has occurred should be suspected of being a terrorist incident.

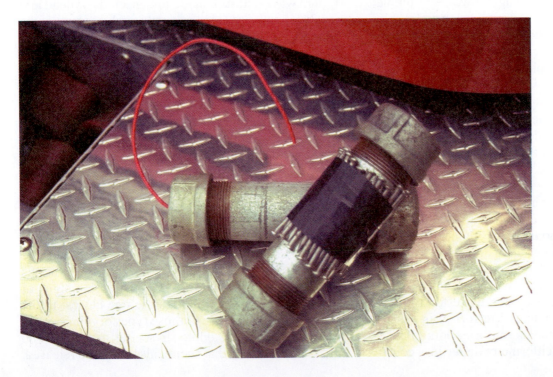

FIGURE 7-5 The most common explosive device is a pipe bomb, and it is very effective. It is a very dangerous device, not only for responders but for the builder as well.

SAFETY

If one explosion has occurred, responders should always assume that there is a second device awaiting their arrival.

In this day and age any suspicious package should be suspected of containing explosives and should be dealt with by a bomb technician. Firefighters should never assume that they can handle the package or remove it from the area. Just like EMS and hazardous materials, the handling of bombs or suspicious packages is a very specialized field and should only be done by a person trained to do that type of job.

The presence of chemicals or lab equipment in an unusual location, such as a home or apartment, is an indication of possible illegal activity. When looking at chemicals or a lab there are three possibilities: drug making, bomb making, or terrorism agent production (chemical or biological). The most likely, based on statistics, is drug making, followed by bomb making. Although very common in some parts of the country, predominantly the West Coast and the Northwest, drug production labs are not commonly located throughout the whole United States. Currently the number of drug labs is increasing in the Midwest, with responders taking down more than twenty a month in some areas. This trend is slowly moving from the West Coast to the East Coast. The most likely scenario is locating someone making explosive devices, as many individuals like to make homemade devices. It is possible, but very unlikely, that firefighters would locate a facility attempting to make a terrorism agent such as sarin. The exception to the biological agents would be for the production of ricin, as the items required to make ricin are easily obtained and the production is just as easy. Fortunately, ricin does not have the potential to easily kill large numbers of people. It is primarily an injection hazard, although it is still very toxic through inhalation or ingestion. A responder in a small community or a rural setting is the most likely to run across any of these types of production areas. Several persons a year are arrested for the possession of ricin with the intent to use it for some type of criminal act. Most of the arrests are occurring in small towns throughout the United States. No matter what agent is located, the responder should immediately isolate the area and call for assistance. The call should go simultaneously to the police, the hazardous materials team, and the bomb squad. This request would apply to all three types of labs—drug, bomb, or warfare agent—as any of these labs should be handled as a cooperative effort.

Another indicator of potential terrorism is the intentional release of chemicals into a building or the environment. Finding a chlorine cylinder in a courthouse would be unusual and should put the responders on alert to the fact that there is a high probability that a terrorist may be at work. Finding chemical containers such as bottles, bags, cylinders, or other containers in unusual locations would also be suspect. In an industrial facility in which chemicals may be common, responders may find that there is the intentional release of a chemical.

NOTE

One of the best indicators of potential terrorism will be a pattern of unexplained illness or injury. If these indicators show up immediately, then the probable cause is a release of a chemical. If these illnesses start showing up a few days to a week later, then the incident is probably biological.

A response to a mall for a seizure patient is not unusual. However, a response to a mall for six people having seizures is very unlikely and could involve a chemical release from a terrorist attack. Seizures, twitching, tightness in the chest, pinpoint pupils, runny nose, nausea, and vomiting are all signs and symptoms of a warfare agent attack. EMS providers will probably be the first group to identify the use of a chemical agent. Imagine arriving at an explosion where there is a large amount of debris and twenty victims. To most the injuries would appear to be blast injuries, as they would in most cases be visible during a quick survey of a patient. Other signs and symptoms, primarily pinpoint pupils, probably would not be identified until the patient is given a more thorough exam. When confronted with victims that are unconscious or dead and now have outward signs of death, such as trauma, the responder must face the possibility that they may have been the victims of a chemical attack. This obviously has to be put in perspective. If responders in a metropolitan area of the Northeast are called to an apartment building in the winter because there are twelve unconscious people, and this happens several times each winter, it is probably carbon monoxide poisoning. However, if responders are called to a mall in the summertime because there are twelve unconscious people, it is probably not due to carbon monoxide but to a chemical release of some type. When distributing a warfare agent the explosive device will usually not have a destructive effect on the building or the surrounding area. In some cases the larger the device the more likely it is that the detonation will consume the agent. Most of the victims will not display signs or symptoms of a blast, although the one closest to the device may suffer some of those types of injuries.

Smelling unusual odors or seeing a vapor cloud may be an indicator that chemicals have been released.

As may be seen with other toxic gases such as chlorine, arriving and finding dead birds or other animals should alert the responder not only to a chemical hazard but also to the potential for terrorism. It would also be unusual to respond to an office building or a house and notice security measures of the type that one would not expect in that occupancy. Items such as extra locks, bars on windows, surveillance cameras, fortified doors, guards, and other unusual protection devices may provide a clue to the responder that something out of the ordinary is at play.

SAFETY

The same hazards that exist in a chemical emergency—thermal, radioactive, asphyxiation, chemical, etiological, mechanical, and psychological—also exist at chemical or biological attacks. In some cases, the risk might be increased, since these incidents are intentional and may be intended specifically to kill or injure responders.

HAZARDOUS MATERIALS CRIMES

Although this chapter covers terrorism, emergency responders may respond to other criminal-related events. Whether the incident is terrorism related or not, the effects on responders can be the same. Criminals are using more weapons than just guns today, and the use of clandestine labs for illicit production is on the rise. These incidents are referred to as hazardous materials crimes, since chemicals are being used in an illegal fashion. Occasionally, there will be a robbery of a convenience store where the weapon is a chemical. The robbery suspect might throw a corrosive liquid on the clerk in order to rob the cash register. There have been other robberies, attacks, or attempted murders using corrosive liquids. Whether or not the law-enforcement community or the prosecutor decides the incident is an act of terrorism, emergency responders need to recognize potentially hazardous situations and have the ability to protect themselves.

One big exposure issue for emergency responders is drug related. Many drug addicts use chemicals as part of the process to get high, and the drugs themselves are usually toxic and may present other hazards. Drug users may use a combination of chemicals to get high, and many of these items are toxic and flammable. Ether is used to assist in the heating process of several types of drugs. This material in pure form is extremely flammable, and the container may eventually become a shock-sensitive explosive. Most drugs are in solid form, which means that they present

little risk unless eaten or touched with bare hands. In some cases the drugs may be stored or used with flammable and toxic liquids.

The most common situation in which emergency responders could be directly affected by drug use is when a person is huffing. While people are huffing they typically use a toxic and/or flammable material. Many common household items such as paints, glues, hair spray, and solvents provide the high that some people desire. From a hazardous materials point of view, these materials are toxic and flammable and airborne. In most cases, users will spray the material into a bag to concentrate the vapors, but when they are done the vapors remain in the room of use.

Clandestine Labs

There are several types of clandestine labs that emergency responders are likely to encounter. The most common is the drug lab, but other possible labs include explosives labs, chemical labs, and biological weapons labs.

SAFETY

All labs have inherent dangers for responders, and all are very dangerous locations to occupy. All may be booby-trapped or be set up to harm responders, and a booby trap does not know if the person coming through the door is a police officer, a firefighter, or an EMT.

For the most part a biological weapons lab, shown in **Figure 7-6,** can run unattended without any major concern and may be shut down in any number of ways without major consequence. A drug, explosives, or chemical lab, on the other hand, should only be shut down by someone trained to do so, as these labs are especially dangerous.

Drug Labs

There are many types of drug labs. Just for methamphetamines, there are more than six common methods of production. In 1973, the Drug Enforcement Administration (DEA) discovered 41 labs; in 1999, they discovered 2,155 labs; in 2001, they raided 12,715 labs; and in 2006, they discovered 6,435 labs. The map shown in **Figure 7-7** provides proof that the labs have moved eastward at a rapid rate; and now all of the United States has to deal with this problem. Due to its popularity, meth production is becoming common but it does involve a dangerous process. The production of drugs requires the use of many chemicals. These chemicals can be purchased outright, stolen, or

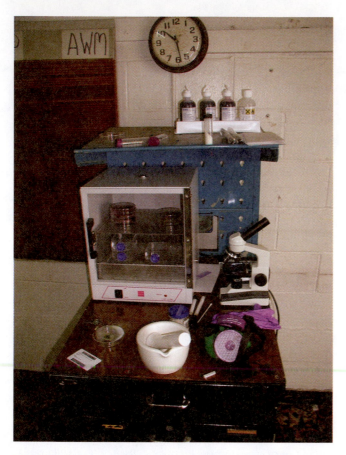

FIGURE 7-6 Example of a possible biological lab. Shown is a microscope and an incubator; both would be used in the production of biological materials.

manufactured using other chemicals. As many drug-producing chemicals are hot listed or cannot be purchased, the producer must resort to innovative methods to produce the chemicals. The "nazi" method of meth production involves the use of anhydrous ammonia, which is usually stolen from a chemical facility. If a 150-pound (68-kg) cylinder of anhydrous ammonia developed a leak or catastrophically failed, people for a considerable distance downwind could be affected, and those in the immediate area would be in grave danger.

Drug labs can be very complicated setups and can be found in any number of locations such as homes, barns, hotels, storage units, and even trucks. Emergency responders routinely encounter these labs inadvertently through other responses. Responders who encounter a drug lab should notify their hazardous materials team, the bomb squad, and the local office of the DEA. The shutting down of a drug lab is a complicated and very dangerous process.

SAFETY

When chemicals are being heated or cooled there is a chance for a violent reaction.

It is this heating and cooling process that indicates the type of lab that may be present. When responders see glassware that is distilling chemicals—in other words, evaporating a certain component—and then rehydrating or condensing another portion of the

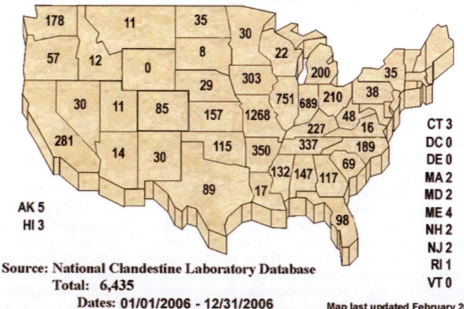

FIGURE 7-7 Methamphetamine lab seizures across the United States. Note the high number of labs in the Midwest. In years past, the largest numbers of lab seizures occurred in the West. The prevalence of methamphetamine labs is moving eastward at a fast pace.

Total of All Meth Clandestine Laboratory Incidents Including Labs, Dumpsites, Chem/Glass/Equipment Calendar Year 2006

178 · 11 · 35 · 30 · 8 · 22 · 57 · 12 · 0 · 303 · 200 · 35 · 29 · 751 · 689 · 210 · 38 · 30 · 11 · 85 · 157 · 1268 · 48 · 16 · 281 · 14 · 30 · 115 · 350 · 227 · 189 · 337 · 69 · 132 · 147 · 117 · 89 · 17 · 98

AK 5
HI 3

CT 3
DC 0
DE 0
MA 2
MD 2
ME 4
NH 2
NJ 2
RI 1
VT 0

Source: National Clandestine Laboratory Database
Total: 6,435
Dates: 01/01/2006 - 12/31/2006

Map last updated February 2007

original chemical, this is indicative of possible drug production. The end result of the process is a solid form, usually a powder. There will be a production line-type formation to the glassware, with some solutions being heated while others are cooled. At some point, gas cylinders may be present and the gas is allowed to mix with some part of the process. Many of the chemicals involved in the production of drugs are flammable, and although many are also toxic the predominant hazard is flammability.

SAFETY

When choosing protective clothing to enter a lab, it is important to protect responders against the predominant hazard.

The materials that are toxic cannot harm the responder as long as SCBA is worn and the materials are not touched with bare hands or eaten. It is highly recommended that personnel from the police, fire, and EMS departments receive training in drug lab awareness, with some receiving specialized training in this area.

Explosives Labs

Although not common, it is possible that emergency responders might encounter an explosives lab, which is the predominant weapon of choice for a terrorist. An explosives lab can be anything from a workbench pipe bomb builder to a full chemical production facility. A person building a pipe bomb does not need much equipment other than some simple hand tools, pipes, caps, and powder. A person making cyclotrimethylenetrinitramine, commonly referred to as RDX, or another more sophisticated explosive will need some more equipment. Depending on the availability of some materials, the bomb maker may have to make some chemical components as opposed to purchasing them. It is when materials are made at home that the danger increases for a responder. The processes to make the chemical components for explosives are very dangerous and present a significant risk to the builder and responders. An explosives lab differs from a drug lab in that most of the processes do not involve heating, cooling, condensing, or distilling. However, some processes used to make the chemicals do perform some of these functions, so there are no black-and-white rules for identification. The major work at an explosives lab is usually mixing of materials, in most cases solids and liquids. If gases are used, they are usually coupled with the explosive device and may be used to increase the heat of the explosion, boosting its efficiency. People making explosives will usually have a large amount of powders in their house.

Some of the other indicators of an explosives lab are the presence of ignition devices, boosting charges, or blasting caps. Many people who manufacture explosives are doing so to make fireworks and will have cardboard tubes for the explosives to be placed into. In one case, it was originally thought that an explosives maker was just making fireworks, but some other explosive devices were found to have BBs and nails taped and glued to the outside of the explosive. Neither of these have an effect on the display ability of the explosive device and are only designed to kill or maim. When an explosives lab is discovered, the local bomb squad and the local hazardous materials team should be called in to assess the materials. As the stability of many of the materials cannot be ensured, it is usually necessary to remotely destroy many of the found materials.

Terrorism Agent Labs

These are the least likely labs to be encountered in emergency response, but responders must be aware of their existence and some of the unique features of these types of labs. The two types of terrorism agent, or **weapons of mass destruction (WMD),** labs are chemical weapons or biological weapons labs. Statistically the most likely lab is a biological toxin lab, which may be used to manufacture ricin, which is shown in **Figure 7-8.** The FBI, on average, arrests a few people a year for the possession of ricin, usually accompanied by a threat. Biological labs may be set up to attempt to make other biological materials. Other than ricin, botulinum toxin, and a few other biological materials, the manufacture of biological weapons is very difficult. The manufacture of ricin is a simple process that only requires a few items, such as castor beans

FIGURE 7-8 Shown is a lab that could be used to produce a biological toxin such as ricin.

and some readily available chemicals. The process to make some of the more advanced biological agents such as anthrax is more difficult and requires a higher level of education, sophisticated equipment, and access to raw materials. The development of biological agents involves culturing the material, usually in a petri dish. These petri or culturing dishes may be placed in an incubator or oven-like device to keep the material warm and at a constant temperature. Depending on the type of agent, there may be grinders, dryers, and sieves present to finish off the product. The major route of entry for many of these products is through touch or ingestion. The only inhalation concern would be during the grinding process, but a simple high-efficiency particulate air (HEPA) mask offers more than enough protection for a responder. Some of the materials used as part of the process may be flammable and have some toxicity, but the chemicals used to clean glassware and tools are usually highly corrosive and in most cases will be sodium hydroxide (lye) and/or bleach. A bioweapons lab will involve some chemicals, but may resemble more of a greenhouse than a chemical lab. Some biological materials are sensitive to light and will be kept in the dark, and the characteristic distilling and condensing glass apparatus will be missing.

Chemical weapons labs use two methods of production: the development of new chemical agents through standard production methods and the synthesis of existing materials. The development of a new chemical agent is the more difficult of the two processes and is nearly impossible except for someone with a chemistry background and access to some chemistry equipment. The recipes for chemical weapons are fairly sophisticated and require access to many raw materials that are hot listed. There have only been two arrests of persons in the United States for attempting to make a chemical weapon such as sarin nerve agent. Both were arrested for ordering the raw materials, and it is thought that the two individuals did not have the educational capacity to manufacture the agent. The production of these agents can be very risky to the producer and, without safety precautions, may result in death. Some of the off gases from the production of chemical warfare agents are extremely dangerous.

The most probable scenario for the development of a chemical warfare agent is the synthesis of an existing product. The criminal would take existing materials, which are usually in diluted form, and synthesize them or reduce them down to a concentrated product. Luckily most, if not all, of the existing products do not present much risk to humans, as they are strictly engineered to harm only insects. This would not stop a terrorist from attempting to use one of these products in an illegal manner, however. There are some pes-

ticides on the market that are applicable to this type of scenario. The standard household pesticide usually has 0.05 to 0.5 percent of pure product mixed with an inert ingredient. A mixture that is used by a farmer may be on the order of 40 to 50 percent pure form and is then diluted in the farmer's tank. In order to be more harmful to humans the pesticide must be concentrated and not diluted. The criminal must devise a method of removing one or more of the inert ingredients. As the inert ingredient is usually flammable or combustible, this is not a difficult task. A mechanism must be devised to off-gas the inert ingredient and then capture the pesticide and collect it.

One form of pesticides, known as technical grade pesticides, is already concentrated. These technical grade pesticides are in pure form and are not diluted. These materials are not in common use but can be found in and around the United States.

INCIDENT ACTIONS

A terrorist incident combines four types of emergency response into a large incident.

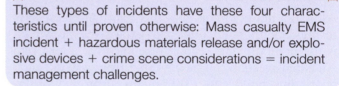

NOTE

These types of incidents have these four characteristics until proven otherwise: Mass casualty EMS incident + hazardous materials release and/or explosive devices + crime scene considerations = incident management challenges.

The first-in companies at these types of incidents can easily be overwhelmed and are going to be committed to basic actions, such as life preservation. The IC will have enormous responsibilities dealing with all of the required actions. All of the components present are difficult to handle individually, and now in this type of incident they are combined and must be handled simultaneously. The handling of a 100-person **mass casualty** incident is difficult and has the potential to overload the IMS, and that may only be one-quarter of the result of a terrorist incident. Responders should examine their response systems and determine how many patients present a concern. Some systems define a mass casualty as an incident involving five victims, while in other systems it may be ten to twelve. Obviously in a system with one EMS unit, more than one patient begins to cause system problems and may involve delays in treatment and transportation.

Imagine responding to an explosion at a mall, with 100 people injured. Such a mass casualty incident (MCI) will have EMS playing a predominant role. The police department, on the other hand, will be concerned with evidence preservation and crime scene considerations. The possibility exists that the perpetrator(s) also used a chemical or biological weapon, and the explosion was the means of distribution. This situation would be an MCI, a crime scene, and a potential chemical release situation with some, if not most, of the patients contaminated. On top of this scenario, add the potential for a secondary explosive device, one that is aimed at the responders! If responders can eliminate the chemical and secondary device issues, then all they have to deal with is the simple 100-person MCI and the crime scene issues. EMS will handle the patients and the police department would handle the criminal element.

Identifying whether an incident is a crime, terrorist act, or just an emergency situation can be difficult. In many cases, in the first few moments, there may be no way to determine a cause. Even a small explosion could be from a number of sources. Differentiating between a chemical or biological attack can sometimes be determined by the immediacy of the symptoms. Chemical agents or toxic industrial chemicals cause immediate reactions and symptoms. Biological agents do not cause immediate effects and may take days to several weeks for symptoms to show. Part of the issue that responders have to deal with is the potential public hysteria regarding biological-threat agents. In some cases, the population may be extremely frightened and may exhibit some psychologically driven medical problems. For some, the symptoms may be real, and for others, they may be suspect.

The other aspect of a suspected terrorism incident will be the massive response from the federal government, even if not requested. Later in this chapter more information is provided about federal resources that are available to respond. In most cases, the minimum response to a suspected terrorism incident would come from the FBI, initially from the local agent. Field offices are located across the country, and almost every major city has a field office. Agent(s) usually arrive one to two hours into the incident. If terrorism is suspected the FBI is the lead agency of the incident as provided by a Presidential Decision Directive known as **PDD 39**. As with hazardous materials incidents it is important to know all involved players before an incident. Knowing who the local FBI agent is can be very important, because meeting for the first time in front of an incident is not conducive to effective scene management.

Working out the "who's in charge" concerns prior to an incident is important. The fire department should liaison with its local police, EMS, and emergency management agency prior to incidents. These and many other agencies are going to be involved and have a variety of responsibilities at an incident.

In general, a unified command is recommended, and although this does not mean command by consensus, input from the various agencies should be considered. The agency with the majority of the tasks to do is generally in charge. In the initial stages the fire/EMS authority would be in charge while rescuing victims, but after the victims are removed and evidence recovery becomes the next priority, the command may switch to the police department.

The magnitude of response to a terrorism incident can be the most difficult part of the incident to manage. The IC will be overtaken by a large number of federal agencies' representatives, who on a regular basis will be replaced by a later arriving supervisor. Response groups consisting of two to seventy responders may arrive uninvited, all trying to assist with the incident. Another group that can also overwhelm the system is the media. In most cases the local media will react as normal, and for the most part will cooperate. Members of the national media do not know, nor do they follow, local protocol, and they will require information. A lack of accurate information can be disastrous to an incident by causing a deterioration of the incident and creating more hysteria than is already present. Media relations are very important in these types of events.

> **NOTE**
> If an incident occurs that involves federal responders, the national media will follow close behind.

One of the most important issues that will arise other than the life safety hazard is that of evidence preservation. As much care as possible should be taken to preserve evidence and make sure it is taken care of appropriately. The collection of evidence is primarily a law-enforcement responsibility. Unless properly trained, fire service personnel should not collect evidence but should alert the police of its presence. A whole host of issues goes along with the collection of evidence, including the chain of custody. This chain of custody, or the paper trail that follows any evidence, is crucial to the successful prosecution of the persons responsible. The failure to follow proper procedure or document the travel of evidence can result in a case being dismissed, regardless of any other evidence. There have been cases, such as the Murrah bombing, in which fire service personnel have collected evidence, but that is a very unusual occurrence. Firefighters who are assigned to collect evidence should be fire investigators or fire marshals because they are typically trained in evidence collection. Another alternative is to double up and use a firefighter and a police officer to collect evidence.

SAMPLING AND EVIDENCE COLLECTION

One of the major concerns of local law enforcement and the FBI regarding the prosecution of a terrorism case is the purity of the evidence. In order for the evidence to be used in a successful prosecution, the evidence must be pure beyond all doubts. If there is any suspicion that the evidence has been tampered with, altered, or contaminated, it could affect the outcome of the prosecution. When responders enter the hazard area, there is great potential for evidence to be destroyed. The best way to proceed is to coordinate your efforts with your local FBI WMD coordinator, who can assist you with the preservation of evidence. Evidence collection is very process driven, but procedures can be implemented that will help preserve any potential evidence. The gold standard for

any courtroom is laboratory analysis, so evidence must be available to be analyzed by a laboratory. This doesn't mean that every incident requires that material be sent off to the laboratory, but steps must be taken to preserve a portion of the evidence in the event that laboratory testing is required.

When there is a large amount of potential evidence, it can be overwhelming, but following a process helps eliminate error and preserve evidence. There is great potential to become involved in a court case, and following the proper procedures every time minimizes your risk in the courtroom. The one time that you don't follow evidentiary procedures will be the time that you are grilled in the courtroom for contaminating the evidence. The FBI and local law enforcement have response teams that collect evidence in these situations, and they can provide great assistance when trying to process a potential crime scene.

A cooperative effort is needed to combat a terrorist attack. The primary functions are rescue/life safety by fire and EMS personnel, hazard identification by the hazardous materials team, identification of possible secondary devices by a bomb technician, and incident management. It is important to communicate the hazards to all personnel, and to limit the response to essential personnel. Instead of having the whole alarm assignment report to the front of the building, it is preferable to use one or two companies to investigate while staging the other companies away from dumpsters, mailboxes, or dead-end streets. A secondary device can be hidden almost anywhere, but the key is to look for something out of the ordinary. When dealing with victims it is essential to isolate them until the cause is identified. The victims can have a large amount of information and should be questioned quickly. Questions to ask include these: What did you see, hear, or smell? Was this coming from one area or was it throughout the building? Did you see anything else suspicious? What type of signs and symptoms do you have? In addition, the police will need to conduct interviews. Documentation and preservation of the evidence are essential to the successful prosecution of the terrorist.

GENERAL GROUPINGS OF WARFARE AGENTS

Terrorists could use any of a number of possible warfare agents. They are classified into three broad areas. Weapons of mass destruction are commonly used

by the military. Some of the regulations that prohibit the making, storing, or using of terrorism agents are called WMD laws or regulations. Any item that has the potential to cause significant harm or damage to a community or a large group of people is considered a WMD. The other two classifications are nuclear, biological, and chemical (NBC) and chemical, biological, radiological, nuclear, and explosive (CBRNE). Both of these are descriptions of the types of materials that could be used in a terrorism attack. Although there are slight variations, they are all used to describe the various types of agents that a terrorist could use. Most of the language differences come from funding legislation or a specific federal agency.

The military has devised a naming system for many of these agents, many of which are listed in **Table 7-1**. Responders should become familiar with these names because much of the literature and help guides refer to these agents by these names. For instance, when using a military detection device, the military name is used. When dealing with terrorism, firefighters are entering another world that has its own language. The fire service has to adopt this new language to survive in this new world. The three groupings mentioned earlier are further subdivided into the categories discussed in the following subsections.

Nerve Agents

Nerve agents are related to organophosphorus pesticides and include tabun, sarin, soman, and V agent. They were designed for one purpose and that is to kill people. Although very toxic, their ability to kill large

SIGNS AND SYMPTOMS OF NERVE AGENTS

All of the nerve agents present the same types of signs and symptoms as organophosphorus pesticides, and in reality the difference is minor. Nerve agents are pesticides for humans and are a stronger, more concentrated version of commercially available pesticides. The signs and symptoms can be generally described using the acronym SLUDGEM, which stands for:

S alivation—excessive drooling
L acrimation—tearing of the eyes
U rination—loss of bladder control
D efecation—loss of bowel control (diarrhea)
G astrointestinal—nausea and vomiting
E mesis—vomiting
M iosis—pinpoint pupils

The term SLUDGEM describes all of the symptoms from the minor ones to the extreme signs and symptoms. A slight exposure to any of the nerve agents will cause pinpoint pupils, a runny nose, and difficulty breathing. A person who has come in contact with the liquid will be experiencing all the SLUDGEM signs in addition to convulsions. A person who is in convulsions needs immediate decontamination and medical treatment in order to survive. This treatment sequence has to occur in less than five to ten minutes. In addition, there must be sufficient medication available on scene to accomplish the treatment. Most paramedic units carry enough medication to treat one or two patients who have severe symptoms.

TABLE 7–1 Military Designations for Agents

Name	Military Designation	UN/DOT Hazard Class
Anthrax	N/A	6.2
Biological agents	N/A	6.2
Chlorine	CL	2.3
Cyanogen chloride	CK	2.3
Distilled mustard	HD	6.1
Hydrogen cyanide	AC	6.1
Lewisite	L	6.1
Mace	CN	6.1
Mustard	H	6.1
Nitrogen mustard	HN	6.1
Pepper Spray	OC	2.2 and 6.1
Phosgene	CG	2.3
Ricin	N/A	6.2
Sarin	GB	6.1
Soman	GD	6.1
Tabun	GA	6.1
Tear gas	CS	6.1
Thickened soman	TGD	6.1
V agent	VX	6.1

numbers of people requires that the dissemination device function correctly and a number of other critical factors be in place to be truly effective. Although several gallons of sarin agent were used in the Tokyo subway attack, the distribution method was ineffective, so out of the twelve people who died, the only people killed by the sarin itself were the two people who actually touched the liquid. The chemical and physical properties of these agents hinder their ability to be effective as a stand-alone killer. To best produce the desired effect, the agents must touch people in liquid form or be breathed in while the materials are in aerosol form. The materials will not stay in aerosol form for very long, and they have a very low vapor pressure and thus will not create vapors as standing liquid. All of the military warfare agents have a vapor pressure less than water, which means they do not evaporate quickly and unless the liquid is touched or placed on the skin it does not present a large hazard.

Incendiary Agents

For the sake of classification, **incendiary agents** are placed into the chemical classification, because chemicals are used in these devices. The most commonly used chemicals are flammable and combustible liquids. The standard Molotov cocktail is an example of an incendiary device that could be used by a terrorist. In some cases arsonists have used a mixture of chemicals, usually oxidizers, to create very fast high-temperature fires.

SIGNS AND SYMPTOMS OF BLISTER AGENT EXPOSURE

One of the biggest risks with the blister agents is that they may present delayed effects. If not detected this could result in victims being released only to later have problems. In general, blister agents are not designed to kill; they were designed to incapacitate the enemy, resulting in troops being assigned to assist with the wounded. It is possible to create scenarios in which fatalities could occur, but these would be unusual cases. The effects from blister agents include irritation of the eyes, burning of the skin, and difficulty in breathing. The more severe exposure results in blisters, which may be delayed. The only real street treatment for these signs and symptoms is decontamination and supportive measures. It is important to have anyone with liquid contact blot the liquid off the skin and avoid spreading the agent. Fortunately, the chemical and physical properties of these agents make them difficult to disseminate, and coming in contact with the liquid would be the primary means of injury.

SIGNS AND SYMPTOMS OF BLOOD AND CHOKING AGENTS

Many of the blood agents are commonly found in industry and may be found at normal chemical facilities. The signs and symptoms of slight exposure to blood agents include dizziness, difficulty in breathing, nausea, and general weakness. With cyanides, the breathing initially will be rapid and deep, followed by respiratory depression and usually death. The two most common choking agents are very common in industrial use, and chlorine can be found in most communities. Signs and symptoms include difficulty in breathing and respiratory distress, eye irritation, and, in higher amounts, skin irritation. Phosgene may present delayed effects, while chlorine's effects are immediate.

Blister (Vesicants)

The category of **vesicants,** or, as they are more commonly called, **blister agents,** includes chemical compounds called mustard, distilled mustard, nitrogen mustard, and lewisite. These materials were never designed to kill. They were designed to incapacitate the enemy so that if one person was affected by one of these agents several more would be needed to care for the affected person. Although at high concentrations these materials can be toxic, their biggest threat is from skin contact which causes severe irritation and blistering. Their chemical and physical properties make them less of a hazard than the nerve agents. One large concern with these agents is the fact that the effects from an exposure can be delayed from fifteen minutes to several hours. Quick identification of a blister agent is key to keeping the victims safe.

Blood and Choking Agents

The four chemicals discussed here in addition to being terrorism agents are also common industrial chemicals. The first category is **blood agents** and includes hydrogen cyanide and cyanogen chloride. Both of these materials are gases that disrupt the body's ability to use the oxygen within the bloodstream. They are also referred to as chemical asphyxiates. The **choking agents,** chlorine and phosgene, are very common in industry. Chlorine is present in almost every town, because it is used for water treatment processes and in swimming pools. Any community that has a water system or swimming pool has some form of chlorine. Chlorine comes in cylinders of 150 pounds (68 kg) to 90-ton (82 metric tons) railcars. It also comes in the tablet form (HTH) that is typically used in residential pools. The release of chlorine from a 90-ton (82 metric tons) railcar could result in several hundred thousand injuries and possibly an equal number of deaths, especially in an urban area. A small amount of chlorine can be very deadly or at a minimum create substantial panic in a community.

Irritants (Riot Control)

The most commonly used materials that are classified as potential terrorism agents are irritants and include mace, pepper spray, and tear gas. An incident that uses an irritant often impacts a large number of people, because the usual target is a school, mall, or other large place of assembly. The use of one small container can affect large numbers of people and

SIGNS AND SYMPTOMS OF IRRITANTS

The signs and symptoms for a slight exposure up to a high dose are the same, with the exception of increasing severity. The signs and symptoms are eye and respiratory irritation. There is no real treatment except removal to fresh air; in fifteen to twenty minutes the symptoms will begin to disappear. Supportive care can be provided.

make them immediately symptomatic. Luckily these materials are not extremely toxic—although they are extremely irritating—and the symptoms will usually disappear after fifteen to twenty minutes of exposure to fresh air. The response to one of these incidents is difficult because patients with real medical problems need treatment and the source of the irritant is often difficult to identify.

Biological Agents and Toxins

Other than explosives, the most likely agents to be used in a terrorism scenario are **biological agents** and **toxins.** Some of the materials in this grouping include anthrax, mycotoxins, smallpox, plague, tularemia, and ricin. Out of all of the agents for a terrorist to make, this grouping is the easiest, especially ricin. The fatal route of entry differs with each of the agents and could occur via skin contact, inhalation, or injection. These agents are difficult to distribute effectively, and in some cases exposure to sunlight may neutralize many of these agents. Some of the potential indicators of biological terrorism include presence of a powder, liquid material, containers associated with biology, and in some cases a threat letter expressing the use of a biological material. The use of a dissemination device or placement so that the material is disseminated is usually more robust than with chemical agents. The visual sighting of a dust or vapor cloud being released in an unusual fashion could be indicative of a biological threat agent release. A small explosion followed by an uncharacteristic dust or vapor cloud indicates a possible dispersal device. Any unusual packages, equipment, or boxes that are out of place could be holding a biological material.

The Centers for Disease Control (CDC) have three categories of biological agents that could be used for weapons, as shown in **Table 7-2.** Category A agents are those that can be easily disseminated, or transmitted person to person, have potential to cause a large-scale public health emergency, might cause public panic and social disruption, and may require special

action for public health preparedness. Category B agents are those that are moderately easy to disseminate, result in moderate to low fatality rates, and require specific enhancements of CDC's diagnostic capacity and enhanced disease surveillance. Category C agents are those that might be emerging pathogens that can be used for mass dissemination in the future because of their availability and ease of production and dissemination, and they have the potential for high fatalities and major health impact. Specific information about the two most popular agents is provided next.

Anthrax

Anthrax is a naturally occurring bacterial disease that is commonly found in dead sheep. It is contagious through skin contact or by inhalation of the anthrax spores. Although relatively easy to obtain, it is more difficult to culture and grow the proper grade of anthrax. To produce fatal effects, the type of anthrax required is called weapons-grade anthrax and is very difficult to produce. Even if developed it must be distributed effectively and under the right conditions.

Ricin

Although ricin is easy to make, the required distribution method leaves a lot to be desired because it must be injected to be truly effective. It is 10,000 times more toxic than the nerve agent sarin. A small amount such as one milligram, about the size of a pinhead, can be fatal. Death usually occurs several days after injection. By other routes of entry, such as inhalation or ingestion, the most likely consequence is that many people would get sick but would eventually recover. After explosives, ricin is the leading choice of domestic terrorists, and several times a year someone is arrested for possession of ricin. Some example ricin cases include:

2002—Kenneth Olsen was arrested for possession of ricin, which he kept in his office cubicle. While at work at a high tech company in Spokane,

TABLE 7-2   Biological Agents

Category A Agents

Name	Notes
Anthrax	Noncontagious bacteria, highly infectious through inhalation
Smallpox	Highly contagious virus, with high potential for fatalities
Botulinum toxin	Produced from *Clostridium botulinum* and one of the worst toxins known
Ebola	Also known as viral hemorrhagic fever, high fatality rates
Plague	Caused by the *Yersinia pestis* bacteria. Easy to manufacture and spread by flea bites. Can be inhaled through the spread of pneumonic plague.
Marburg	A viral hemorrhagic fever, with high fatality potential
Tularemia	Caused by the *Francisella tularensis* bacteria. Also known as rabbit fever. Incapacitates as opposed to causing fatalities.

Category B Agents

Name	Notes
Brucellosis	
Epsilon toxin	*Clostridium perfringens*
Food Safety Threats	*Salmonella spp., E. coli* o157:H7, *Shigella*
Glanders	*Burkholder mallei*
Melioidosis	*Burkholder pseudomallei*
Psittacosis	*Chlamydia psittaci*
Q fever	*Caxiella burnetii*
Ricin toxin	*Ricinus Communis*
Staphylococcal enterotoxin B	
Typhus	*Rickettsia prowazekii*
Viral encephalitis	Alphaviruses, Venezuelan equine encephalitis, eastern and western equine encephalitis
Water supply threats	*Vibrio cholerae, Cryptosporidium parvum*

Category C Agents

Name	Notes
Nipah virus	N/A
Hantavirus	N/A
Multidrug-resistant tuberculosis	N/A

Washington, he researched other poisons and explosives and left this information throughout the office.

2003—A letter containing ricin was mailed to the White House. It was one of two letters from the "fallen angel" who was protesting the change of work hours being mandated by the Department of Transportation (DOT). The second letter was located at an airport postal facility in Greenville, South Carolina.

2003—FBI agents arrested Bertier Ray Riddle of Omaha, Arkansas, for sending a letter supposedly containing ricin to the local FBI office in Little Rock.

2004—Ricin was located in Senator Bill Frist's office in the Dirksen Senate Office building. It is not

ANTHRAX SCARE 2001

The FBI is still investigating the anthrax attacks of October 2001 that started in Boca Raton, Florida, at the American Media building. It is suspected that a letter was sent and opened at this facility, which publishes the National Enquirer. One person died and several others fell ill at this building, but a letter was never recovered from this facility. The letters that were sent targeted the media and members of Congress. The NBC studios, the New York Post, and ABC all received letters. Two members of Congress, Senator Daschle and Senator Leahy, received letters. The letters killed five people and made eighteen others ill. Some of the deaths and illnesses involved people at various post offices. It is thought that the mail handling process enabled anthrax to aerosolize and get into the air. The letters contained small amounts of anthrax, which has been tested to be the Ames strain of anthrax that was produced in the United States. A lot has been learned about anthrax and its ability to be used as a weapon. The theoretical dose of anthrax in each of the recovered letters was estimated to be able to kill several hundred thousand people. The reality is that each letter typically killed one person. The other major factor in the deaths was a delay in treatment and in some cases misdiagnoses. The persons who did not seek quick treatment or who were misdiagnosed had difficult or unsuccessful recoveries. Those who sought quick medical treatment, during which the agent was recognized as anthrax, survived. The CDC reported that there were ten cases of inhalational anthrax and twelve confirmed or suspected cases of cutaneous anthrax. Of the ten inhalational cases, seven occurred to postal workers in New Jersey and Washington, D.C., at mail sorting facilities. In the American Media case, one person who received the letter died, and the person sorting the mail fell ill. Six of the ten individuals with inhalational cases of anthrax survived after treatment, which is higher than originally anticipated.

The signs and symptoms of inhalational anthrax are a one- to four-day period of malaise, fatigue, fever, muscle tenderness, and a nonproductive cough followed by a rapid onset of respiratory distress, cyanosis, and sweating. The recent cases also had profound, often drenching, sweating, along with nausea and vomiting.

known what the source of the ricin was since no letter was found. Frist from Tennessee was the majority leader at the time. There is a possibility that the ricin is connected to several other ricin letters from the "fallen angel."

2004—Richard Alberg was arrested for possession of ricin after he placed a large order of castor beans, which are the main component required to make ricin. Although no specific attack was determined, Alberg fantasized about the use of biological agents to kill.

2005—Six persons were arrested in London for conspiring to commit terrorist acts. Inside their apartment, the police found ricin and materials that can be used to manufacture ricin. The Islamic group that was arrested had ties to Chechnya and to al Qaeda terrorists who plotted the millenium bombing attempt in the United States. The establishment of an al Qaeda and Chechnya relationship is very troublesome. The Chechnyan region of Russia is responsible for well-coordinated terrorist attacks in Russia. Their large-scale attacks have killed thousands of Russian citizens, and they are the group responsible for the Beslam school attack, the Moscow theater takeover, the downing of two planes, and a subway bombing, among many other attacks.

2006—Ricin along with pipe bombs and blasting caps were found in the home of William Matthews of Nashville, Tennessee. Matthews reportedly had problems at work, and his wife had sought a protection order due to his substance abuse. It is not known what he intended to do with the ricin, which was found in a baby food jar.

Radioactive Agents

Nuclear agents, unfortunately, have to be put back on the list of possibilities that a terrorist could use. There are two types of radiation events, nuclear detonation and **radiological dispersion devices (RDDs).** The use of an actual nuclear detonation device is very unlikely given the security these materials have. The amount required and the specific type also make the use unlikely. Although there is current speculation that there are some small nuclear devices missing from Russia, this has never been substantiated. There is the potential, however, for some nuclear material that could be used in an RDD coming from Eastern Europe. An RDD is a device that disperses radiological materials, usually through an explosive device. One example would be the use of a pharmaceutical grade radioactive material attached to a pipe bomb, which could cause a large amount of radioactive material to be distributed. The

RESPONSE TO ANTHRAX HOAXES

Emergency responders should establish a plan for response for noncredible anthrax events. In 2001 and 2002, these events kept hazardous materials and other emergency response teams very busy. From the outset it would be difficult for someone to manufacture and successfully pull off an anthrax threat. The use of real anthrax is most likely an isolated case. One cannot say with absolute certainty, but a new attack is not likely and is not likely to kill or injure mass numbers of people. In the hazardous materials and WMD business one should avoid the use of the words never, always, and best. The credibility factor and the technological difficulty make weaponized anthrax an unlikely candidate for use as a weapon. In many of the hoax incidents, letters proclaimed that the recipients had been exposed to anthrax. In some of these cases the only thing present in the envelope was a letter, which meant that there was no other material present. Anthrax is not invisible, and in order to improve the likelihood that the attack may be successful a quantity of the material must be present. In order to cause health effects the material must be inhaled, and in order to be inhaled the material needs to be distributed to put it into the air. The material, since it is a solid, requires a device to put it into the air. If this device is not present the risk to any persons in a building is very small. The only action required is to double- or triple-bag the envelope, package, or material in bags suitable for evidence collection. Procedures for the collection of evidence should be followed, and the local police as well as the FBI should be consulted. The persons who touched the material should be instructed to wash their hands with soap and water. They do not require full body decontamination nor are special solutions required such as a bleach and water mixture. The military advises that water is more than sufficient for humans. Anyone who was in the immediate vicinity (i.e., several feet) of the person does not require decontamination. The only reason for a full body wash would be if the material was splashed on a person from head to toe, but it is not required to be done immediately. The person can be taken to a shower or be provided privacy to take a shower. The person who did open the envelope should be entered into the health care system, advised of the signs and symptoms of exposure, and provided with emergency contact information. There is no need to start prophylactic antibiotics just because a person opened a letter with a powder in it. There is sufficient time to do lab analysis on the material before medical treatment is required. The threat from a solid material (no matter how toxic) is from inhaling the dust or touching the material. Gloves and SCBA are adequate protection for the collection of evidence. The FBI has a number of labs around the country set up to assist with the identification of WMD agents, and the local FBI office should be contacted for assistance.

strength of the radiation source dictates how harmful an RDD would be. In many cases the RDD would present more of an investigative issue than a significant health issue. There are some radiation sources that would present a risk, but these are not in common use. The other factor is the explosive device and its limitations. Technologically an RDD is viable for a small area such as one-half acre of land. Attempting to distribute a significant amount of radioactive material takes a significant explosive device, not to mention the radiation source. The larger the RDD the more danger there is to the terrorist who has to assemble, transport, and detonate the device. There is an advantage to a noncredible RDD or a small RDD, and that is the public's reaction. The perception to the public and to many responders is that this would be a radioactive disaster. Radiation meters would indicate that there were radioactive materials spread around. The reality is that the amount of radiation would not be dangerous. As time passed, the danger would lessen as the radioactive material became less hazardous. Still, there would likely be tremendous concern in the community. Radiation causes fear because it is a big unknown, making it a prime weapon for a terrorist. Education on the hazards of radiation and the effective use of radiation monitors can ease this fear and allow responders to make an informed response.

Other Terrorism Agents

Although considerable emphasis is placed on warfare agents, in reality many other common industrial or household materials can be just as deadly, if not more so.

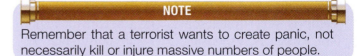

NOTE

Remember that a terrorist wants to create panic, not necessarily kill or injure massive numbers of people.

A likely scenario is a pipe bomb, followed by the use of ricin. Responders should not be lulled into a false sense of security if they do not find a "warfare" agent or if the device is only a small pipe bomb and

FBI AND CDC LABORATORY RESPONSE NETWORK

When discussing WMD detection, it is important to note the laboratory system that the FBI and CDC have set up. The Laboratory Response Network, known as the LRN, handles the initial analysis of potential WMD materials. The LRNs have been in existence since 1999 but really came to light in 2001 after the anthrax attacks. The FBI partnered with the CDC to ensure that standardized protocols were implemented and that both public health and evidentiary concerns were addressed. The WMD coordinator can assist in getting evidence or samples to an LRN facility. The gold standard for analysis is laboratory testing, which is the desire of the FBI. Without laboratory testing, they can't make their case, and the suspect may be allowed to go free. They do not place much emphasis on street detection methods and generally discourage messing with the potential evidence. Prior to shipment they want the evidence screened for fire, corrosive, toxic, and radioactive hazards. The FBI has a number of hazardous materials response teams throughout the United States and maintains a highly technical response group known as the **Hazardous Materials Response Unit (HMRU),** who can provide assistance when potential WMD materials are found or suspected. **Figure 7-9** shows one their response vehicles.

There are 140 LRN labs set up for biological threat agents throughout the United States, and generally the local or state health departments run the labs. For chemical attacks, there are 62 laboratories that can conduct chemical threat agent analysis. These are the labs that most responders would send their samples to, through the FBI WMD coordinator.

FIGURE 7-9 The FBI's Hazardous Materials Response Unit (HMRU) has responsibility to assist in the collection of evidence at high hazard crime scenes. These hazards typically are WMD materials or other hazardous materials. They are also equipped to handle building collapses, trench scenes, and confined spaces, much like a hazardous materials or USAR team.

not a moving truck filled full of ammonium nitrate and fuel oil. Nerve agents (or any other chemical warfare agent) have never been used in the United States, nor manufactured by anyone other than the military. The FBI Bomb Data Center reports that out of the 3,000 bombings that occur each year, there have been only two large bombings, but these "small" bombs kill an average of 32 and injure 277 people each year.

DETECTION OF TERRORISM AGENTS

The detection of terrorism agents is difficult, and given the potential circumstances is done under severe conditions. The response to terrorism incidents has changed how many hazardous materials teams operate and has increased their capability to handle other situations. The confirmation that terrorist agents have been used is difficult because they are extremely hard to detect. The exceptions are the standard indus-

trial materials such as chlorine, which is easy to detect and confirm its presence.

The detection of terrorism agents addresses three major categories of hazards: chemical agents, radiological materials, and biological agents. There are a number of devices that detect chemical warfare agents. They range from inexpensive paper strip tests (M-8 and M-9) to sophisticated electronic devices costing more than $125,000. The most common are direct reading devices, which detect nerve and blister agents, **Figure 7-10.** Some devices will also detect drugs, explosives, mace and pepper spray, and industrial chemicals. The radiological detection field has dosimeters that track the dose of radiation that one is receiving to handheld instruments. Radiation pagers and pager/dosimeters are very small. Pager-sized devices are available and can detect levels of radiation and alert the user to potential danger. Typical radiation monitors detect the dose a human is absorbing, while some newer models will determine and identify the exact type of radiation source present. The detection of biological agents is more difficult, and

FIGURE 7-10 This detection device has the capability to detect chemical warfare agents such as sarin nerve agent and several toxic industrial chemicals.

there are limited choices. One detection device that is new to the emergency response world is a polymerase chain reaction (PCR) unit, which detects the DNA of the sample and compares it to other DNA samples loaded in the library. It is the same method that is used in the laboratory to detect biological agents. The only other choice is handheld bioassays, which, depending on the brand, function with varying degrees of accuracy. Some work well and have accuracy rates in the 90th percentile, while others have accuracy rates in the 30th percentile. The major issue with any biological agent detection is that the sample collection method has to follow exacting standards without any deviation. The detection of any terrorism agent is difficult, and the hazardous materials team should take responsibility for doing the testing.

In some cases standard civilian detection devices such as photo-ionization detectors also play a role in the detection of terrorism agents. Other tests are used by hazardous materials teams that have applicability in the detection of terrorism agents. Due to the many mitigating factors the detection of these agents is going to be difficult, and to be conclusive will probably require lab tests.

FEDERAL ASSISTANCE

The federal government has established roles and responsibilities in the event an act of terrorism occurs, and these roles are provided in PDD 39. Per this document, the FBI is designated the lead agency during the emergency (crisis management) stage of an incident.

> **NOTE**
>
> At the end of November 2002, the Department of Homeland Security (DHS) was formed. This organization is designed to protect the United States against terrorist attacks and to respond to natural disasters. Its goal is to prevent, deter, and respond to terrorism and disaster situations. A number of existing federal agencies were absorbed into DHS. The one agency within DHS that has direct interaction with state and local responders is the Federal Emergency Management Agency.

The Federal Emergency Management Agency (FEMA) becomes the lead when the incident is no longer in the emergency phase (consequence management). The FBI has a Hazardous Materials Response Unit (HMRU) that provides identification, mitigation assistance, and evidence collection for potential terrorist incidents. The HMRU is a multifaceted group that not only responds to terrorism incidents, but also responds to incidents involving explosives, drug labs/incidents, and environmental crimes. The FBI HMRU has responsibility to collect evidence from potentially hazardous environments. A response from FEMA will vary with the incident, as will the number of FEMA personnel, but the response will be similar to any other disaster; FEMA assists in restoration and recovery issues.

The Urban Search and Rescue (USAR) teams fall under FEMA, and are activated by following the emergency management chain from the local level to the state level and then to FEMA for activation. At this time there are twenty-eight USAR teams across the country. They provide expertise in heavy rescue operations such as building collapses. This team of seventy people is composed of rescue specialists, dog search teams, medical specialists, hazardous materials WMD specialists, communications specialists, and an engineering and rigging component. The majority of victims are rescued by the first responders

BASIC INCIDENT PRIORITIES

When dealing with an incident that involves terrorism, first responders should follow these guidelines:

- To rescue live victims, use full protective clothing including SCBA. All of the agents listed in this chapter are predominantly hazardous through inhalation. Avoid touching any unknown liquids or solids, because most are also toxic through skin contact.

- Use a quick in/quick out approach: Do not treat victims, remove them from the area. Keep in mind the terrorist may be among the injured. Watch for secondary devices.

- Request hazardous materials and the police bomb squad. The sooner responders eliminate the potential for chemical agents or a secondary device, the better off they will be. The hazardous materials team may have the ability to detect the agents listed in this chapter, and most hazardous materials teams are working with their bomb squads on a more frequent basis.

- Limit personnel operating in the hazard area.

- Establish multiple staging areas, out of the line of sight.

- Notify the local emergency management agency so that they can mobilize the state and federal resources.

- If a building has collapsed or there is potential for a building collapse, request assistance from a tactical rescue team or a USAR team.

- Isolate all victims, separating contaminated from clean victims.

- Establish a safe triage, treatment, and transport area away from the impact or hot zone.

- Notify all area hospitals of the incident.

- Remember that the incident is a crime scene and make provisions to preserve as much evidence as possible.

- If there is reason to suspect the presence of chemical agents, use the DOT ERG or other reference sources such as the Medical Management of Chemical Casualties Handbook to suggest safety precautions and patient treatments.

and other local specialized resources. But in some cases trapped victims may be alive for many days and may require the expertise and equipment that a USAR team has available. The unfortunate thing is that the USAR teams have a delayed response, because it takes them several hours to become airborne, not to mention travel time. If there is a possibility they may be needed, their assistance should be requested early. Like many of the teams, it is not uncommon for an advance party to arrive hours prior to the arrival of the remainder of the team.

A number of other agencies may be involved in the event of an incident. The military has a couple of units that have terrorism response capabilities and responsibilities. The National Guard has Civil Support Teams (CST) that are set up like a local hazardous material response team. They are comprised of 22 persons and carry detection devices, protective clothing, state of the art communications abilities, and decontamination equipment. There are fifty-five such teams across the United States, which are activated by the states, even though they are federally funded. The army has the Technical Escort Unit (TEU), headquartered at the Aberdeen Proving Grounds, Maryland. There are other units at Dugway Proving Grounds, Utah; Pine Bluff Arsenal, Arkansas; and Fort Belvoir, Virginia. The TEU is assigned to provide escort service for warfare agents and to be the troubleshooting group in the event of an incident involving warfare agents or explosives. When such an incident occurs, the TEU will respond to assist with identification and the mitigation of the incident. Team members handle chemical, biological, and explosive materials as well as other hazardous materials. TEU is a self-contained unit and has the lab resources of the Research Development and Engineering Command (RDECOM) at the Aberdeen Proving Grounds available to assist them.

The United States Marines have a unit known as the Chemical and Biological Incident Response Force (CBIRF), which comes from Indian Head, Maryland. This unit responds to acts of terrorism across the world and has three main components: decontamination, medical, and security. It is a self-contained unit and requires only a water source for conducting long-term operations. CBIRF is able to respond nationwide upon request to terrorist incidents and can provide detection and mitigation as well. All of the federal resources including CBIRF, CST, and TEU can be and have been pre-positioned for certain events, such as the Olympics, political conventions, and visits by dignitaries. These resources integrate within the local system, usually hidden away from the public, and are immediately available.

Within DHS, there are a number of agencies within FEMA that have been reorganized. As part of this reorganization, major national preparedness components and functions to include The Office of Grants and Training, the United States Fire Administration, National Capital Region Coordination, Chemical Stockpile Emergency Preparedness, and the Radiological Emergency Preparedness Program transferred to FEMA, effective April 1, 2007. Other federal programs include training under the Nunn-Lugar-Domenici Legislation, which was passed in September 1996 (P.L. 104-201). This law, known as the Domestic Preparedness Training Initiative, mandated that the Department of Defense provide train-ing to the 157 major cities and counties across the United States. It is intended to enhance the capability of the local, state, and federal response to incidents involving NBC materials. Part of this process is an assessment to determine the cities' capabilities after the training. Also set up around the country are Metropolitan Medical Response Teams (MMRTs), which are a group of 129 people on each team. They are trained and equipped to handle the medical component of a terrorism event. The plan is to have them in the 120 largest metropolitan areas in the country.

LESSONS LEARNED

Within each community there exist multiple agencies that would respond to a terrorist attack. One of the major issues would be the coordination of those resources. The response to a potential terrorism incident can be very challenging, and every responder must be alert to the possibility of such an incident. To be safe, responders should wear all PPE, not linger in the environment, and relocate to a safe area once the live victims are out. It is important to be aware of the potential for secondary devices and request hazardous materials and bomb squad assistance quickly. One to thousands of victims may be injured or killed. Scenarios involve tremendous loss of life, including a large number of responders. One of the major challenges involves the evidence collection process and collection of potentially hazardous evidence. There exists the possibility that responders may lose and the terrorist will win, a situation that can be avoided by training, planning, and preparing for such an incident.

There are a number of Federal resources that can assist with the response to a terrorist attack, or even a suspected attack. Working with and coordinating this response is an important consideration. Special response companies or teams should know who their local contacts are and should develop a relationship with the Federal liaisons. As the FBI has the responsibility and is the lead agency to investigate an act of terrorism, responders should know that the FBI has WMD coordinators who are responsible for interacting with local responders to help coordinate the overall response.

KEY TERMS

Anthrax   A biological material that is naturally occurring and is severely toxic to humans. It is commonly used in hoax incidents.

Biological Agents   Microorganisms that cause disease in humans, plants, and animals; they also cause the victims' health to deteriorate. Biological agents have been designed for warfare purposes.

Blister Agents   A group of chemical agents that cause blistering and irritation of the skin. Sometimes referred to as vesicants.

Blood Agents   Chemicals that affect the body's ability to use oxygen. If they prevent the body from using oxygen, fatalities result.

Butyric Acid   A fairly common lab acid that has been used in many attacks on abortion clinics. Although not extremely hazardous, it has a character-istic stench that permeates the entire area where it is spilled.

Choking Agent   Agent that causes a person to cough and have difficulty breathing. The terrorism agents that are considered choking agents are chlorine and phosgene, both very toxic gases.

Hazardous Materials Crime   A criminal act that uses or threatens the use of hazardous materials as a weapon.

Hazardous Materials Response Unit (HMRU)   A specialized response group within the FBI Laboratory division that responds to WMD and other potentially hazardous crime scenes.

Incendiary Agents   Chemicals that are used to start fires, the most common being a Molotov cocktail.

Mass Casualty   An incident in which the number of patients exceeds the capability of the EMS to manage the incident effectively. In some jurisdictions this can be two patients, while in others it may take ten to make the incident a mass casualty.

Nerve Agents   Chemicals that are designed to kill humans, specifically in warfare. They are chemically similar to organophosphorus pesticides and cause the same medical reaction in humans.

PDD 39   Presidential Decision Directive 39, which established the FBI as the lead agency in terrorism incidents responsible for crisis management. It also established FEMA as the lead for consequence management.

Radiological Dispersion Device (RDD)   An explosive device that spreads radioactive material throughout an area.

Ricin   A biological toxin that can be used by a terrorist or other person attempting to kill or injure someone. It is the easiest terrorist agent to produce and one of the most common.

Toxins   Disease-causing materials that are extremely toxic and in some cases more toxic than other warfare agents such as nerve agents.

Vesicants   A group of chemical agents that cause blistering and irritation of the skin. Commonly referred to as blister agents.

Weapons of Mass Destruction (WMD)   A term that is used to describe explosive, chemical, biological, and radiological weapons used for terrorism and mass destruction.

REVIEW QUESTIONS

1. Describe four potential targets of terrorism.
2. Describe four indicators of potential terrorist activity.
3. What are the most readily available agents that could be used in a terrorist attack?
4. Which three main local/regional agencies or groups should be notified immediately of a suspected terrorist attack?
5. Describe the process of requesting federal assistance.
6. Explain which of the CBRNE agents is the most likely to be found at an incident.
7. Describe the second most likely agent to be found at an incident.
8. Describe which agent is designed to kill immediately.
9. What are the immediate signs and symptoms of sarin exposure?
10. Who is the lead agency to investigate acts of terrorism?
11. What is significance of category A biological agents?
12. Which biological toxin is used most of the time in the United States?
13. Which is more toxic, ricin or sarin?
14. What are the outward indicator differences between a chemical attack and a biological attack?
15. Which federal agency should be contacted to assist with the collection of evidence in a hazardous environment?

FOR FURTHER REVIEW

For additional review of the content covered in this chapter, including activities, games, and study materials to prepare for the certification exam, please refer to the following resources:

Firefighter's Handbook Online Companion
Click on our Web site at **http://www.delmarfire.cengage.com** for FREE access to games, quizzes, tips for studying for the certification exam, safety information, links to additional resources and more!

Firefighter's Handbook Study Guide
Order#: 978-1-4180-7322-0
An essential tool for review and exam preparation, this Study Guide combines various types of questions and exercises to evaluate your knowledge of the important concepts presented in each chapter of *Firefighter's Handbook*.

ENDNOTE

1. The Roe v. Wade decision was the Supreme Court case that allowed legalized abortions, and is the case that has created a lot of the turmoil between pro-choice and pro-life forces.

ADDITIONAL RESOURCES

Bevelacqua, Armando, *Hazardous Materials Chemistry,* 2nd ed. Thomson Delmar Learning, Clifton Park, NY, 2006.

Bevelacqua, Armando and Richard Stilp, *Hazardous Materials Field Guide,* 2nd ed. Thomson Delmar Learning, Clifton Park, NY, 2006.

Bevelacqua, Armando and Richard Stilp, *Terrorism Handbook for Operational Responders,* 3rd ed. Delmar, Cengage Learning, Clifton Park, NY, 2003.

Buck, George, *Preparing for Biological Terrorism: An Emergency Services Guide*. Delmar Thomson Learning, Albany, NY, 2002.

Buck, George, *Preparing for Terrorism: An Emergency Services Guide.* Delmar Publishers, Albany, NY, 1997.

Buck, George, Lori Buck, and Barry Mogil, *Preparing for Terrorism: The Public Safety Communicator's Guide.* Thomson Delmar Learning, Clifton Park, NY, 2003.

Hawley, Chris, *Hazardous Material Air Monitoring and Detection Devices,* 2nd ed. Thomson Delmar Learning, Clifton Park, NY, 2006.

Hawley, Chris, *Hazardous Materials Incidents,* 3rd ed. Thomson Delmar Learning, Clifton Park, NY, 2007.

Hawley, Chris, Michael Hildebrand, and Gregory Noll, *Special Operations: Response to Terrorism and HazMat Crimes.* Red Hat Publishing, Chester, MD, 2002.

Henry, Timothy V., *Decontamination for Hazardous Materials Emergencies*. Delmar Publishers, Albany, NY, 1998.

Medical Management of Biological Casualties. U.S. Army Medical Research Institute of Infectious Diseases, Ft. Dietrick, Fredrick, MD, 1996.

Medical Management of Chemical Casualties Handbook. U.S. Army Chemical Casualty Care Office, Medical Research Institute of Chemical Defense, Aberdeen Proving Ground, MD, Sept. 1995.

Pickett, Mike, *Explosives Identification Guide,* 2nd ed. Thomson Delmar Learning, Clifton Park, NY, 2004.

Schnepp, Rob and Paul Gantt, *Hazardous Materials: Regulations, Response and Site Operations.* Delmar Publishing, Albany, NY, 1998.

Smelby, L. Charles, Jr. (ed.), *Hazardous Materials Response Handbook,* 3rd ed. National Fire Protection Association, Quincy, MA, 1997.

Stilp, Richard and Armando Bevelacqua, *Citizen's Guide to Terrorism Preparedness.* Thomson Delmar Learning, Clifton Park, NY, 2002.

Stilp, Richard, and Armando Bevelacqua, *Emergency Medical Response to Hazardous Materials Incidents.* Delmar Publishers, Albany, NY, 1996.

Acronyms

ACGIH   American Conference of Governmental Industrial Hygienists

ACS   American Chemical Society

ALARA   as low as reasonably achievable

ALOHA   aerial location of hazardous atmospheres

ALS   advanced life support

ANFO   ammonium nitrate and fuel oil

APR   air-purifying respirators

AST   aboveground storage tanks

ASTM   American Society for Testing and Materials

ATF   (Bureau of) Alcohol, Tobacco, Firearms (and Explosives)

BLEVE   boiling liquid expanding vapor explosion

BLS   basic life support

BOCA   Building Officials Conference Association

CAAA   Clean Air Act Amendment

CAMEO   Computer-Aided Management for Emergency Operations

CANUTEC   Canadian Transportation Emergency Center

CAS   Chemical Abstracts Service

CBIRF   Chemical and Biological Incident Response Force (Marines)

CBRNE   chemical, biological, radiological, nuclear, and explosive

CDC   Centers for Disease Control

CERCLA   Comprehensive Environmental Response, Compensation, and Liability Act

CFR   *Code of Federal Regulations*

CGI   combustible gas indicators

CHEMTREC   Chemical Transportation Emergency Center

CHRIS   Chemical Hazards Risk Information System

CMS   chip measurement system

CNS   central nervous system

CO   carbon monoxide

CST   (National Guard) Civil Support Team

DCM   dangerous cargo manifest

DECIDE   detect, estimate, choose, identify, do the best, evaluate

DHS   Department of Homeland Security

DNR   Department of Natural Resources

DOT   Department of Transportation (U.S.)

EHS   extremely hazardous substance

EMS   emergency medical services

EPA   Environmental Protection Agency

EPCRA   Emergency Planning and Community Right to Know Act

ERG   Emergency Response Guidebook

ERP   emergency response planning

FBI   Federal Bureau of Investigation

FD   fire department

FEMA   Federal Emergency Management Agency

FFPC   firefighter protective clothing

FFPE   firefighter protective ensemble

GEDAPER   gather information, estimate potential, determine goals, assess tactical options, plan, evaluate, review

gpm   gallons per minute

Gy   gray

HAZMAT   hazardous materials

HAZWOPER   hazardous waste operations and emergency response

HCS   hazard communication standard

HMIS   Hazardous Materials Information System

HMRU   Hazardous Materials Response Unit (FBI)

HMTD   hexamethylene triperoxide diamine

HVAC   heating, ventilation, and air conditioning

IC   incident commander

ICt$_{50}$ incapacitating concentration to 50 percent of the population, with "t" representing time, usually expressed in minutes

IDLH   immediately dangerous to life and health

IDR   instant detection and response

IM   intermodal

IMS   (National Fire Service) Incident Management System

IRS   Internal Revenue Service

ISC   Industrial Scientific Corporation

ISEA   International Safety Equipment Association

LC$_{50}$ lethal concentration to 50 percent of the population

LCt$_{50}$ lethal concentration to 50 percent of the population, with "t" representing time, usually expressed in minutes

LD$_{50}$ lethal dose to 50 percent of the population

LEL   lower explosive (flammable) limit

LEPC   Local Emergency Planning Committee

LOX   liquid oxygen

LRN   Laboratory Response Network (of the CDC and FBI)

LSA   low specific activity (radiation)

LUST   leaking underground storage tank

MCI   mass casualty incident

MOS   metal oxide sensor

mph   miles per hour

MSDS   Material Safety Data Sheet

NBC   nuclear, biological, and chemical

NFPA   National Fire Protection Association

NIMS   National Incident Management System

NIOSH   National Institute for Occupational Safety and Health

NOS   not otherwise specified

NPL national priority list

NRC Nuclear Regulatory Commission

OPA Oil Pollution Act

OPP organophosphate pesticide

ORM-D other regulated material, Class D

OSHA Occupational Safety and Health Administration

P polymerization hazard

PASS personal accountability safety system

PDD 39 Presidential Decision Directive 39

PEL permissible exposure limit

PG I, II, or III packing group I, II, or III

PID photo-ionization detector

PIH poison inhalation hazard

PIO public information officer

PNS peripheral nervous system

ppb parts per billion

PPE personal protective equipment

ppm parts per million

psi pounds per square inch

psig pounds per square inch gauge

PVC polyvinyl chloride

R&I recognition and identification

rad radiation absorbed dose

RDECOM Research, Development, and Engineering Command

REL recommended exposure limit

REM roentgen equivalent man; also abbreviated as R

RPG rocket-propelled grenade

RQ reportable quantity

SADT self-accelerating decomposition temperature

SAR supplied air respirator

SARA Superfund Amendments and Reauthorization Act

SBCCOM Soldiers Biological and Chemical Command

SBIMAP South Baltimore Industrial Mutual Aid Plan

SCA surface contaminated articles (radiation)

SCBA self-contained breathing apparatus

SERC State Emergency Response Committee

SETIQ Emergency Transportation System for the Chemical Industry

SHB suicide/homicide bomber

SOP standard operating procedure

SPE solid polymer electrolyte (sensor)

STCC Standard Transportation Commodity Code

STEL short-term exposure limit

TATP triacetone triperoxide

TEU Technical Escort Unit

TIA Tentative Interim Amendment

TLV threshold limit value

TOFC trailers on flat cars

TRACEMP thermal, radiation, asphyxiation, chemical, etiological, mechanical hazards, and potential psychological harm

UEL upper explosive (flammable) limit

UN/NA United Nations/North America

UPS United Parcel Service

USAR Urban Search and Rescue (team)

UST underground storage tank

UV ultraviolet

VTR violent tank rupture

WMD weapons of mass destruction

Glossary

8-Step Process A management system used to organize the response to a chemical incident. The elements are site management and control, identifying the problem, hazard and risk evaluation, selecting PPE and equipment, information management and resource coordination, implementing response objectives, decon and cleanup operations, and terminating the incident.

Abortifacient A chemical or material that can cause abortions.

Aboveground Storage Tank (AST) Tank that is stored above the ground in a horizontal or vertical position. Smaller quantities of fuels are often stored in this fashion.

Absorbed Dose A measure of the amount of radiation transferred to a material.

Absorption A defensive method of controlling a spill by applying a material that absorbs the spilled chemical.

Activity A measure of the number of decays per second that occur with the source.

Acute A quick one-time exposure to a chemical.

ADR/RID Abbreviations used by the international shipping community and established by a European Agreement by the United Nations Economic and Social Council Committee of Experts on Transporting Dangerous Goods. The U.S. DOT is a voting member of this shipping body, which governs Regulations Concerning the International Transport of Dangerous Goods by Rail (RID) and Road (ADR).

Air Bill The term used to describe the shipping papers used in air transportation.

Air Monitoring Devices Used to determine oxygen, explosive, or toxic levels of gases in air.

Air-Purifying Respirator (APR) Respiratory protection that filters contaminants out of the air, using filter cartridges. Requires the atmosphere to have sufficient oxygen, in addition to other regulatory requirements.

Allergen A material that causes a reaction by the body's immune system.

ANFO The acronym that is used for ammonium nitrate fuel oil mixture, which is a common explosive. ANFO was used in the Oklahoma City bombing incident.

Anthrax A biological material that is naturally occurring and is severely toxic to humans. It is commonly used in hoax incidents.

Awareness Level The basic level of training for emergency response to an incident involving hazardous materials/weapons of mass destruction (WMDs), the basis of which is the firefighters' ability to recognize a hazardous situation, protect themselves and call for trained assistance, and secure the scene.

Becquerel (Bq) A measure of radioactivity; the metric version of curies.

Biological Agents Microorganisms that cause disease in humans, plants, and animals; they also cause the victims' health to deteriorate. Biological agents have been designed for warfare purposes.

Biomimetic A form of gas sensor that is used to determine levels of carbon monoxide. It is of the type of sensors used in home CO detectors. It closely re-creates the body's reaction to CO and activates an alarm.

Blister Agents A group of chemical agents that cause blistering and irritation of the skin. Sometimes referred to as vesicants.

Blood Agents Chemicals that affect the body's ability to use oxygen. If they prevent the body from using oxygen, fatalities result.

Boiling Point The temperature where a liquid will convert to a gas at a vapor pressure equal to or greater than atmospheric pressure.

Building Officials Conference Association (BOCA) A group that establishes minimum building and fire safety standards.

Bulk Tank A large transportable tank, comparable to a tote, but considered to be the larger of the two.

Bump Test Used to determine if an air monitor is working. It will alarm if a toxic gas is present. It is a quick check to make sure the instrument responds to a sample of gas.

Butyric Acid A fairly common lab acid that has been used in many attacks on abortion clinics. Although not extremely hazardous, it has a characteristic stench that permeates the entire area where it is spilled.

Calibration Used to set the air monitor and to ensure that it reads correctly. When calibrating a monitor, it is exposed to a known quantity of gas to make sure it reads the values correctly.

Carcinogen A material that is capable of causing cancer in humans.

Catalytic Bead The most common type of combustible gas sensor that uses two heated beads of metal to determine the presence of flammable gases.

Ceiling Level The highest exposure a person can receive without suffering any ill effects. It is combined with the PEL, TLV, or REL as a maximum exposure.

Chemtrec The Chemical Transportation Emergency Center, which provides technical assistance and guidance in the event of a chemical emergency; a network of chemical manufacturers that provide emergency information and response teams if necessary.

Chip Measurement System (CMS) A form of colorimetric air sampling in which the gas sample passes through a tube. If the correct color change occurs, the monitor interprets the amount of change and indicates a level of the gas on an LCD screen.

Choking Agent Agent that causes a person to cough and have difficulty breathing. The terrorism agents that are considered choking agents are chlorine and phosgene, both very toxic gases.

Chronic A continual or repeated exposure to a hazardous material.

Clandestine Drug Labs Illegal labs set up to manufacture street drugs.

Colorimetric Tubes Crystal-filled tubes that change colors in the presence of the intended gases. These tubes are made for the detection of known and unknown gases.

Computer-Aided Management for Emergency Operations (CAMEO) Program A computer program that combines a chemical information database with emergency planning software. It is commonly used by hazardous materials teams to determine chemical information.

Consist The shipping papers that list the cargo of a train. The listing is by railcar, and the consist lists all of the cars.

Convulsant A chemical that has the ability to cause seizure-like activity.

Cryogenic Gas Any gas that exists as a liquid at a very cold temperature, always below −150°F (−101°C).

Curies (Ci) The measure of activity level for radiation sources.

Damming The stopping of a body of water, which at the same time stops the spread of the spilled material.

Dangerous Cargo Manifest (DCM) The shipping papers for a ship, which list the hazardous materials on board.

DECIDE Process A management system used to organize the response to a hazardous materials incident. The factors of DECIDE are detect, estimate, choose, identify, do the best, and evaluate.

Decontamination The physical removal of contaminants (chemicals) from people, equipment, or the environment. Most often used to describe the process of cleaning to remove chemicals from a person. Often shortened to "decon."

Deflagrates Rapid burning, which in reality with regard to explosions can be considered a slow explosion, but is traveling at a lesser speed than a detonation.

Diking A defensive method of stopping a spill. A common dike is constructed of dirt or sand and is used to hold a spilled product. In some facilities, a dike may be preconstructed, such as around a tank farm.

Dilution The addition of a material to the spilled material to make it less hazardous. In most cases water is used to dilute a spilled material, although other chemicals could be used.

Diverting Using materials to divert a spill around an item. For instance, several shovels full of dirt can be used to divert a running spill around a storm drain.

Emergency Decontamination The rapid removal of a material from a person when that person (or responder) has become contaminated and needs immediate cleaning. Most emergency decon setups use a single hoseline to perform a quick gross decon of a person with water.

Emergency Planning and Community Right to Know Act (EPCRA) The portion of SARA that specifically outlines how industries report their chemical inventory to the community.

Emergency Response Guidebook (ERG) Book provided by the DOT that assists the first responder in making decisions primarily at transportation-related hazardous materials incidents.

Emergency Response Planning (ERP) Levels that are used for planning purposes and are usually associated with the preplanning for evacuation zones.

Encapsulated Suit A chemical suit that covers the responder, including the breathing apparatus. Usually associated with Level A clothing, which is gas and liquid tight, but there are some Level B styles that are fully encapsulated but not gas or liquid tight.

Endothermic Reaction Chemical reactions that absorb heat or require heat to bond atoms or molecules.

Etiological A form of a hazard that includes biological, viral, and other disease-causing materials.

Evacuation The movement of people from an area, usually their homes, to another area that is considered to be safe. People are evacuated when they are no longer safe in their current area.

Excepted Packaging Type of packaging used to transport low-risk radioactive materials; the packaging is exempted from DOT regulations.

Exothermic Reaction A chemical reaction that releases heat, such as when two chemicals are mixed and the resulting mixture is hot.

External Floating Roof Tank Tank with the roof that covers the liquid within the tank exposed on the outside. The roof floats on the top of the liquid, which does not allow for vapors to build up.

Extremely Hazardous Substances (EHS) A list of 366 substances that the EPA has determined present an extreme risk to the community if released.

Fine Decontamination The most detailed of the types of decontamination. Usually performed at a hospital that has trained staff and is equipped to perform fine-decon procedures.

Firefighters Protective Clothing (FFPC) Protective clothing that firefighters use for firefighting, rescue, and other first-response activities; also known as turnout or bunker gear.

First Responders A group designated by the community as those who may be the first to arrive at a chemical incident. This group is usually composed of police officers, EMS providers, and firefighters.

Fissile Isotopes that can be induced into fission, which creates a release of energy.

Fit Testing A test that ensures the respiratory protection fits the face and offers maximum protection.

Frangible Disk A type of pressure-relieving device that actually ruptures in order to vent the excess pressure. Once opened, the disk remains open; it does not close after the pressure is released.

Freezing Point The temperature at which liquids become solids.

Gas A state of matter in which the material moves freely about and is difficult to control. Steam is an example.

GEDAPER Process A management system used to organize the response to a chemical incident. The factors are gather information, estimate potential, determine goals, assess tactical options, plan, evaluate, and review.

Gross Negligence Occurs when an individual disregards training and continues to act in a manner without regard for others.

Half-Life The amount of time for a given radiation source to decay by one half, which means it is emitting radioactivity.

Hazardous Materials Crime A criminal act that uses or threatens the use of hazardous materials as a weapon.

Hazardous Materials Response Unit (HMRU) A specialized response group within the FBI Laboratory division that responds to WMD and other potentially hazardous crime scenes.

Hazardous Waste Operations and Emergency Response (HAZWOPER) The OSHA regulation that covers safety and health issues at hazardous waste sites, as well as response to hazardous materials incidents.

Hyperbaric Chamber A chamber that is usually used to treat scuba divers who ascended too quickly and need extra oxygen to survive. The chamber re-creates the high-pressure atmosphere of diving and forces oxygen into the body. It is also successful in the treating of carbon monoxide poisoning and smoke inhalation, because both of these problems require high amounts of oxygen to assist with the patient's recovery.

ICt$_{50}$ The incapacitating level for time to 50 percent of the exposed group. It is a military term that is often used in conjunction with LCt$_{50}$.

Incendiary Agents Chemicals that are used to start fires, the most common being a Molotov cocktail.

Incident Commander Level A training level that encompasses the operations level with the addition of incident command training. Intended to be the person who may command a chemical incident.

Infrared Sensor A sensor that uses infrared light to determine the presence of flammable gases. The light is emitted in the sensor housing and the gas passes through the light. If the gas is flammable the sensor will indicate the presence of the gas.

Intermodal Containers These are constructed in a fashion so that they can be transported by highway, rail, or ship. Intermodal containers exist for solids, liquids, and gases.

Internal Floating Roof Tank Tank with a roof that floats on the surface of the stored liquid, but also has a cover on top of the tank to protect the top of the floating roof.

Ionization Potential (IP) The ability of a gas or vapor to be ionized. It is most commonly used to determine whether a photoionization detector can detect a gas or vapor.

Irritant A material that is irritating to humans, but usually does not cause any long-term adverse health effects.

Isolation Area An area that is set up by responders and is intended to keep people, both citizens and responders, out. May later become the hot zone/sector as the incident evolves. Is the minimum area that should be established at any chemical spill.

Isotope A material that has a different form due to the number of neutrons that are in the nucleus.

Laws Legislation that is passed by the House and Senate, and signed by the president.

LCt$_{50}$ The lethal concentration for time to 50 percent of the group. Same as the LC$_{50}$, but adds the element of time. It is a military term.

Leaking Underground Storage Tank (LUST) Describes a leaking tank that is underground.

Lethal Concentration (LC$_{50}$) A value for gases that provides the amount of chemical that could kill 50 percent of the exposed group.

Lethal Dose (LD$_{50}$) A value for solids and liquids that provides the amount of a chemical that could kill 50 percent of the exposed group.

Level A Ensemble (of Protective Clothing) Fully encapsulated chemical protective clothing. It is gas and liquid tight and offers protection against chemical attack.

Level B Ensemble (of Protective Clothing) A level of protective clothing that is usually associated with splash protection. Level B requires the use of SCBA. Various clothing styles are considered Level B.

Liability The possibility of being held responsible for individual actions.

Liquid A state of matter that implies fluidity, which means a material has the ability to move as water would. There are varying states of being a liquid from moving very quickly to very slowly. Water is an example.

Local Emergency Planning Committee (LEPC) A group composed of members of the community, industry, and emergency responders to plan for a chemical incident and to ensure that local resources are adequate to handle an incident.

Lower Explosive Limit (LEL) The lower part of the flammable range, and is the minimum required to have a fire or explosion.

Low Specific Activity (LSA) Designation that indicates that a material is emitting low levels of radiation.

Mass Casualty An incident in which the number of patients exceeds the capability of the EMS to manage the incident effectively. In some jurisdictions this can be two patients, while in others it may take ten to make the incident a mass casualty.

Mass Decontamination A process used to decontaminate large numbers of contaminated victims.

Material Safety Data Sheet (MSDS) Information sheet for employees that provides specific information about a chemical, with attention to health effects, handling, and emergency procedures.

Melting Point The temperature at which solids become liquids.

Metal Oxide Sensor (MOS) A coiled piece of wire that is heated to determine the presence of flammable gases. Also called tin dioxide sensor.

Mission-Specific Competencies The knowledge, skills, and abilities to perform the mission of the organization. Examples are provided in NFPA 472.

Multi-Gas Detector A term used to describe an air monitor that measures oxygen levels, explosive (flammable) levels, and one or two toxic gases such as carbon monoxide or hydrogen sulfide.

National Response Center (NRC) The location that must be called to report a spill if it is in excess of the reportable quantity.

Negligence Acting in an irresponsible manner or different from the way in which someone was trained; that is, differing from the standard of care.

Nerve Agents Chemicals that are designed to kill humans, specifically in warfare. They are chemically similar to organophosphorus pesticides and cause the same medical reaction in humans.

Nuclear Regulatory Commission (NRC) The government regulatory agency responsible for oversight of nuclear materials.

Operations Level The next level of training above awareness that provides the foundation which allows for the responder to perform defensive activities at a hazardous materials incident.

Ordinary Tank A horizontal or vertical tank that usually contains combustible or other less hazardous chemicals. Flammable materials and other hazardous chemicals may be stored in smaller quantities in these types of tanks.

Overpacked A response action that involves the placing of a leaking drum (or container) into another drum. There are drums made specifically to be used as overpack drums in that they are oversized to handle a normal-sized drum.

Oxidizer Materials that readily release oxygen; by yielding oxygen, an oxidizer can easily cause or enhance the combustion of other materials. Oxidizers can dramatically increase the rate of burning when the combustible material is ignited.

Paragraph q The paragraph within HAZWOPER that outlines the regulations that govern emergency response to hazardous materials incidents.

PDD 39 Presidential Decision Directive 39, which established the FBI as the lead agency in terrorism incidents responsible for crisis management. It also established FEMA as the lead for consequence management.

Permeation The movement of chemicals through chemical protective clothing on a molecular level; does not cause visual damage to the clothing.

Permissible Exposure Limit (PEL) An OSHA value that regulates the amount of a chemical that a person can be exposed to during an eight-hour day.

Persistence An indication of the time that a material will remain as a liquid, and is related to vapor pressure.

Photoionization Detector (PID) An air monitoring device used by hazardous materials teams to determine the amount of toxic materials in the air.

Polymerize A chain reaction in which the material quickly duplicates itself and, if contained, can be very explosive.

Psychological Decontamination The process performed when persons who have been involved in a situation think they have been contaminated and want to be decontaminated. Responders who have identified that the persons have not been contaminated should still consider what can be done to make them feel better.

Radioactive Decay Process whereby as a material emits radiation, it decays, changing its form; for example, uranium eventually becomes lead after it decays.

Radioisotope A radioactive form of an element unstable due to the number of neutrons.

Radiological Dispersion Device (RDD) An explosive device that spreads radioactive material throughout an area.

Radon A radioactive gas that is emitted from the earth, sometimes collecting in basement areas.

Recommended Exposure Limit (REL) An exposure value established by NIOSH for a ten-hour day, forty-hour workweek. Similar to the PEL and TLV.

Regulations Developed and issued by a governmental agency and have the weight of law.

Relief Valve A device designed to vent pressure in a tank, so that the tank itself does not rupture due to an increase in pressure. In most cases these devices are spring loaded so that when the pressure decreases the valve shuts, keeping the chemical inside the tank.

Remote Shutoffs Valves that can be used to shut off the flow of a chemical. The term remote is used to denote valves that are located away from the spill.

Reportable Quantity (RQ) Both the EPA and DOT use the term. It is a quantity of chemicals that may require some type of action, such as reporting an inventory or reporting an accident involving a certain amount of the chemical.

Retention The digging of a hole in which to collect a spill. Can be used to contain a running spill or collect a spill from the water.

Ricin A biological toxin that can be used by a terrorist or other person attempting to kill or injure someone. It is the easiest terrorist agent to produce and one of the most common.

Risk-Based Response An approach to responding to a chemical incident by categorizing a chemical into a fire, corrosive, or toxic risk. Use of a risk-based approach can assist the responder in making tactical, evacuation, and PPE decisions.

Sea Containers Shipping boxes that were designed to be stacked on a ship, then placed onto a truck or railcar.

Sector An area established and identified for a specific reason, typically because a hazard exists within the sector. The sectors are usually referred to as hot, warm, and cold sectors and provide an indication of the expected hazard in each sector. Sometimes referred to as a zone.

Self-Accelerating Decomposition Temperature (SADT) Temperature at which a material will ignite itself without an ignition source present. Can be compared to ignition temperature.

Sensitizer A chemical that after repeated exposures may cause an allergic-type effect on some people.

Shelter in Place A form of isolation that provides a level of protection while leaving people in place, usually in their homes. People are usually sheltered in place when they may be placed in further danger by an evacuation.

Short-Term Exposure Limit (STEL) A fifteen-minute exposure to a chemical followed by a one-hour break between exposures. Only allowed four times a day.

Sick Building A term that is associated with indoor air quality. A building that has an air quality problem is referred to as a sick building. In a sick building, occupants become ill as a result of chemicals in and around the building.

Sick Building Chemical When a building is referred to as a sick building, certain chemicals exist within that cause health problems for the occupants. These chemicals are referred to as sick building chemicals.

Solid A state of matter that describes materials that may exist in chunks, blocks, chips, crystals, powders, dusts, and other types. Ice is an example.

Specialist Level A level of training that provides for a specific type of training, such as railcar specialist; someone who has a higher level of training than a technician.

Specification (Spec) Plates All trucks and tanks have a specification plate that outlines the type of tank, capacity, construction, and testing information.

Spent Nuclear Fuel Radioactive fuel that was used in a nuclear reactor.

Standards Usually developed by consensus groups establishing a recommended practice or standard to follow.

Standard Transportation Commodity Code (STCC) A number assigned to chemicals that travel by rail.

State Emergency Response Committee (SERC) A group that ensures that the state has adequate training and resources to respond to a chemical incident.

States of Matter Describe in what form matter exists, such as solids, liquids, or gases.

Sublimation The ability of a solid to go to the gas phase without being liquid.

Superfund Amendments and Reauthorization Act (SARA) A law that regulates a number of environmental issues, but is primarily for chemical inventory reporting by industry to the local community.

Supplied Air Respirator (SAR) Respiratory protection that provides a face mask, air hose connected to a large air supply, and an escape bottle. Typically used for waste sites or confined spaces.

Surface Contaminated Object (SCO) Materials that may be contaminated with radioactive waste and are usually low hazard.

Technical Decontamination The process used to perform decontamination of persons or equipment.

Technician Level A high level of training that allows specific offensive activities to take place, to stop or handle a hazardous materials incident.

Threshold Limit Value (TLV) An exposure value that is similar to the PEL, but is issued by the ACGIH. It is based on an eight-hour day.

Tote A large tank usually 250 to 500 gallons (946–1893 liters), constructed to be transported to a facility and dropped for use.

Toxins Disease-causing materials that are extremely toxic and in some cases more toxic than other warfare agents such as nerve agents.

TRACEMP An acronym for the types of hazards that exist at a chemical incident: thermal, radiation, asphyxiation, chemical, etiological, mechanical, and potential psychological harm.

Type A Container A container designed to hold low-risk radioactive material and that offers moderate protection against accidents.

Type B Container A container designed to hold high-risk radioactive materials and is a hardened transport case capable of withstanding severe accidents.

Underground Storage Tank (UST) Tank that is buried under the ground. The most common are gasoline and other fuel tanks.

Upper Explosive Limit (UEL) The upper part of the flammable range. Above the UEL, fire or an explosion cannot occur because there is too much fuel and not enough oxygen.

Vapor Dispersion The intentional movement of vapors to another area, usually by the use of master streams or hoselines.

Vapor Pressure The amount of force that is pushing vapors from a liquid. The higher the force the more vapors (gas) being put into the air.

Vesicants A group of chemical agents that cause blistering and irritation of the skin. Commonly referred to as blister agents.

Waybill A term that may be used in conjunction with consist, but is a description of what is on a specific railcar.

Weapons of Mass Destruction (WMD) A term that is used to describe explosive, chemical, biological, and radiological weapons used for terrorism and mass destruction.

Zone (zoning) An area established and identified for a specific reason, typically because a hazard exists within the zone. The zones are usually referred to as hot, warm, and cold zones and provide an indication of the expected hazard in each zone. Sometimes referred to as a sector.

In addition to *Firefighter's Handbook,* Delmar, Cengage Learning, publishes a wide variety of training resources and textbooks for the fire service. To request a copy of our latest catalog: e-mail delmar.fire@cengage.com or visit delmarfire.cengage.com

For descriptions of all the offerings in the *Firefighter's Handbook,* Third Edition suite of products, please see the preface of this book.

Firefighter

Firefighter Online Course and Firefighter Course on CD-ROM

A comprehensive, blended electronic solution for Firefighter I & II and Hazardous Materials Awareness and Operations education. Contact us to receive a demo CD or 30-day trial of the online course.

Order #: Online Course 978-1-4180-3970-7
CD-ROM Courseware: 978-1-4180-3972-1

Firefighter Job Performance Requirements CD-ROM

With gaming-quality animations that present realism and detail, this CD demonstrates the Job Performance Requirements (JPRs) required by NFPA Standard 1001 and introduced in the Firefighter's Handbook—taking classroom training to a new level.

Order #: 978-1-4180-3973-8

Exam Preparation for Firefighter I and II/Walter, Rutledge, and Hawley

Order #: 978-1-4018-9923-3

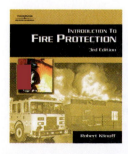

Introduction to Fire Protection, 3rd ed./Klinoff

This book offers a complete introduction to the field of fire protection, technology, and the wide range of services provided by both public and private fire departments of today. It covers fighting fires and the provisions of other emergency services, hazardous materials control, fire prevention, and public education.

Order #: 978-1-4180-0177-3

Rapid Intervention Company Operations/Mason and Pindelski

This book serves as a "one stop" reference for all aspects of rapid intervention operations, outfitting fire service personnel with the techniques and procedures they need to effectively take charge of rescue situations.

Order #: 978-1-4018-9503-7

Occupational Health and Safety in the Emergency Services, 2nd ed./Angle

A comprehensive approach to program management for fire and emergency service occupational safety and health is provided in this practical book. Safety officers and fire department and EMS managers will make good use of this one-stop resource, recently revised.

Order #: 978-1-4018-5903-9

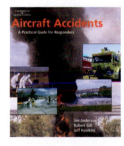

Aircraft Accidents: A Practical Guide for Responders/Anderson, Hawkins, and Gill

This straightforward guidebook prepares emergency responders for the successful response and management of aircraft accidents in their communities, enabling them to communicate and work effectively with the various agencies involved.

Order #: 978-1-4018-7910-5

Principles of Fire Behavior/Quintiere

While explaining the science of fire with a precision found nowhere else, this text presents an ideal introduction to the scientific principles behind fire behavior. New edition available in 2009.

Order #: 978-0-8273-7732-5

Wildland Firefighting Practices/Lowe

The reader will learn in detail all aspects of wildland firefighting with this new, well-illustrated text. Written in a clear, how-to style by a seasoned wildland fire officer, it provides a comprehensive explanation of all the skills a firefighter needs to operate effectively against any type of wildland blaze.

Order #: 978-0-7668-0147-9

Firefighters from the Heart: True Stories and Lessons Learned/Chikerotis

In more than 45 exciting true short stories recounted by real firefighters and fire officers, walk alongside the participants as they fight fires, face career-altering decisions, and learn critical lessons that forever change their perception on what it means to be a firefighter.

Order #: 978-1-4180-1423-0

Firefighter Reference

Encyclopedia of Fire Protection, **2nd ed./Nolan**

Full of insightful information, this edition remains current with the ever changing fire protection industry. Current terminology, codes, organizations, and other need-to-know information is presented in a reader-friendly, engaging format. Excellent desk reference for any firefighter.

Order #: 978-1-4180-2014-9

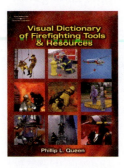

Visual Dictionary of Firefighting Tools and Resources/Queen

This is a one-stop resource containing definitions for various tools utilized in search and rescue, as well as other resources critical to incident response. Photos and graphics accompany the terms and definitions for easy identification.

Order #: 978-1-4018-9790-1

Mastering the CPAT: A Comprehensive Guide/Wasser and Kimble

This exceptional resource provides aspiring firefighters with the information needed to pass the Candidate Physical Ability Test. It includes background on the test, implementing an effective training program, and various aspects of successful CPAT preparation.

Order #: 978-1-4180-1229-8

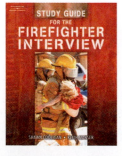

Study Guide for the Firefighter Interview/Cooligan and Manser

This insightful book helps prepare aspiring firefighters for effective and professional fire department entrance interviews. Practical topics prepare you for presenting in an interview, while other sections help you face the challenges of the question-and-answer session.

Order #: 978-1-4180-5072-6

Fire Department Interview Tactics/Mahoney

Focused on both the entrance and promotional interview, this training manual features a thorough explanation of the fire department oral interview process and how to prepare for it—from the examination announcement and eligibility list to resumé guidelines and post-interview procedures.

Order #: 978-1-4180-3004-9

Driver/Operator

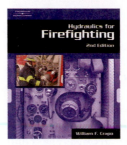

Hydraulics for Firefighting, **2nd ed./Crapo**

This book leads readers throughout the principles, theory, and practical application of fire service hydraulics. It is written in a format that will help guide the new firefighter through even the most technical hydraulic principles and complex laws of physics.

Order #: 978-1-4180-6402-0

Introduction to Fire Pump Operations, **2nd ed./Sturtevant**

This book offers the updated knowledge required to efficiently, effectively, and safely operate and maintain fire pumps. With an emphasis on NFPA standards and safety, the book is logically presented in three sections: Pump Construction/Peripherals, Pump Procedures, and Water Flow Calculations.

Order #: 978-0-7668-5452-9

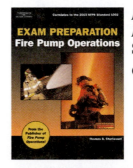

Exam Preparation for Fire Pump Operations/ **Sturtevant**

Order #: 978-1-4180-2088-0

Practical Problems in Mathematics for Emergency Services/Sturtevant

This is the only math-related text specifically written for the emergency services field. This book may be used as a preparation for certification and promotional exams with math components, as well as a quick reference for the seasoned professional.

Order #: 978-0-7668-0420-3

Fire Officer

***Firefighting Strategies and Tactics,* 2nd ed./Angle, Gala, Harlow, Lombardo, and Maciuba**

A complete source for learning firefighting strategies and tactics, from standard company responsibilities and assignments to specialized situational strategies and tactics. The reader will progress from basic concepts to the application of tactics and situational strategies for particular occupancies or types of fires.

Order #: 978-1-4180-4893-8

***Company Officer,* 2nd ed./Smoke**

Any firefighter that wants to gain certification as a Fire Officer will find this practical guide an excellent resource. Based on the latest information and requirements outlined in NFPA 1021, the book gives the user the information necessary to meet NFPA Standard competencies for certification.

Order #: 978-1-4018-2605-5

***Exam Preparation for Fire Officer I and II*/Cline**

Order #: 978-1-4018-9922-6

***Going for the Gold*/Coleman**

Author Ronny Coleman offers a unique, must-have resource for the thousands of individuals who hope to carry the fire chief's badge. The book provides a realistic appraisal of what it takes to aspire to, achieve, and then succeed as fire chief.

Order #: 978-0-7668-0868-3

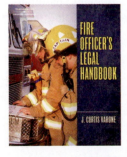

***Fire Officer's Legal Handbook*/Varone**

This all-inclusive legal desk reference includes discussions, cases, and examples that will truly speak to a variety of fire chiefs and officers, municipal officials, emergency managers, and attorneys. The book includes explanations of laws as they relate to firefighting and covers a broad range of legal topics facing fire departments today.

Order #: 978-1-4180-4113-7

Delmar's Fire Officer I and II DVD

This four-part program focuses on the "soft skills" essential to Fire Officers at Levels I and II. The Fire Officer I and II DVD offers challenges that an Officer may face in relation to a variety of topics, includes relevant scenarios, and encourages discussion.

Order #: 978-1-4018-8293-8

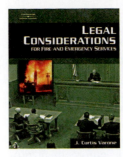

***Legal Considerations for Fire and Emergency Services*/Varone**

Written by a lawyer who is also an experienced firefighter, this book examines the most challenging legal issues confronting firefighters and emergency service personnel today. Explores such major legal concerns as fire service liability issues, the jurisdiction of OSHA over fire departments, search and seizure, employment discrimination, and more.

Order #: 978-1-4018-6571-9

***Fire Department Incident Safety Officer,* 2nd ed./Dodson**

This is the only book that provides a clear, focused, and detailed approach to making a difference as an incident safety officer. Company officers, battalion chiefs, safety officers, and incident commanders will benefit from the foundation material and the incident safety officer action model presented in this book.

Order #: 978-1-4180-0942-7

Instructor/Educator

Training Officer's Resource Kit CD-ROM

This customizable CD-ROM curriculum package contains 48 lesson plans, Powerpoint® presentations, Skill Sheets, and tests based on the following texts: *Firefighter's Handbook,* Second Edition, *Company Officer,* Second Edition, *Introduction to Fire Pump Operations,* Second Edition, and *First Responder: Fire Service Edition.* The straightforward, lesson-oriented approach and customizable electronic format allows training officers to easily adapt the materials to suit specific training needs.

Order #: 978-1-4018-9964-6

***A Practical Guide to Teaching in the Fire Service*/Morse**

Designed for the person without formal training or a degree in education who is found teaching in front of a classroom, this how-to book is the perfect resource. Its no-nonsense approach covers the day-to-day information that is needed to conduct a successful class in fire service.

Order #: 978-0-7668-0432-6

Inspection/Investigation

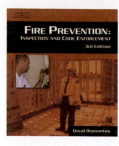

Fire Prevention: Inspection and Code Enforcement, 3rd ed./Diamantes

This is a vital resource for the application of building and fire prevention codes in the inspection of buildings and facilities and for compliance through the code enforcement process. Issues such as enforcement authority, determining inspection priorities, maintenance of rated assemblies, and much more are covered in depth in this comprehensive guide.

Order #: 978-1-4180-0944-1

Principles of Fire Prevention/Diamantes

This book helps readers understand the value of fire prevention, protection, and associated programs. The book includes the origins of our prevention programs, elements of plan review, inspection and investigation, as well as the logistics of staffing and financial management of fire prevention.

Order #: 978-1-4018-2611-6

2006 International Fire Code

The 2006 International Fire Code, coordinated with the 2006 International Building Code®, references national standards to comprehensively address fire safety in new and existing buildings.

Order #: 978-1-58001-255-3

2006 International Building Code

The 2006 International Building Code® addresses the design and installation of building systems through requirements that emphasize performance.

Order #: 978-1-58001-251-5

Significant Changes to the International Fire Code: 2006 Edition/Shapiro

This resource identifies the significant changes that occurred between the 2003 and 2006 editions of the International Fire Code. Coverage focuses squarely on those provisions that have special significance, are utilized frequently, or have had a change on the Code's application.

Order #: 978-1-4180-5301-7

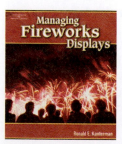

Managing Fireworks Displays/Kanterman

From planning and pre-event activities to post-operations procedures, this resource covers the entire process of safely and effectively executing display operations. This innovative book expands upon the basics of the 2006 edition of NFPA 1123

Code for Fireworks Displays with a step-by-step approach that includes clear explanations, practical examples, and comprehensive coverage.

Order #: 978-1-4180-7284-1

Fire Protection Systems/Jones

A practical understanding of fire protection systems is essential to effective management of a fire scene. *Fire Protection Systems* focuses on the operational characteristics and abilities of different types of systems and equipment that are used during fire department operations to access a water source, apply a suppression agent to control a particular type of fire, provide information concerning the location of a fire, and more.

Order #: 978-1-4018-6262-6

Design of Water-Based Fire Protection Systems/Gagnon

A vital reference for every inspector and designer of fire protection, sprinkler, architectural, or engineering systems, this book is a must. Hydraulic calculations for the most commonly encountered water-based fire protection systems are covered in detail.

Order #: 978-0-8273-7883-4

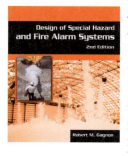

Design of Special Hazard and Fire Alarm Systems, 2nd ed./Gagnon

As the most current guide to the design of state-of-the-art special hazard and fire protection systems, this book is essential to architects, engineers, layout technicians, plumbers, mechanical contractors, and sprinkler firms.

Order #: 978-1-4180-3950-9

Hazardous Materials/WMD/Terrorism

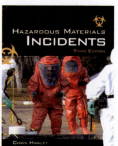

Hazardous Materials Incidents, 3rd ed./Hawley

Hazardous Materials Incidents is an invaluable procedural manual and all-inclusive information resource for emergency services professionals. Easy-to-read and perfect for use in hazardous materials awareness and operations training courses.

Order #: 978-1-4283-1796-3

Hazardous Materials Incidents,
Spanish ed./Hawley

Order #: 978-1-4180-1156-7

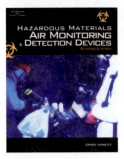

Hazardous Materials Air Monitoring and Detection Devices, **2nd ed./Hawley**

This book provides hazardous materials teams with a thorough guide to effective air monitoring in emergency response situations. Each type of air monitoring device available for emergency services is described in detail, including operating guidelines and sampling strategies.

Order #: 978-1-4180-3831-1

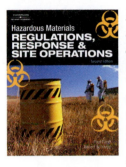

Hazardous Materials Regulations, Response, and Site Operations, **2nd ed./Gantt and Schnepp**

This essential guide provides the student with a practical approach to the concepts of handling hazardous materials. Based on OSHA "HAZWOPER" regulations, this invaluable text addresses the specific competencies required of persons responding to a hazardous materials emergency.

Order #: 978-1-4180-4992-8

Hazardous Materials Chemistry,
2nd ed./Bevelacqua

Hazardous Materials Chemistry covers the basic concepts of chemistry, emphasizing the decision-making process so that appropriate strategies and tactics will be chosen.

Order #: 978-1-4018-8089-7

Hazardous Materials Chemistry Field Operations Guide/Bevelacqua

Order #: 978-1-4018-8090-3

Hazardous Materials Field Guide, **2nd ed./Bevelacqua and Stilp**

Whether the incident involves hazardous materials, a clandestine laboratory, terrorism, or a confined space operation, this user-friendly resource includes information that is consistent with the mission of all agencies. The guidebook's easy-access format allows rapid identification of placards, labels, silhouettes, and common commodities that move on roadways and railways.

Order #: 978-1-4180-3828-1

HotZone Log: Personal Hazmat Record/Toy

This pocket-sized personal log enables individuals to maintain a historical record of personal performance following the completion of training events while also providing opportunities to document experiences gained in response to actual hazardous materials incidents.

Order #: 978-1-4018-7229-8

Emergency Decontamination for Hazardous Materials Responders/Henry

This one-of-a-kind book focuses entirely on decontamination, a crucial aspect of hazardous materials emergency response. The book brings together facts about chemical contamination gathered over the last ten years and presents them in a simple, streetwise way.

Order #: 978-0-7668-0693-1

Emergency Medical Response to Hazardous Materials Incidents/Stilp and Bevelacqua

Medical aspects of hazardous materials response, including the initial response, chemical and toxicological information, and effects on body systems, are explained in this book.

Order #: 978-0-8273-7829-2

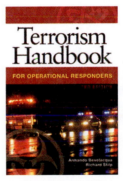

Terrorism Handbook for Operational Responders, **3rd ed./Bevelacqua and Stilp**

This updated book is a guide to the most significant points that surround the emergency response processes needed to cope with terrorism incidents. It highlights new equipment and strategies that can enhance a responder's detection, monitoring, and protection capabilities against chemical and biological agents.

Order #: 978-1-4283-1145-9

Bioterrorism: A Field guide for First Responders, **2nd ed./Imaginatics**

An effective pocket guide specifically designed to help first responders deal with the unique nature of a biological terrorist attack, from personal protection to emergency actions to the guidelines for the twelve most probable biological agents.

Order #: 978-0-9740-6322-5

Chemical/Nuclear Terrorism: A Field Guide for First Responders, **2nd ed./Imaginatics**

An ideal pocket guide for first responders during the aftermath of a chemical or nuclear terrorist attack. From evacuation distances to emergency actions to the guidelines for nerve, blister, and choking agents, this book is ideal for field use.

Order #: 978-0-9740-6323-2

Preparing for Terrorism: The Public Safety Communicator's Guide/Buck, Buck, and Mogil

Emphasis throughout the book is on how to prepare communications center staff and their families for a terrorist event by providing them with well-thought-out employee emergency plans and contingencies. Solutions to communications problems, such as cellular and landline telephone overload situations, are addressed as well.

Order #: 978-1-4018-7131-4

Citizen's Guide to Terrorism Preparedness/**Stilp and Bevelacqua**

This book provides readers with facts, figures, and practical guidelines to follow as they go about their daily lives. It is designed specifically for average citizens who want to take all of the steps they can to prepare themselves for a terrorist act in their state, city, or neighborhood.

Order #: 978-1-4018-1474-8

Preparing for Biological Terrorism: An Emergency Services Guide/Buck

This book contains vitally important information to guide local agencies in their efforts to secure and coordinate the influx of state and federal resources before, during, and after an attack. This resource walks through the fundamental concepts of emergency planning.

Order #: 978-1-4018-0987-4

Explosives Identification Guide, **2nd ed.**/**Pickett**

With this reference guide, through color photographs and short descriptions, the student can identify explosives by general type and learn the appropriate way to treat each of them.

Order #: 978-1-4018-7821-4

Confined Space Rescue/**Browne and Crist**

This clearly written book identifies confined space rescue challenges, showing users how to address them when rescuing a victim. With this book, responders can learn a simple set of skills that will provide a foundation for growth into advanced rescue operations.

Order #: 978-0-8273-8559-7

Engineering Practical Rope Rescue Systems/**Brown**

A practical look at rope rescue systems from the point of view of an experienced professional, each chapter features exciting stories and real-life situations. The book provides a complete review of team integrity and development issues, as well as team efficiency concepts, that create a superior survival profile.

Order #: 978-0-7668-0197-4

Index